Current Knowledge in Thyroid Cancer—From Bench to Bedside

Special Issue Editor
Daniela Gabriele Grimm

MDPI • Basel • Beijing • Wuhan • Barcelona • Belgrade

MDPI

Special Issue Editor
Daniela Gabriele Grimm
Aarhus University
Denmark

Editorial Office
MDPI AG
St. Alban-Anlage 66
Basel, Switzerland

This edition is a reprint of the Special Issue published online in the open access journal *International Journal of Molecular Sciences* (ISSN ISSN 1422-0067) from 2016–2017 (available at:
http://www.mdpi.com/journal/ijms/special_issues/thyroid_cancer).

For citation purposes, cite each article independently as indicated on the article page online and as indicated below:

Author 1; Author 2. Article title. *Journal Name*. **Year**. Article number/page range.

First Edition 2017

ISBN 978-3-03842-476-5 (Pbk)
ISBN 978-3-03842-477-2 (PDF)

Table of Contents

About the Special Issue Editor

Daniela Grimm, MD, is currently Professor of Space Medicine at the Department of Biomedicine, Aarhus University, Denmark. She studied Human Medicine at the University of Würzburg, Germany and worked at the Institute of Pathology (1989–1992) and Clinic for Internal Medicine II (1992–1999) of the University Regenburg, Germany. From 1999 to 2008, she worked at the Institute of Toxicology and Clinical Pharmacology, Charité-University-Medicine Berlin, Germany. She is a specialist of Internal Medicine and Clinical Pharmacology. Since 2008, she has been working at the Department of Biomedicine, Aarhus University, Denmark. Since 2016, she has been Guest Professor of Gravitational Biology and Translational Regenerative Medicine at the Otto-von-Guericke-University Magdeburg, Germany.

Preface to "Current Knowledge in Thyroid Cancer— From Bench to Bedside"

In recent years, studies in the field of thyroid cancer have been performed in order to identify and verify thyroid specific biomarkers, as well as cancer-specific changes in gene expression patterns and alterations of the protein content. Furthermore, new drugs, small molecules and antibodies were developed and tested in vitro and in vivo. Trials investigated the ratio between therapeutic and adverse effects. Tyrosine kinase inhibitors (TKI) have become a new therapeutic option of both differentiated thyroid cancer and medullary thyroid cancer. In the last few years, new substances for targeted systemic therapy have been approved after their efficacy was demonstrated in Phase III trials. Most of them show a moderate response. However, adverse effects are common. TKI are used in patients with advanced metastatic thyroid cancer that is radioiodine (RAI)-refractory.

In this Special Issue, original studies on the pathophysiology, diagnosis, and therapy of thyroid cancer, including genetics, proteomics, metabolomics, molecular and cell biology, will be published. It will also cover reports on patients, providing novel mechanistic insights into the underlying pathogenesis or new aspects that may impact clinical therapy, and recent study results in order to review the current status of new therapy options in thyroid cancer.

Daniela Grimm
Special Issue Editor

International Journal of
Molecular Sciences

MDPI

Review

Molecular Signature of Indeterminate Thyroid Lesions: Current Methods to Improve Fine Needle Aspiration Cytology (FNAC) Diagnosis

Silvia Cantara *, Carlotta Marzocchi, Tania Pilli, Sandro Cardinale, Raffaella Forleo, Maria Grazia Castagna and Furio Pacini

Department of Medical, Surgical and Neurological Sciences, University of Siena, 53100 Siena, Italy; carlottamarzocchi@libero.it (C.M.); t.pilli.e@ao-siena.toscana.it (T.P.); sandro.cardinale@gmail.com (S.C.); forleo.r@gmail.com (R.F.); m.g.castagna@ao-siena.toscana.it (M.G.C.); furio.pacini@unisi.it (F.P.)
* Correspondence: cantara@unisi.it; Tel.: +39-0577-585243

Academic Editor: Daniela Gabriele Grimm
Received: 15 March 2017; Accepted: 3 April 2017; Published: 6 April 2017

Abstract: Fine needle aspiration cytology (FNAC) represents the gold standard for determining the nature of thyroid nodules. It is a reliable method with good sensitivity and specificity. However, indeterminate lesions remain a diagnostic challenge and researchers have contributed molecular markers to search for in cytological material to refine FNAC diagnosis and avoid unnecessary surgeries. Nowadays, several "home-made" methods as well as commercial tests are available to investigate the molecular signature of an aspirate. Moreover, other markers (i.e., microRNA, and circulating tumor cells) have been proposed to discriminate benign from malignant thyroid lesions. Here, we review the literature and provide data from our laboratory on mutational analysis of FNAC material and circulating microRNA expression obtained in the last 6 years.

Keywords: fine needle aspiration cytology (FNAC); indeterminate lesions; next generation sequencing; gene expression classifier; microRNAs (miRNAs)

---

1. Thyroid Nodules

In countries where iodine deficiency has been corrected by iodine prophylaxis, thyroid nodules are found in approximately 4–7% of the population [1]. However, in countries affected by moderate or severe iodine deficiency, the prevalence is even greater [2]. Subclinical nodules, detected by thyroid ultrasound, are found in over 50% of women older than 60 years, a number similar to that reported in autopsy series.

Any type of nodule may be found as a single lump in an otherwise normal thyroid gland or in the context of a multinodular goitre. Regardless of the presentation, the large majority are benign hyperplastic nodules, frequently an expression of underlying nodular goitre or autoimmune thyroiditis. Thyroid cancer is found in less than 10% of hypo-functioning nodules that are solid or mixed on thyroid ultrasound (US) and more than 80% of them are differentiated thyroid cancer of the follicular epithelium.

Surgical treatment of thyroid nodules without selection would expose millions of people annually to surgery. Since only a small proportion of these nodules finally result malignant at histology, this approach would imply a tremendous number of unnecessary surgeries and high financial costs. Thyroid nodules must therefore undergo rigorous selection based on a rational diagnostic protocol.

2. Diagnostic Evaluation of Thyroid Nodules: Fine-Needle Aspiration Cytology (FNAC)

The ultimate objective of the diagnostic protocol is to differentiate between benign and malignant nodules. Nowadays, the problem has largely been solved by fine-needle aspirate cytology. In expert hands, FNAC has an overall accuracy of 95%. The sensitivity is between 43% and 98% and the specificity is between 72% and 100%, with positive and negative predictive values of 89–98% and 94–99%, respectively [3]. False positive and false negative results are between 1–11% and 0–7%, respectively.

The Bethesda Classification System [4], a six diagnostic category system, is at present the most widespread reporting system for thyroid fine needle aspiration (FNA) cytology. The categories are: (I) non-diagnostic/unsatisfactory; (II) benign; (III) atypia of undetermined significance/follicular lesion of undetermined significance (AUS/FLUS); (IV) follicular neoplasm/suspicious for follicular neoplasm (FN); (V) suspicious for malignancy (SUSP) and (VI) malignant.

Category I refers to samples where inadequate or insufficient material is present for a diagnosis or the interpretation is precluded by technical artefacts. In different series, the rate of inadequate cytologies varies between 15% and 20% and in these cases it is recommended to repeat the procedure after some weeks or months [5–8]. Around 70% (range 53–90%) of aspirates are classified as category II, meaning that the features are consistent with a nodular goitre or thyroiditis. Meanwhile 4% (1–10%) are classified as category VI when unequivocal features of papillary, medullary or anaplastic carcinoma or lymphoma are present, and 10% (5–23%) are classified as category V. A particular issue is represented by category III and IV, representing nearly 20% of FNACs. In this case, the aspirate is represented by a monotonous population of follicular cells arranged on cohesive groups, whose cellular and nuclear features are similar whether the nodule is benign or malignant. The distinction is based on the presence of vascular and capsular invasion, which is detectable only at histology. In a meta-analysis of 25,445 thyroid FNAC [4] cases reported from eight studies using the Bethesda System, 9.6% of all samples were diagnosed as AUS/FLUS (category III) and 10.1% were diagnosed as follicular neoplasm/suspicious for follicular neoplasm (FN/SFN) (category IV) with an average cancer risk at final histology of 15.9% and 26.1%, respectively. It is evident that, both the AUS/FLUS and FN/SFN have a cancer risk that cannot be ignored. However, at final histology, only about 25% of the lesions result malignant, so the risk of cancer is not high enough to definitely support surgery as treatment of all indeterminate lesions.

3. Protein-Based Assays to Increase FNAC Performance

To avoid unnecessary surgeries and to increase FNAC performance especially for Categories III and IV, several markers have been proposed in the past years. Among those studied by immunocytochemistry, galectin-3 is one of the most reliable. Galectins are carbohydrate-binding proteins that are members of the β-galactoside binding lectin family. Galectin-3, (Gal-3) appears to be necessary for the maintenance of transformed thyroid papillary cancer (PTC) cell lines in vitro [9]. The use of Gal-3 in the detection of thyroid malignancy in indeterminate or suspicious FNA has a sensitivity that ranges from 20% to 100% and a specificity ranging from 62% to 100% [10–17]. In case of indeterminate FNAC with a positive staining for Gal-3, surgery is strongly recommended, however no specific suggestions can be made in case of Gal-3 negative staining [14]. Similar results have been described with the Hector Battiflora Mesothelial-1 (HBME-1), a monoclonal antibody developed against the microvillous surface of mesothelial cells, which has shown a sensitivity of 79–87% and a specificity of 83–96% [12,13,18–20] in Bethesda categories III and IV. Another proposed marker is CD44v6, a polymorphic family of immunologically related cell-surface glycoproteins, which have a functional role in regulating cell–cell and cell–matrix interactions, cell migration, tumor growth and progression [21–24]. The combined use of CD44v6 and Galectin-3 in indeterminate lesions showed 88% sensitivity, 98% specificity with a positive predictive value (PPV) of 91%, and a diagnostic accuracy of 97% [15].

Although the usage of different markers has been tested to improve the diagnostic efficacy in FNAC, so far none of the tested molecules has provided sufficient sensitivity and specificity to advocate its use in routine practice (not recommended by the America Thyroid Association (ATA).

4. Use of Molecular Markers in the Differential Diagnosis of Thyroid Nodules

The discovery of genetic alterations specific for differentiated thyroid cancer have provided molecular markers to be searched for in the material obtained by FNA, thus increasing the diagnostic accuracy of traditional cytology. The need to search for genetic alterations in FNAC sample should be considered especially for Bethesda categories III and IV. However, the revised guidelines for the management of thyroid cancer published by ATA in 2015 [25] do not provide strong recommendation in support of the use of molecular markers to help the management of patients with indeterminate cytology.

The most frequent genetic alterations detected in papillary and follicular thyroid carcinoma (PTC and FTC, respectively) are B-Raf proto-oncogene, serine/threonine kinase (BRAF), rat sarcoma (N-H-KRAS) point mutations, and REarranged during Transfection proto-oncogene (RET)/PTC, and Paired box 8/Peroxisome proliferator activated receptor gamma (PAX8/PPARγ) rearrangements [26,27]. Telomerase reverse transcriptase (TERT) promoter and Tumor protein 53 (TP53) mutations are more frequent in less differentiated carcinomas [28–30]. Nowadays, there are three diffuse approaches to investigating the molecular profile of FNA: (1) the seven genes panel; (2) the Afirma classifier and (3) next generation sequencing (NGS) assays.

4.1. Seven Genes Mutational Panel

The first study analyzing the contribution of molecular testing to thyroid fine-needle aspiration cytology was published in 2010 [31]. In this work, the authors considered BRAF and RAS gene mutations, as well as RET/PTC, and PAX8/PPAR-γ gene rearrangements in 117 indeterminate cytologies. Among these, 35 (29.9%) cases had a neoplastic outcome and 20 (17.1%) cases were found to be carcinoma. Positive molecular results were found in 12 cases, all of which were PTC. The authors found that the cancer probability for AUS/FLUS and FN/SFN with molecular alteration was 100%, while the probability for AUS/FLUS and FN/SFN without molecular alteration was 7.6%.

In the same year, another study [32] analyzed 174 consecutive FNAC (all categories) for BRAF, RAS, RET, TRK, and PPAR-γ alterations. Mutations were found in 67/235 (28.5%) cytological samples. Of the 67 mutated samples, 23 (34.3%) were mutated by RAS, 33 (49.3%) by BRAF, and 11 (16.4%) by RET/PTC. The presence of mutations at cytology was associated with cancer in 91.1% of the cases and with follicular adenoma in 8.9% of the time. The accuracy of molecular analysis was 90.2%, with a sensitivity of 78.2%, specificity of 96.2%, PPV of 91% and negative predictive value (NPV) of 89.9%. Considering only categories III and IV (*n* = 41), the authors found that 7/41 (17%) samples were mutated (2 BRAF, 2 RET-PTC, 3 RAS). At final histology, all but one (follicular adenoma) were PTC. Of the 34 samples with no mutation, 33 were benign lesions and only one was PTC. Specificity was 97%, sensitivity was 85% and accuracy 95%.

The most complete work aimed to disclose the clinical utility of molecular testing of thyroid FNA samples with indeterminate cytology was published in 2011 [33]. Nikiforov and co-workers analyzed the presence of BRAF, N-H and K-RAS point mutations and RET/PTC1-3, PAX8/PPARγ rearrangements in 1056 consecutive thyroid FNA samples with indeterminate cytology. In 967/1056 (92%) cytologies, the material was adequate for molecular analysis. They found 87 mutations including 62 RAS (71.3%), 19 BRAF (21.8%), 1 RET/PTC (1.1%) and 5 PAX8/PPARγ rearrangements (5.8%). In the AUS/FLUS category, sensitivity was 63%, specificity 99%, PPV 88%, NPV 94% and accuracy 94%. For the FN/SFN group, sensitivity was 57%, specificity 97%, PPV 87%, NPV 86% and accuracy 86%. In AUS/FLUS, FN/SFN categories the detection of any mutation conferred the risk of histological malignancy of 88 and 87%, respectively. The risk of cancer in mutation-negative nodules was 6%, 14%, and 28%, respectively.

In conclusion, mutation panels intended to identify malignancies in indeterminate lesions must include at least BRAF and RAS point mutations (H, K and NRAS), and RET/PTC, PAX8/PPAR-γ rearrangements. Several "homemade" methods comprising PCR with final Sanger sequencing and some commercial kits are available to screen for these alterations with the limitation that they cannot rule out malignancy with a NPV > 95%.

Since the publication of our previous work [32], we applied molecular testing in clinical routine, especially for FNAC categories III and IV. We collected 197 consecutive indeterminate samples and searched for BRAF, RAS (H, K and NRAS), and TERT point mutations, and RET/PTC1-3 and PAX8/PPAR-γ rearrangements. End point PCR, real time PCR, denaturing high performance liquid chromatography (DHPLC) and direct sequencing were used for the analysis [32]. The exam was performed on 176/197 (89.4%) of the sample as in 21/197 (10.6%) the collected material was inadequate for the investigation. We found 17 mutations (9.6%) including 3 BRAF, 2 HRAS, 5 NRAS, 1 KRAS and 6 RET/PTCs. These 17 patients were subjected to surgery and 15/17 (88.2%) were confirmed malignant at final histology (3 FTC, 5 PTC and 7 follicular variant PTC) whereas 2/17 (11.7%) were follicular adenoma (1 NRAS and 1 RET/PTC). Among the 159 nodules negative for mutations, 23 underwent surgery for other reasons (i.e., ultrasound characteristics, patient's decision, increased nodule size over time) and 21/23 (91.3%) were confirmed benign lesions at histology whereas 2/23 (8.6%) were malignant (2 microcarcinomas). The PPV was 88.2% and the NPV was 91.3%, with an accuracy of 90% (Table 1). One-hundred and thirty-six nodules/176 (77.2%) negative for mutation and not subjected to surgery are still under follow up. In a period of time from 1 up to 6 years, no increase in nodule size or changes in ultrasound features were observed. Twenty-two/136 (16.2%) samples repeated a second FNAC and a category II was found for these lesions confirming the results of molecular test. Despite the encouraging results, the method of the "seven genes" has the limitation that collected material can be inadequate to perform the complete panel, thus increasing the number of false negative results.

Table 1. Results from mutation analysis on indeterminate lesions treated with surgery.

Atypia of Undetermined Significance/Follicular Lesions of Undetermined Significance (AUS/FLUS)			
Follicular neoplasma/suspicious for follicular neoplasma (FN/SFN) (n = 40)			
	Histology Malignant	**Histology Benign**	
Mutation positive (n = 17)	7 RAS (6 FVPTC, 1 FTC) 3 BRAF (3 PTC) 5 RET/PTC (1 FVPTC, 2 PTC, 2 FTC)	1 NRAS (FA) 1 RET/PTC (FA)	Sensitivity 88.2% Specificity 91.3% PPV 88.2% NPV 91.3% Accuracy 90%
Mutation negative (n = 23)	2 microcarcinoma	21 (9 FA, 12 HN)	

PTC = papillary thyroid cancer; FTC = follicular thyroid cancer; FVPTC = follicular variant of PTC; FA = follicular adenoma; HN = hyperplastic nodules; PPV = positive predictive value; NPV = negative predictive value.

4.2. Afirma Classifier

The Afirma test is a gene expression classifier (GEC) [34] which uses the expression of 142 genes to categorize thyroid nodules into benign or suspicious (rule out method). The test was validated in a multi-institutional (for a total of 49 clinical sites) prospective double-blind study funded by industry (Veracyte) in indeterminate nodules [35]. Authors obtained 577 cytologically-indeterminate aspirates, 413 of which had corresponding histopathological specimens from excised lesions. After inclusion criteria were met, only 265 aspirated were allocated to GEC and were included in the final analysis [35]. Of these 265, 85 (32%) were confirmed to be malignant at histology. In the 265 indeterminate cytology nodules, the sensitivity of the Afirma test was 92% (95% confidence interval (CI), 84%, 97%, 78/85) and the specificity was 52% (95% CI, 44%, 59%, 93/180). In another study by same authors [36] on 339 cytologically-indeterminate nodules (165 AUS/FLUS; 161 FN; 13 suspicious for malignancy), 174/339 (51%) were GEC benign and 148/339 (44%) were GEC suspicious. Among GEC-suspicious nodules, 121 were surgically removed and 53 (44%) were malignant, confirming the previous study in terms of sensitivity and specificity. Recent studies, have shown results from the GEC classifier in indeterminate cytologies obtaining high sensitivity but lower specificity compared to previous reports thus stressing the need of additional, independent, non-industry supported studies to establish the performance of the classifier [37–39]. In summary, based on the above studies, Afirma test sensitivity has been reported to range from 83% to 100% and specificity from 7 to 52%, where the prevalence of malignancy in histopathologically confirmed study populations has ranged from 17% to 51% [35,37,38,40].

4.3. Thyroseq and Other NGS Platform

Targeted next generation sequencing (NGS) is a promising method to simultaneously examine multiple genes with high sensitivity potentially achieving not only high PPV but also high negative predictive value (NPV) [41] and with low input of starting material (5 to 10 ng).

ThyroSeq is a NGS-based gene mutation and fusion panel initially designed to target 12 cancer genes with 284 mutational hot spots [41], showing 100% accuracy with a sensitivity of 3–5% of mutant alleles. In the first work reporting data from Thyroseq, the authors analyzed 229 thyroid neoplastic and non-neoplastic samples and found mutation in 70% of PTCs (19/27), 83% of papillary thyroid cancer follicular variant (PTCFV) (25/30), 59% of FTC (21/36), 30% of poorly differentiated thyroid carcinoma (3/10), 74% of anaplastic thyroid cancer (ATC) (20/27) and in 73% of medullary thyroid carcinomas (11/15). The majority of samples were mutated for BRAF and RAS. Other studies confirmed the high PPV of the ThyroSeq (88% and 87% for AUS/FLUS and FN, respectively) [42,43] and indicate that the test could potentially be used as a "rule in" test.

In 2014, results from ThyroSeq v2, an enhanced version of the test, on AUS/FLUS and FN cytologies were published [44]. ThyroSeq v2 allowed the analysis of 14 genes (more than 1000 mutations) and RNA alterations (approximately 42 fusions) reaching a sensitivity and specificity of 90% and 93%, respectively, a PPV of 83%, an NPV of 96%, and accuracy of 92% [44]. These results suggested that ThyroSeq v2 may potentially works as both "rule out" and "rule in" test for nodules with indeterminate cytology. Finally, owing to the limited studies and data from literature, the value of ThyroSeq v2 needs further investigation. Moreover, clinical validation results are not available yet, whereas data for lung and other tumors show that next generation sequencing is as robust as Sanger sequencing in routine diagnostics and, in addition, is able to reveal mutations in low percentage and screen the mutational status of different critical samples offering innovative diagnostic opportunities [45–48]. Furthermore, methodological problems like result interpretation (e.g., for unknown mutations), definition of cut offs for mutation calling, and bioinformatics analysis need to be solved and common standard operation procedures (SOPs) need to be defined. On the matter of bioinformatics analysis, a recent report describes a solution called "SeqReport" [49]. This module automatically imports patient data and related NGS run information and allows comprehensive review of all variants by users linking to both COSMIC and dbSNP databases and manual review of variants. In addition, the program automatically locates variants with low frequency or coverage and compares the status of Sanger sequencing confirmation. In this method, the cut off values are determined for each multigene panel during validation. Human errors are minimized and the creation of clinical report is automatic also with appropriate clinical comments.

Le Mercier and colleagues [50] performed a pilot study with a commercially available NGS-based 50-gene panel kit (Ion AmpliSeq Cancer Hotspot Panel version 2; Thermo Fisher Scientific, Gent, Belgium) to evaluate 34 indeterminate FNA samples. The panel is designed to amplify 207 amplicons covering approximately 2800 COSMIC mutations from 50 oncogenes and tumor suppressor genes. The authors identified cytologies with a "molecular test negative" (including patients carrying germline polymorphisms, mutations of unknown clinical significance, or no mutation) or "molecular test positive" for patients carrying pathogenic mutations reaching a sensitivity and specificity of 71% and 89%, respectively. The PPV and NPV were 63% and 92%, respectively, with an accuracy of 85%.

ThyroSeq v2 actually has shown the best results in terms of sensitivity, specificity, PPV and NPV but further studies including a larger number of cases are required for the Ion AmpliSeq Panel.

5. Role of miRNAs in the Differential Diagnosis of Thyroid Lesions

MicroRNA (miRNAs) are small molecules of RNA (approximately 22 nt), non-encoding for protein which negatively regulate gene expression-targeting specific mRNAs [51]. miRNAs can be detected in plasma and serum as they circulate in the blood in a stable, cell-free form [52]. Furthermore, tumor cells have been shown to release miRNAs into the circulation [52] and profiles of miRNAs in plasma and serum have been found to be altered in cancer and other disease states [53–55]. Larger scale miRNA analysis has proven that miRNA expression enables the distinction of benign tissues from their malignant counterparts [56,57]. The mechanisms of miRNA implication in cancer development are linked to downregulation of tumor suppressor genes or upregulation of oncogenes.

Several studies have demonstrated a different miRNA signature between benign and malignant thyroid tissues [58–65] unfortunately using different detection systems (microarray and/or Quantative RT-PCR (Q-RT-PCR) and producing inconsistent results in terms of selected miRNAs, sensitivity and specificity. However, all authors concluded that a limited set of miRNAs can be used for the differential diagnosis between benign and malignant lesions in the surgical samples with high accuracy, implying the potential role of miRNAs in differentiating the nature of thyroid nodules in FNAC. Again, the most important question is question is whether the analysis of miRNAs in cytological samples can improve FNAC results, particularly for indeterminate lesions [61,66–74]. All the studies which addressed this issue obtained a similar diagnostic odds ratio (mean 20.3) and concluded that a set of multiple miRNAs seems to be more sensitive (sensitivity of 87%) than a single miRNA (sensitivity of 71%) although there is discrepancy in terms of set of miRNA proposed. Pooling together the results from these studies, however, a relative small set of 15 miRNAs emerge as the more powerful diagnostic panel for indeterminate lesions. The panel is composed of miRNA7, -146, -146b, -155, -221, -222, -21, -31, -187, -30a-3p, -30d, -146b-5p, -199b-5p, -328 and miRNA197. Future prospective and retrospective research are recommended on a large cohort of indeterminate lesions to validate the diagnostic value of this panel.

As pointed out previously, FNAC represents the gold standard for the differential diagnosis of thyroid nodules, however it is an invasive technique compared to blood sampling. Thus, the idea is to use miRNAs as a serological marker for thyroid cancer (TC) diagnosis from the moment that TC releases miRNAs into the bloodstream. Three studies addressed this issue [75–77], two of them in the Chinese population [77,78] and only one study in the Caucasian population [77]. Although these studies found different set of miRNAs, the preliminary results are promising for future research showing a good sensitivity (ranging from 61.4 to 94%) and specificity (ranging from 57.9% to 98.7%).

We performed a study on serum miRNA expression (miRNA95 and miRNA190) on 982 consecutive patients undergoing FNAC at our institute. We collected serum from 114/982 (11.6%) subjects with a Bethesda III and IV FNAC result. Seventy-five/114 (65.7%) underwent surgery and at final histology we had 11 follicular adenomas (FA), 32 hyperplastic nodules (HN), 27 PTCs, 4 FTCs and 1 Hurthle cell carcinoma (HC). miRNAs were extracted from 200 μL of serum using the miRNeasy Serum/Plasma Kit (Qiagen, Milan, Italy) and retro-transcribed by miScript II RT Kit (Qiagen). Two μL of cDNA was used as template for real time PCR (RT-PCR) to measure miRNA expression levels with the miScript SYBR Green PCR Kit (Qiagen) with specific primers for miRNA-95 and -190 (Qiagen). RT-PCR was performed in duplicate on the Rotor-gene Q MDx (Qiagen) under the following cycling conditions: 95 °C for 15 min, 40 cycles at 94 °C for 15 s, 55 °C for 30 s and 70 °C for 30 s. Relative expression levels were calculated using the $2^{-\Delta\Delta Ct}$ method with the miRNA-16 as endogenous control. miRNA expression correctly identified 38/43 (specificity of 88.4%) of the benign lesions and 27/32 (sensitivity of 84.3%) of malignant sample. We had 5 false positive (FP) (5 HNs) and 5 false negative (FN) (4 PTCs, 1 FTC) results, obtaining a final accuracy of 86.7% (Table 2). Despite the promising results, molecular analysis on FNAC has a better performance and further studies are required to identify the optimal set of circulating miRNAs specific for indeterminate lesions.

Table 2. Results from microRNA (miRNA) expression analysis on indeterminate lesions treated with surgery.

Histology	miRNA Expression Negative for Malignancy	miRNA Expression Positive for Malignancy	Performance
Benign at histology	38 (11 FA, 27 HN)	5 (5 HN)	Sensitivity 84.3% Specificity 88.4%
Malignant at histology	5 (1 FTC, 4 PTC)	27 (23 PTC, 3 FTC, 1 HC)	PPV 88.4% NPV 84.3% Accuracy 86.7%

PTC = papillary thyroid cancer, FTC = follicular thyroid cancer, FA = follicular adenoma, HN = hyperplastic nodules, HC = Hurtlhe cell carcinoma, PPV = positive predictive value, NPV = negative predictive value.

Serum normally contains low amounts of total RNA, of which miRNAs only constitute 0.4–0.5%. In addition, serum samples may be affected by technical problems, such as hemolysis and it is not known whether circulating serum expression can be influenced by other comorbidities. In this view, the analysis on FNAC may be preferable.

In 2016 two studies [78,79] were published on clinical validation of the RosettaGX Reveal test, a miRNA-based assay which evaluates a set of 24 miRNAs (by real time PCR) specific for cytologically indeterminate thyroid nodules. The assay can be used directly on FNA smears and it is able to categorize benign or suspicious nodules even when as little as 1% of thyroid cells is present or less than 5 ng RNA are extracted. The overall NPV reported was 99%, with sensitivity of 98% and specificity of 78%.

6. Proteomics: An Interesting Alternative Approach to Stratify Thyroid FNAC

Proteomics is the large-scale study of proteins and it is widely used to discover cancer biomarkers. In the field of thyroid cancer, proteomics has been initially applied on thyroid tissue specimens and cancer cell lines [80–89], using different techniques such as surface-enhanced laser desorption/ionization-time-of-flight-mass spectrometry (SELDI-TOF-MS), liquid chromatography–mass spectrometry (LC/MS) and MS alone. All these studies ended by identifying specific protein signatures for malignant and benign lesions with a final selection of clusters of proteins with discriminating abilities. In particular, proteins involved in oxidative stress, metabolic pathways, nuclear stability, turnover of thyroglobulin, and kinase signaling are those more represented in thyroid cancer. Techniques such as matrix-assisted laser desorption/ionization (MALDI)-TOF-MS and MALDI-imaging mass spectrometry (MALDI-IMS) have been applied in several studies [90–95] to cytological thyroid specimens. Most of these studies used ex vivo FNA [90,91,93–95] and one study [94] used pre-surgical FNAC obtaining an overall sensitivity of 87% and specificity of 94% in discriminating benign from suspicious samples, with a good reproducibility among studies. Proteomics could serve to improve the preoperative diagnosis of indeterminate lesions, but some aspects such as the limitation in the availability of these technologies and the lack of uniformity among techniques, need to be addressed before its introduction in clinical practice.

7. Conclusions

In summary, the purpose of thyroid molecular testing is to discriminate the nature of thyroid nodules and reduce the diagnostic uncertainty of cytologically indeterminate lesions prior to surgery. Mutation panels intended to identify malignancies must include at least BRAF, and RAS point mutations as well as RET/PTC, NTRK, and PAX8/PPARγ rearrangements. Several "home-made" methods and some commercial kits are available to screen for these alterations with the limitation that they cannot rule out malignancy with an NPV >95%. GEC recognizes benign lesions on the basis of an expression pattern of mRNA extracted from one or two dedicated FNA needle passes. A negative result in the Afirma test has resulted in a major decrease in the number of surgeries performed in samples classified as Bethesda categories III and IV. However, Afirma shows a low PPV. On the other hand, the risk of malignancy calculated by ThyroSeq or other NGS platforms is superior to that of the Afirma, reaching an NPV of 95% or more, with good sensitivity and high PPV. The identification of new biomarkers (i.e., miRNA, proteomic profiles) in the thyroid needs to be corroborated in larger studies with final histology as a gold standard and adequate follow up before use in the clinical routine [96]. Therefore, molecular testing must be always performed in specialized laboratories and results interpreted within the context of the clinical, radiographic, and cytological findings. In addition, clinicians may take into account that the interpretation of molecular testing and its utility are strongly influenced by the prevalence of cancer in each cytological category [96] which can differ among centers. Due to this aspect, molecular test performance may vary significantly.

Acknowledgments: Italian Ministry of Health RF-2011-02350673.

Author Contributions: Silvia Cantara performed molecular analysis, designed and performed experiments on miRNA and wrote the paper. Carlotta Marzocchi performed experiments on miRNA expression. Tania Pilli and Sandro Cardinale review data on miRNA expression. Raffaella Forleo and Maria Grazia Castagna identified indeterminate lesions for molecular analysis and review the data base with results. Furio Pacini review the manuscript.

Conflicts of Interest: The authors declare no conflict of interest.

References

1. Aschebrook-Kilfoy, B.; Ward, M.H.; Sabra, M.M.; Devesa, S.S. Thyroid cancer incidence patterns in the United States by histologic type, 1992–2006. *Thyroid* **2011**, *21*, 125–134. [CrossRef] [PubMed]
2. Belfiore, A.; La Rosa, G.L.; La Porta, G.A.; Giuffrida, D.; Milazzo, G.; Lupo, L.; Regalbuto, C.; Vigneri, R. Cancer risk in patients with cold thyroid nodules: Relevance of iodine intake, sex, age, and multinodularity. *Am. J. Med.* **1992**, *93*, 363–369. [CrossRef]
3. Gharib, H.; Goellner, J.R.; Johnson, D.A. Fine needle aspiration of the thyroid. A 12 year experience with 11,000 biopsies. *Clin. Lab. Med.* **1993**, *13*, 699–709. [PubMed]
4. Bongiovanni, M.; Spitale, A.; Faquin, W.C.; Mazzucchelli, L.; Baloch, Z.W. The Bethesda System for Reporting Thyroid Cytopathology: A meta-analysis. *Acta Cytol.* **2012**, *56*, 333–339. [CrossRef] [PubMed]
5. Hamming, J.F.; Goslings, B.M.; Van Steenis, G.J.; van Ravenswaay Claasen, H.; Hermans, J.; van de Velde, C.J.H. The value of one-needle aspiration biopsy in patients with nodular thyroid disease divided into groups of suspicion of malignant neoplasms on clinical grounds. *Arch. Intern. Med.* **1990**, *150*, 113–116. [CrossRef] [PubMed]
6. Caruso, D.; Mazzaferri, E.L. Fine-needle aspiration in the management of thyroid nodules. *Endocrinologist* **1991**, *1*, 194–202. [CrossRef]
7. Caplan, R.H.; Kisken, W.A.; Strutt, P.J.; Wester, S.M. Fine-needle aspiration biopsy of thyroid nodules: A cost-effective diagnostic plan. *Postgrad. Med.* **1991**, *90*, 183–190. [CrossRef] [PubMed]
8. Hamburger, J.I. Extensive personal experience. Diagnosis of thyroid nodules by fine needle biopsy: Use and abuse. *JCEM* **1994**, *79*, 335–339. [PubMed]
9. Yoshii, T.; Inohara, H.; Takenaka, Y.; Honjo, Y.; Akahani, S.; Nomura, T.; Raz, A.; Kubo, T. Galectin-3 maintains the transformed phenotype of thyroid papillary carcinoma cells. *Int. J. Oncol.* **2001**, *18*, 787–792. [CrossRef] [PubMed]
10. Sapio, M.R.; Guerra, A.; Posca, D.; Limone, P.P.; Deandrea, M.; Motta, M.; Troncone, G.; Caleo, A.; Vallefuoco, P.; Rossi, G.; et al. Combined analysis of galectin-3 and BRAFV600E improves the accuracy of fine-needle aspiration biopsy with cytological findings suspicious for papillary thyroid carcinoma. *Endocr. Relat. Cancer* **2007**, *14*, 1089–1097. [CrossRef] [PubMed]
11. Bryson, P.C.; Shores, C.G.; Hart, C.; Thorne, L.; Patel, M.R.; Richey, L.; Farag, A.; Zanation, A.M. Immunohistochemical distinction of follicular thyroid adenomas and follicular carcinomas. *Arch. Otolaryngol. Head Neck Surg.* **2008**, *134*, 581–586. [CrossRef] [PubMed]

12. Torregrossa, L.; Faviana, P.; Filice, M.E.; Materazzi, G.; Miccoli, P.; Vitti, P.; Fontanini, G.; Melillo, R.M.; Santoro, M.; Basolo, F. CXC chemokine receptor 4 immunodetection in the follicular variant of papillary thyroid carcinoma: Comparison to galectin-3 and hector battifora mesothelial cell-1. *Thyroid* **2010**, *20*, 495–504. [CrossRef] [PubMed]

13. Saggiorato, E.; de Pompa, R.; Volante, M.; Cappia, S.; Arecco, F.; Dei Tos, A.P.; Orlandi, F.; Papotti, M. Characterization of thyroid 'follicular neoplasms' in fine-needle aspiration cytological specimens using a panel of immunohistochemical markers: A proposal for clinical application. *Endocr. Relat. Cancer* **2005**, *12*, 305–317. [CrossRef] [PubMed]

14. Raggio, E.; Camandona, M.; Solerio, D.; Martino, P.; Franchello, A.; Orlandi, F.; Gasparri, G. The diagnostic accuracy of the immunocytochemical markers in the pre-operative evaluation of follicular thyroid lesions. *J. Endocrinol. Investig.* **2010**, *33*, 378–381. [CrossRef]

15. Bartolazzi, A.; Gasbarri, A.; Papotti, M.; Bussolati, G.; Lucante, T.; Khan, A.; Inohara, H.; Marandino, F.; Orlandi, F.; Nardi, F.; et al. Application of an immunodiagnostic method for improving preoperative diagnosis of nodular thyroid lesions. *Lancet* **2001**, *357*, 1644–1650. [CrossRef]

16. Zhang, L.; Krausz, T.; DeMay, R.M. A pilot study of galectin-3, HBME-1, and p27 triple immunostaining pattern for diagnosis of indeterminate thyroid nodules in cytology with correlation to histology. *Appl. Immunohistochem. Mol. Morphol.* **2015**, *23*, 481–490. [CrossRef] [PubMed]

17. Carpi, A.; Naccarato, A.G.; Iervasi, G.; Nicolini, A.; Bevilacqua, G.; Viacava, P.; Collecchi, P.; Lavra, L.; Marchetti, C.; Sciacchitano, S.; et al. Large needle aspiration biopsy and galectin-3 determination in selected thyroid nodules with indeterminate FNA-cytology. *Br. J. Cancer* **2006**, *95*, 204–209. [CrossRef] [PubMed]

18. Franco, C.; Martínez, V.; Allamand, J.P.; Medina, F.; Glasinovic, A.; Osorio, M.; Schachter, D. Molecular markers in thyroid fine-needle aspiration biopsy: A prospective study. *Appl. Immunohistochem. Mol. Morphol.* **2009**, *17*, 211–215. [CrossRef] [PubMed]

19. Das, D.K.; Al-Waheeb, S.K.; George, S.S.; Haji, B.I.; Mallik, MK. Contribution of immunocytochemical stainings for galectin-3, CD44, and HBME1 to fine-needle aspiration cytology diagnosis of papillary thyroid carcinoma. *Diagn. Cytopathol.* **2014**, *42*, 498–505. [CrossRef] [PubMed]

20. Trimboli, P.; Guidobaldi, L.; Amendola, S.; Nasrollah, N.; Romanelli, F.; Attanasio, D.; Ramacciato, G.; Saggiorato, E.; Valabrega, S.; Crescenzi, A. Galectin-3 and HBME-1 improve the accuracy of core biopsy in indeterminate thyroid nodules. *Endocrine* **2016**, *52*, 39–45. [CrossRef] [PubMed]

21. Naor, D.; Sionov, R.V.; Ish-Shalom, D. CD44: Structure, function, and association with the malignant process. *Adv. Cancer Res.* **1997**, *71*, 241–319. [PubMed]

22. Günthert, U.; Hofmann, M.; Rudy, W.; Reber, S.; Zöller, M.; Haussmann, I.; Matzku, S.; Wenzel, A.; Ponta, H.; Herrlich, P. A new variant of glycoprotein CD44 confers metastatic potential to rat carcinoma cells. *Cell* **1991**, *65*, 13–24. [CrossRef]

23. Matesa, N.; Samija, I.; Kusić, Z. Accuracy of fine needle aspiration biopsy with and without the use of tumor markers in cytologically indeterminate thyroid lesions. *Coll. Antropol.* **2010**, *34*, 53–57. [PubMed]

24. Maruta, J.; Hashimoto, H.; Yamashita, H.; Yamashita, H.; Noguchi, S. Immunostaining of galectin-3 and CD44v6 using fine-needle aspiration for distinguishing follicular carcinoma from adenoma. *Diagn Cytopathol.* **2004**, *31*, 392–396. [CrossRef] [PubMed]

25. Haugen, B.R.; Alexander, E.K.; Bible, K.C.; Doherty, G.M.; Mandel, S.J.; Nikiforov, Y.E.; Pacini, F.; Randolph, G.W.; Sawka, A.M.; Schlumberger, M.; et al. 2015 American thyroid association management guidelines for adult patients with thyroid nodules and differentiated thyroid cancer: The American thyroid association guidelines task force on thyroid nodules and differentiated thyroid cancer. *Thyroid* **2016**, *26*, 1–133. [CrossRef] [PubMed]

26. Sobrinho-Simões, M.; Máximo, V.; Rocha, A.S.; Trovisco, V.; Castro, P.; Preto, A.; Lima, J.; Soares, P. Intragenic mutations in thyroid cancer. *Endocrinol. Metab. Clin. N. Am.* **2008**, *37*, 333–362. [CrossRef] [PubMed]

27. Xing, M. Molecular pathogenesis and mechanisms of thyroid cancer. *Nat. Rev. Cancer* **2013**, *13*, 184–199. [CrossRef] [PubMed]

28. Nikiforova, M.N.; Kimura, E.T.; Gandhi, M.; Biddinger, P.W.; Knauf, J.A.; Basolo, F.; Zhu, Z.; Giannini, R.; Salvatore, G.; Fusco, A.; et al. BRAF mutations in thyroid tumors are restricted to papillary carcinomas and anaplastic or poorly differentiated carcinomas arising from papillary carcinomas. *J. Clin. Endocrinol. Metab.* **2003**, *88*, 5399–5404. [CrossRef] [PubMed]

29. Vinagre, J.; Almeida, A.; Pópulo, H.; Batista, R.; Lyra, J.; Pinto, V.; Coelho, R.; Celestino, R.; Prazeres, H.; Lima, L.; et al. Frequency of TERT promoter mutations in human cancers. *Nat. Commun.* **2013**, *4*, 2185. [CrossRef] [PubMed]

30. Melo, M.; da Rocha, A.G.; Vinagre, J.; Batista, R.; Peixoto, J.; Tavares, C.; Celestino, R.; Almeida, A.; Salgado, C.; Eloy, C.; et al. TERT promoter mutations are a major indicator of poor outcome in differentiated thyroid carcinomas. *J. Clin. Endocrinol. Metab.* **2014**, *99*, E754–E765. [CrossRef] [PubMed]

31. Ohori, N.P.; Nikiforova, M.N.; Schoedel, K.E.; LeBeau, S.O.; Hodak, S.P.; Seethala, R.R.; Carty, S.E.; Ogilvie, J.B.; Yip, L.; Nikiforov, Y.E. Contribution of molecular testing to thyroid fine-needle aspiration cytology of "follicular lesion of undetermined significance/atypia of undetermined significance". *Cancer Cytopathol.* **2010**, *118*, 17–23. [CrossRef] [PubMed]

32. Cantara, S.; Capezzone, M.; Marchisotta, S.; Capuano, S.; Busonero, G.; Toti, P.; Di Santo, A.; Caruso, G., Carli, A.F.; Brilli, L.; et al. Impact of proto-oncogene mutation detection in cytological specimens from thyroid nodules improves the diagnostic accuracy of cytology. *JCEM* **2010**, *95*, 1365–1369. [CrossRef] [PubMed]

33. Nikiforov, Y.E.; Ohori, N.P.; Hodak, S.P.; Carty, S.E.; LeBeau, S.O.; Ferris, R.L.; Yip, L.; Seethala, R.R.; Tublin, M.E.; Stang, M.T.; et al. Impact of mutational testing on the diagnosis and management of patients with cytologically indeterminate thyroid nodules: A prospective analysis of 1056 FNA samples. *JCEM* **2011**, *96*, 3390–3397. [CrossRef] [PubMed]

34. Chudova, D.; Wilde, J.I.; Wang, E.T.; Wang, H.; Rabbee, N.; Egidio, C.M.; Reynolds, J.; Tom, E.; Pagan, M.; Rigl, C.T.; et al. Molecular classification of thyroid nodules using high-dimensionality genomic data. *JCEM* **2010**, *95*, 5296–5304. [CrossRef] [PubMed]

35. Alexander, E.K.; Kennedy, G.C.; Baloch, Z.W.; Cibas, E.S.; Chudova, D.; Diggans, J.; Friedman, L.; Kloos, R.T.; LiVolsi, V.A.; Mandel, S.J.; et al. Preoperative diagnosis of benign thyroid nodules with indeterminate cytology. *N. Eng. J. Med.* **2012**, *367*, 705–715. [CrossRef] [PubMed]

36. Alexander, E.K.; Schorr, M.; Klopper, J.; Kim, C.; Sipos, J.; Nabhan, F.; Parker, C.; Steward, D.L.; Mandel, S.J.; Haugen, B.R. Multicenter clinical experience with the Afirma gene expression classifier. *JCEM* **2014**, *99*, 119–125. [CrossRef] [PubMed]

37. Harrell, R.M.; Bimston, D.N. Surgical utility of Afirma: Effects of high cancer prevalence and oncocytic cell types in patients with indeterminate thyroid cytology. *Endocr. Pract.* **2014**, *20*, 364–369. [CrossRef] [PubMed]

38. Lastra, R.R.; Pramick, M.R.; Crammer, C.J.; LiVolsi, V.A.; Baloch, Z.W. Implications of a suspicious afirma test result in thyroid fine-needle aspiration cytology: An institutional experience. *Cancer Cytopathol.* **2014**, *122*, 737–744. [CrossRef] [PubMed]

39. Marti, J.L.; Avadhani, V.; Donatelli, L.A.; Niyogi, S.; Wang, B.; Wong, R.J.; Shaha, A.R.; Ghossein, R.A.; Lin, O.; Morris, L.G.; et al. Wide inter-institutional variation in performance of a molecular classifier for indeterminate thyroid nodules. *Ann. Surg. Oncol.* **2015**, *22*, 3996–4001. [CrossRef] [PubMed]

40. McIver, B.; Castro, M.R.; Morris, J.C.; Bernet, V.; Smallridge, R.; Henry, M.; Kosok, L.; Reddi, H. An independent study of a gene expression classifier (Afirma) in the evaluation of cytologically indeterminate thyroid nodules. *JCEM* **2014**, *99*, 4069–4077. [CrossRef] [PubMed]

41. Nikiforova, M.N.; Wald, A.I.; Roy, S.; Durso, M.B.; Nikiforov, Y.E. Targeted next-generation sequencing panel (ThyroSeq) for detection of mutations in thyroid cancer. *JCEM* **2013**, *98*, E1852–E1860. [CrossRef] [PubMed]

42. Beaudenon-Huibregtse, S.; Alexander, E.K.; Guttler, R.B.; Hershman, J.M.; Babu, V.; Blevins, T.C.; Moore, P.; Andruss, B.; Labourier, E. Centralized molecular testing for oncogenic gene mutations complements the local cytopathologic diagnosis of thyroid nodules. *Thyroid* **2014**, *24*, 1479–1487. [CrossRef] [PubMed]

43. Eszlinger, M.; Krogdahl, A.; Münz, S.; Rehfeld, C.; Precht Jensen, E.M.; Ferraz, C.; Bösenberg, E.; Drieschner, N.; Scholz, M.; Hegedüs, L.; et al. Impact of molecular screening for point mutations and rearrangements in routine air-dried fine-needle aspiration samples of thyroid nodules. *Thyroid* **2014**, *24*, 305–313. [CrossRef] [PubMed]

44. Nikiforov, Y.E.; Carty, S.E.; Chiosea, S.I.; Coyne, C.; Duvvuri, U.; Ferris, R.L.; Gooding, W.E.; Hodak, S.P.; LeBeau, S.O.; Ohori, N.P.; et al. Highly accurate diagnosis of cancer in thyroid nodules with follicular neoplasm/suspicious for a follicular neoplasm cytology by ThyroSeq v2 next-generation sequencing assay. *Cancer* **2014**, *120*, 3627–3634. [CrossRef] [PubMed]

45. Coco, S.; Truini, A.; Vanni, I.; Dal Bello, M.G.; Alama, A.; Rijavec, E.; Genova, C.; Barletta, G.; Sini, C.; Burrafato, G.; et al. Next generation sequencing in non-small cell lung cancer: New avenues toward the personalized medicine. *Curr. Drug Targets* **2015**, *16*, 47–59. [CrossRef] [PubMed]

46. Malapelle, U.; Vigliar, E.; Sgariglia, R.; Bellevicine, C.; Colarossi, L.; Vitale, D.; Pallante, P.; Troncone, G. Ion Torrent next-generation sequencing for routine identification of clinically relevant mutations in colorectal cancer patients. *J. Clin. Pathol.* **2015**, *68*, 64–68. [CrossRef] [PubMed]

47. Chevrier, S.; Arnould, L.; Ghiringhelli, F.; Coudert, B.; Fumoleau, P.; Boidot, R. Next-generation sequencing analysis of lung and colon carcinomas reveals a variety of genetic alterations. *Int. J. Oncol.* **2014**, *45*, 1167–1174. [CrossRef] [PubMed]

48. Ross, J.S.; Badve, S.; Wang, K.; Sheehan, C.E.; Boguniewicz, A.B.; Otto, G.A.; Yelensky, R.; Lipson, D.; Ali, S.; Morosini, D.; et al. Genomic profiling of advanced-stage, metaplastic breast carcinoma by next-generation sequencing reveals frequent, targetable genomic abnormalities and potential new treatment options. *Arch. Pathol. Lab. Med.* **2015**, *139*, 642–649. [CrossRef] [PubMed]

49. Roy, S.; Durso, M.B.; Wald, A.; Nikiforov, Y.E.; Nikiforova, M.N. SeqReporter: Automating next-generation sequencing result interpretation and reporting workflow in a clinical laboratory. *J. Mol. Diagn.* **2014**, *16*, 11–22. [CrossRef] [PubMed]

50. Le Mercier, M.; D'Haene, N.; de Nève, N.; Blanchard, O.; Degand, C.; Rorive, S.; Salmon, I. Next-generation sequencing improves the diagnosis of thyroid FNA specimens with indeterminate cytology. *Histopathology* **2015**, *66*, 215–224. [CrossRef] [PubMed]

51. Carthew, R.W.; Sontheimer, E.J. Origins and Mechanisms of miRNAs and siRNAs. *Cell* **2009**, *136*, 642–655. [CrossRef] [PubMed]

52. Mitchell, P.S.; Parkin, R.K.; Kroh, E.M.; Fritz, B.R.; Wyman, S.K.; Pogosova-Agadjanyan, E.L.; Peterson, A.; Noteboom, J.; O'Briant, K.C.; Allen, A.; et al. Circulating microRNAs as stable blood-based markers for cancer detection. *Proc. Natl. Acad. Sci. USA* **2008**, *105*, 10513–10518. [CrossRef] [PubMed]

53. Chen, X.; Ba, Y.; Ma, L.; Cai, X.; Yin, Y.; Wang, K.; Guo, J.; Zhang, Y.; Chen, J.; Guo, X.; et al. Characterization of microRNAs in serum: A novel class of biomarkers for diagnosis of cancer and other diseases. *Cell Res.* **2008**, *18*, 997–1006. [CrossRef] [PubMed]

54. Lawrie, C.H.; Gal, S.; Dunlop, H.M.; Pushkaran, B.; Liggins, A.P.; Pulford, K.; Banham, A.H.; Pezzella, F.; Boultwood, J.; Wainscoat, J.S.; et al. Detection of elevated levels of tumour-associated microRNAs in serum of patients with diffuse large B-cell lymphoma. *Br. J. Haematol.* **2008**, *141*, 672–675. [CrossRef] [PubMed]

55. Taylor, D.D.; Gercel-Taylor, C. MicroRNA signatures of tumor-derived exosomes as diagnostic biomarkers of ovarian cancer. *Gynecol. Oncol.* **2008**, *110*, 13–21. [CrossRef] [PubMed]

56. Lu, J.; Getz, G.; Miska, E.A.; Alvarez-Saavedra, E.; Lamb, J.; Peck, D.; Sweet-Cordero, A.; Ebert, B.L.; Mak, R.H.; Ferrando, A.A.; et al. MicroRNA expression profiles classify human cancers. *Nature* **2005**, *435*, 834–838. [CrossRef] [PubMed]

57. Volinia, S.; Calin, G.A.; Liu, C.G.; Ambs, S.; Cimmino, A.; Petrocca, F.; Visone, R.; Iorio, M.; Roldo, C.; Ferracin, M.; et al. A microRNA expression signature of human solid tumors defines cancer gene targets. *Proc. Natl. Acad. Sci. USA* **2006**, *103*, 2257–2261. [CrossRef] [PubMed]

58. Tetzlaff, M.T.; Liu, A.; Xu, X.; Master, S.R.; Baldwin, D.A.; Tobias, J.W.; Livolsi, V.A.; Baloch, Z.W. Differential expression of miRNAs in papillary thyroid carcinoma compared to multinodular goiter using formalin fixed paraffin embedded tissues. *Endocr. Pathol.* **2007**, *18*, 163–173. [CrossRef] [PubMed]

59. Chen, Y.T.; Kitabayashi, N.; Zhou, X.K.; Fahey, T.J., 3rd; Scognamiglio, T. MicroRNA analysis as a potential diagnostic tool for papillary thyroid carcinoma. *Mod. Pathol.* **2008**, *21*, 1139–1146. [CrossRef] [PubMed]

60. He, H.; Jazdzewski, K.; Li, W.; Liyanarachchi, S.; Nagy, R.; Volinia, S.; Calin, G.A.; Liu, C.G.; Franssila, K.; Suster, S.; et al. The role of microRNA genes in papillary thyroid carcinoma. *Proc. Natl. Acad. Sci. USA* **2005**, *102*, 19075–19080. [CrossRef] [PubMed]

61. Nikifororva, M.N.; Tseng, G.C.; Steward, D.; Diorio, D.; Nikiforov, Y. MicroRNA expression profiling of thyroid tumors: Biological significance and diagnostic utility. *JCEM* **2008**, *93*, 1600–1608.

62. Swierniak, M.; Wojcicka, A.; Czetwertynska, M.; Stachlewska, E.; Maciag, M.; Wiechno, W.; Gornicka, B.; Bogdanska, M.; Koperski, L.; de la Chapelle, A.; et al. In-depth characterization of the microRNA transcriptome in normal thyroid and papillary thyroid carcinoma. *JCEM* **2013**, *98*, E1401–E1409. [CrossRef] [PubMed]

63. Mancikova, V.; Castelblanco, E.; Pineiro-Yanez, E.; Perales-Paton, J.; de Cubas, A.A.; Inglada-Perez, L.; Matias-Guiu, X.; Capel, I.; Bella, M.; Lerma, E.; et al. MicroRNA deep-sequencing reveals master regulators of follicular and papillary thyroid tumors. *Mod. Pathol.* **2015**, *28*, 748–757. [CrossRef] [PubMed]

64. Kitano, M.; Rahbari, R.; Patterson, E.E.; Xiong, Y.; Prasad, N.B.; Wang, Y.; Zeiger, M.A.; Kebebew, E. Expression profiling of difficult-to-diagnose thyroid histologic subtypes shows distinct expression profiles and identify candidate diagnostic microRNAs. *Ann. Surg. Oncol.* **2011**, *18*, 3443–3452. [CrossRef] [PubMed]

65. Yip, L.; Kelly, L.; Shuai, Y.; Armstrong, M.J.; Nikiforov, Y.E.; Carty, S.E.; Nikiforova, M.N. MicroRNA signature distinguishes the degree of aggressiveness of papillary thyroid carcinoma. *Ann. Surg. Oncol.* **2011**, *18*, 2035–2041. [CrossRef] [PubMed]

66. Mazeh, H.; Levy, Y.; Mizrahi, I.; Appelbaum, L.; Ilyayev, N.; Halle, D.; Freund, H.R.; Nissan, A. Differentiating benign from malignant thyroid nodules using micro ribonucleic acid amplification in residual cells obtained by fine needle aspiration biopsy. *J. Surg. Res.* **2013**, *180*, 216–221. [CrossRef] [PubMed]

67. Agretti, P.; Ferrarini, E.; Rago, T.; Candelieri, A.; de Marco, G.; Dimida, A.; Niccolai, F.; Molinaro, A.; Di Coscio, G.; Pinchera, A.; et al. MicroRNA expression profile helps to distinguish benign nodules from papillary thyroid carcinomas starting from cells of fine-needle aspiration. *Eur. J. Endocrinol.* **2012**, *167*, 393–400. [CrossRef] [PubMed]

68. Wei, W.J.; Shen, C.T.; Song, H.J.; Qiu, Z.L.; Luo, Q.Y. MicroRNAs as a potential tool in the differential diagnosis of thyroid cancer: A systematic review and meta-analysis. *Clin. Endocrinol.* **2016**, *84*, 127–133. [CrossRef] [PubMed]

69. Kitano, M.; Rahbari, R.; Patterson, E.E.; Steinberg, S.M.; Prasad, N.B.; Wang, Y.; Zeiger, M.A.; Kebebew, E. Evaluation of candidate diagnostic microRNAs in thyroid fine-needle aspiration biopsy samples. *Thyroid* **2012**, *22*, 285–291. [CrossRef] [PubMed]

70. Keutgen, X.M.; Filicori, F.; Crowley, M.J.; Wang, Y.; Scognamiglio, T.; Hoda, R.; Buitrago, D.; Cooper, D.; Zeiger, M.A.; Zarnegar, R.; et al. A panel of four miRNAs accurately differentiates malignant from benign indeterminate thyroid lesions on fine needle aspiration. *Clin. Cancer Res.* **2012**, *18*, 2032–2038. [CrossRef] [PubMed]

71. Mazeh, H.; Mizrahi, I.; Halle, D.; Ilyayev, N.; Stojadinovic, A.; Trink, B.; Mitrani-Rosenbaum, S.; Roistacher, M.; Ariel, I.; Eid, A.; et al. Development of a microRNA-based molecular assay for the detection of papillary thyroid carcinoma in aspiration biopsy samples. *Thyroid* **2011**, *21*, 111–118. [CrossRef] [PubMed]

72. Panebianco, F.; Mazzanti, C.; Tomei, S.; Aretini, P.; Franceschi, S.; Lessi, F.; Di Coscio, G.; Bevilacqua, G.; Marchetti, I. The combination of four molecular markers improves thyroid cancer cytologic diagnosis and patient management. *BMC Cancer* **2015**, *19*, 918. [CrossRef] [PubMed]

73. Vriens, M.R.; Weng, J.; Suh, I.; Huynh, N.; Guerrero, M.A.; Shen, W.T.; Duh, Q.Y.; Clark, O.H.; Kebebew, E. MicroRNA expression profiling is a potential diagnostic tool for thyroid cancer. *Cancer* **2012**, *118*, 3426–3432. [CrossRef] [PubMed]

74. Ludvíková, M.; Kalfeřt, D.; Kholová, I. Pathobiology of MicroRNAs and Their Emerging Role in Thyroid Fine-Needle Aspiration. *Acta Cytol.* **2015**, *59*, 435–444. [CrossRef] [PubMed]

75. Lee, Y.S.; Lim, Y.S.; Lee, J.C.; Wang, S.G.; Park, H.Y.; Kim, S.Y.; Lee, B.J. Differential expression levels of plasma-derived miR-146b and miR-155 in papillary thyroid cancer. *Oral Oncol.* **2015**, *51*, 77–83. [CrossRef] [PubMed]

76. Yu, S.; Liu, Y.; Wang, J.; Guo, Z.; Zhang, Q.; Yu, F.; Zhang, Y.; Huang, K.; Li, Y.; Song, E.; et al. Circulating microRNA profiles as potential biomarkers for diagnosis of papillary thyroid carcinoma. *JCEM* **2012**, *97*, 2084–2092. [CrossRef] [PubMed]

77. Cantara, S.; Pilli, T.; Sebastiani, G.; Cevenini, G.; Busonero, G.; Cardinale, S.; Dotta, F.; Pacini, F. Circulating miRNA95 and miRNA190 are sensitive markers for the differential diagnosis of thyroid nodules in a Caucasian population. *JCEM* **2014**, *99*, 4190–4198. [CrossRef] [PubMed]

78. Lithwick-Yanai, G.; Dromi, N.; Shtabsky, A.; Morgenstern, S.; Strenov, Y.; Feinmesser, M.; Kravtsov, V.; Leon, M.; Hajdúch, M.; Ali, S.Z.; et al. Multicentre validation of a microRNA-based assay for diagnosing indeterminate thyroid nodules utilising fine needle aspirate smears. *J. Clin. Pathol.* **2016**. [CrossRef] [PubMed]

79. Benjamin, H.; Schnitzer-Perlman, T.; Shtabsky, A.; VandenBussche, C.J.; Ali, S.Z.; Kolar, Z.; Pagni, F.; Rosetta Genomics Group; Bar, D.; Meiri, E. Analytical validity of a microRNA-based assay for diagnosing indeterminate thyroid FNA smears from routinely prepared cytology slides. *Cancer Cytopathol.* **2016**, *124*, 711–721. [CrossRef] [PubMed]

80. Suriano, R.; Lin, Y.; Ashok, B.T.; Schaefer, S.D.; Schantz, S.P.; Geliebter, J.; Tiwari, R.K. Pilot study using SELDI-TOF-MS based proteomic profile for the identification of diagnostic biomarkers of thyroid proliferative diseases. *J. Proteome Res.* **2006**, *5*, 856–861. [CrossRef] [PubMed]

81. Torres-Cabala, C.; Bibbo, M.; Panizo-Santos, A.; Barazi, H.; Krutzsch, H.; Roberts, D.D.; Merino, M.J. Proteomic identification of new biomarkers and application in thyroid cytology. *Acta Cytol.* **2006**, *50*, 518–528. [CrossRef] [PubMed]

82. Brown, L.M.; Helmke, S.M.; Hunsucker, S.W.; Netea-Maier, R.T.; Chiang, S.A.; Heinz, D.E.; Shroyer, K.R.; Duncan, M.W.; Haugen, B.R. Quantitative and qualitative differences in protein expression between papillary thyroid carcinoma and normal thyroid tissue. *Mol. Carcinog.* **2006**, *45*, 613–626. [CrossRef] [PubMed]

83. Krause, K.; Karger, S.; Schierhorn, A.; Poncin, S.; Many, M.C.; Fuhrer, D. Proteomic profiling of cold thyroid nodules. *Endocrinology* **2007**, *148*, 1754–1763. [CrossRef] [PubMed]

84. Puxeddu, E.; Susta, F.; Orvietani, P.L.; Chiasserini, D.; Barbi, F.; Moretti, S.; Cavaliere, A.; Santeusanio, F.; Avenia, N.; Binaglia, L. Identification of differentially expressed proteins in papillary thyroid carcinomas with V600E mutation of BRAF. *Proteom. Clin. Appl.* **2007**, *1*, 672–680. [CrossRef] [PubMed]

85. Netea-Maier, R.T.; Hunsucker, S.W.; Hoevenaars, B.M.; Helmke, S.M.; Slootweg, P.J.; Hermus, A.R.; Haugen, B.R.; Duncan, M.W. Discovery and validation of protein abundance differences between follicular thyroid neoplasms. *Cancer Res.* **2008**, *68*, 1572–1580. [CrossRef] [PubMed]

86. Musso, R.; di Cara, G.; Albanese, N.N.; Marabeti, M.R.; Cancemi, P.; Martini, D.; Orsini, E.; Giordano, C.; Pucci-Minafra, I. Differential proteomic and phenotypic behaviour of papillary and anaplastic thyroid cell lines. *J. Proteom.* **2013**, *90*, 115–125. [CrossRef] [PubMed]

87. Chaker, S.; Kashat, L.; Voisin, S.; Kaur, J.; Kak, I.; MacMillan, C.; Ozcelik, H.; Siu, K.W.; Ralhan, R.; Walfish, P.G. Secretome proteins as candidate biomarkers for aggressive thyroid carcinomas. *Proteomics* **2013**, *13*, 771–787. [CrossRef] [PubMed]

88. Pagni, F.; L'Imperio, V.; Bono, F.; Garancini, M.; Roversi, G.; de Sio, G.; Galli, M.; Smith, A.J.; Chinello, C.; Magni, F. Proteome analysis in thyroid pathology. *Expert Rev. Proteom.* **2015**, *12*, 375–390. [CrossRef] [PubMed]

89. Galli, M.; Pagni, F.; de Sio, G.; Smith, A.; Chinello, C.; Stella, M.; L'Imperio, V.; Manzoni, M.; Garancini, M.; Massimini, D.; et al. Proteomic profiles of thyroid tumors by mass spectrometry-imaging on tissue microarrays. *Biochim. Biophys. Acta* **2016**. [CrossRef] [PubMed]

90. Giusti, L.; Iacconi, P.; Ciregia, F.; Giannaccini, G.; Donatini, G.L.; Basolo, F.; Miccoli, P.; Pinchera, A.; Lucacchini, A. Fine-needle aspiration of thyroid nodules: Proteomic analysis to identify cancer biomarkers. *J. Proteome Res.* **2008**, *7*, 4079–4088. [CrossRef] [PubMed]

91. Giusti, L.; Iacconi, P.; Ciregia, F.; Giannaccini, G.; Basolo, F.; Donatini, G.; Miccoli, P.; Lucacchini, A. Proteomic analysis of human thyroid fine needle aspiration fluid I. *J. Endocrinol. Nvestig.* **2007**, *30*, 865–869. [CrossRef] [PubMed]

92. Ciregia, F.; Giusti, L.; Molinaro, A.; Niccolai, F.; Agretti, P.; Rago, T.; Di Coscio, G.; Vitti, P.; Basolo, F.; Iacconi, P.; et al. Presence in the pre-surgical fine-needle aspiration of potential thyroid biomarkers previously identified in the post-surgical one. *PLoS ONE* **2013**, *8*, e72911. [CrossRef] [PubMed]

93. Mainini, V.; Pagni, F.; Garancini, M.; Giardini, V.; de Sio, G.; Cusi, C.; Arosio, C.; Roversi, G.; Chinello, C.; Caria, P.; et al. An alternative approach in endocrine pathology research: MALDI-IMS in papillary thyroid carcinoma. *Endocr. Pathol.* **2013**, *24*, 250–253. [CrossRef] [PubMed]

94. Pagni, F.; Mainini, V.; Garancini, M.; Bono, F.; Vanzati, A.; Giardini, V.; Scardilli, M.; Goffredo, P.; Smith, A.J.; Galli, M.; et al. Proteomics for the diagnosis of thyroid lesions: Preliminary report. *Cytopathology* **2015**, *26*, 318–324. [CrossRef] [PubMed]

95. Pagni, F.; de Sio, G.; Garancini, M.; Scardilli, M.; Chinello, C.; Smith, A.J.; Bono, F.; Leni, D.; Magni, F. Proteomics in thyroid cytopathology: Relevance of MALDI-imaging in distinguishing malignant from benign lesions. *Proteomics* **2016**, *16*, 1775–1784. [CrossRef] [PubMed]

96. Ferris, R.L.; Baloch, Z.; Bernet, V.; Chen, A.; Fahey, T.J., 3rd; Ganly, I.; Hodak, S.P.; Kebebew, E.; Patel, K.N.; Shaha, A.; et al. American thyroid association statement on surgical application of molecular profiling for thyroid nodules: Current impact on perioperative decision making. *Thyroid* **2015**, *25*, 760–768. [CrossRef] [PubMed]

International Journal of
Molecular Sciences

MDPI

Article

Diagnostic Limitation of Fine-Needle Aspiration (FNA) on Indeterminate Thyroid Nodules Can Be Partially Overcome by Preoperative Molecular Analysis: Assessment of *RET/PTC1* Rearrangement in *BRAF* and *RAS* Wild-Type Routine Air-Dried FNA Specimens

Young Sin Ko [1,2], Tae Sook Hwang [2,3,*], Ja Yeon Kim [3], Yoon-La Choi [4], Seung Eun Lee [5], Hye Seung Han [2,3], Wan Seop Kim [2,3], Suk Kyeong Kim [6] and Kyoung Sik Park [7]

[1]   Diagnostic Pathology Center, Seegene Medical Foundation, Seoul KS013, Korea; noteasy@mf.seegene.com
[2]   Molecular Genetics and Pathology, Department of Medicine, Graduate School of Konkuk University, Seoul KS013, Korea; aphsh@kuh.ac.kr (H.S.H.); wskim@kuh.ac.kr (W.S.K.)
[3]   Department of Pathology, Konkuk University School of Medicine, Seoul KS013, Korea; 78jykim@hanmail.net
[4]   Department of Pathology and Translational Genomics, Samsung Medical Center, Sungkyunkwan University School of Medicine, Seoul KS013, Korea; yla.choi@samsung.com
[5]   Department of Pathology, Konkuk University Medical Center, Seoul KS013, Korea; 20150063@kuh.ac.kr
[6]   Department of Internal Medicine, Konkuk University School of Medicine, Seoul KS013, Korea; endolife@kuh.ac.kr
[7]   Department of Surgery, Konkuk University School of Medicine, Seoul KS013, Korea; kspark@kuh.ac.kr
*   Correspondence: tshwang@kuh.ac.kr; Tel.: +82-2-2030-5641; Fax: +82-2-2030-5629

Academic Editor: Daniela Gabriele Grimm
Received: 21 February 2017; Accepted: 7 April 2017; Published: 12 April 2017

Abstract: Molecular markers are helpful diagnostic tools, particularly for cytologically indeterminate thyroid nodules. Preoperative *RET/PTC1* rearrangement analysis in *BRAF* and *RAS* wild-type indeterminate thyroid nodules would permit the formulation of an unambiguous surgical plan. Cycle threshold values according to the cell count for detection of the *RET/PTC1* rearrangement by real-time reverse transcription-polymerase chain reaction (RT-PCR) using fresh and routine air-dried TPC1 cells were evaluated. The correlation of *RET/PTC1* rearrangement between fine-needle aspiration (FNA) and paired formalin-fixed paraffin-embedded (FFPE) specimens was analyzed. *RET/PTC1* rearrangements of 76 resected *BRAF* and *RAS* wild-type classical PTCs were also analyzed. Results of RT-PCR and the Nanostring were compared. When 100 fresh and air-dried TPC1 cells were used, expression of *RET/PTC1* rearrangement was detectable after 35 and 33 PCR cycles, respectively. The results of *RET/PTC1* rearrangement in 10 FNA and paired FFPE papillary thyroid carcinoma (PTC) specimens showed complete correlation. Twenty-nine (38.2%) of 76 *BRAF* and *RAS* wild-type classical PTCs had *RET/PTC1* rearrangement. Comparison of *RET/PTC1* rearrangement analysis between RT-PCR and the Nanostring showed moderate agreement with a κ value of 0.56 ($p = 0.002$). The *RET/PTC1* rearrangement analysis by RT-PCR using routine air-dried FNA specimen was confirmed to be technically applicable. A significant proportion (38.2%) of the *BRAF* and *RAS* wild-type PTCs harbored *RET/PTC1* rearrangements.

Keywords: *RET/PTC* gene rearrangement; air-dried FNA specimen; RT-PCR; Nanostring

1. Introduction

The evaluation of a thyroid nodule is a very common clinical problem. Epidemiologic studies have shown the prevalence of palpable thyroid nodules to be approximately 5% in women and 1% in men living in iodine-sufficient parts of the world [1,2]. In contrast, high-resolution ultrasound (US) can detect thyroid nodules in 19–68% of randomly selected individuals, with higher frequencies in women and the elderly [3,4]. The clinical importance of thyroid nodules rests with the need to exclude thyroid cancer, which occurs in 7–15% of cases depending on age, sex, radiation exposure history, family history, and other factors [5,6]. Differentiated thyroid cancer (DTC) includes papillary and follicular cancer, and comprises the vast majority (>90%) of all thyroid cancers [7]. In the United States, approximately 63,000 new cases of thyroid cancer were predicted to be diagnosed in 2014 [8] compared with 37,200 in 2009 when the last ATA guidelines were published. The yearly incidence has nearly tripled from 4.9 per 100,000 in 1975 to 14.3 per 100,000 in 2009 [9].

The most prevalent type of thyroid malignancy in Korea is papillary thyroid carcinoma (PTC), which constitutes more than 97% of the cases, followed by follicular thyroid carcinoma (FTC), comprising 1.5% of the thyroid cancer [10]. Compared to Western countries, the prevalence of PTC is much higher. Therefore, the evaluation of a thyroid nodule in Korea is primarily a search for PTC.

Fine-needle aspiration (FNA) is the safest and most reliable test that can provide a definitive preoperative diagnosis of malignancy [11]. The sensitivity and specificity of FNA are reported to be 68–98% and 56–100%, respectively [12]. However, 15–30% of thyroid FNA diagnoses are "atypia of undetermined significance (AUS)/follicular lesion of undetermined significance (FLUS)", "follicular neoplasm or suspicious for follicular neoplasm (FN/SFN)", and "suspicious for malignancy" [13]. This leads to an increased rate of unnecessary surgery, as only about 25% of the indeterminate cases will receive a postoperative malignant diagnosis by histological examination [11]. Moreover, patients with a diagnosis of indeterminate category usually undergo hemithyroidectomy, and about 25% of the patients need to have a second stage completion thyroidectomy in most centers [12]. Two-stage surgery has higher morbidity than initial total thyroidectomy undertaken with a definitive malignant diagnosis on FNA. Preoperative molecular analysis using a panel of genetic alterations would overcome the limitation of FNA diagnosis. The most common genetic alteration in thyroid cancer is the activation of the mitogen-activated protein kinase pathway. Activation of this pathway occurs through mutually exclusive mutations of the *BRAF* and *RAS* genes and rearrangements of the *RET/PTC* and *NTRK*. The overall prevalence of the *BRAF* mutations is approximately 45% (range, 27.3–87.1%) [14,15], with a significantly higher prevalence in Asia—especially Korea—relative to Western countries [15–17]. The mutations of the *RAS* genes are the second most common genetic alterations in thyroid tumors, and are mostly present in follicular-patterned lesions. The prevalence of *RAS* mutations in follicular variant of papillary thyroid carcinoma (FVPTC) varies from 26.5% to 33.3% in Korea, where most of the follicular patterned thyroid malignancy is FVPTC [18,19].

RET proto-oncogene rearrangements are commonly seen in PTC. These rearrangements play a role in pathogenesis of PTC, and derive from the fusion of the *RET* tyrosine kinase domain sequence with 50 sequences of heterologous genes. The resulting chimeric oncogenes are termed *RET/PTCs* [20–24]. *RET/PTC* rearrangements are typically common in tumors from patients with a history of radiation exposure (50–80%) and PTC of children and young adults (40–70%) [25,26]. The distribution of *RET/PTC* rearrangements within this tumor is quite heterogeneous, and varies from the involvement of almost all neoplastic cells to presence in only a small fraction of the tumor cells [27,28]. To date, 13 different types of *RET/PTC* rearrangements have been reported; *RET/PTC1* and *RET/PTC3* account for more than 90% of all rearrangements.

The prevalence of the *RET/PTC* rearrangements in PTC varies widely in different populations (range, 0–86.8%) [29–31], with significant variability in mutational frequency—even within the same geographical regions. Rates of 0–54.5% have been reported in Asia [30–32], 2.4–72.0% in the United States [17,33], and 8.1–42.9% in Europe [34,35]. The marked variations may reflect the small size of the studies, geographic variability, or different sensitivities of the detection methods [36,37].

When this variability is considered, the prevalence of *RET/PTC* rearrangements in Asia is generally low [29–32,38,39]. The subclonal occurrence of *RET/PTC* rearrangement in PTC can influence the sensitivity of some methods, and might explain why the reported prevalence of *RET/PTC* rearrangements in PTCs varies in different studies. Very recent studies demonstrated that *RET/PTC* rearrangements in benign thyroid nodules are not an uncommon occurrence, and suggested that its presence could be associated with a faster nodular enlargement [40–42]. A variety of methods have been used to identify *RET/PTC* rearrangements. These include real-time reverse transcription-polymerase chain reaction (RT-PCR), Southern blot analysis, fluorescence in situ hybridization, and NanoString nCounter Gene Expression Assay.

Most preoperative detection of these rearrangements has been performed in fresh FNA material. Recently, detection of the *PAX8/PPARG* and *RET/PTC* rearrangements in routine air-dried FNA samples was reported [43–48]. The FNA approach suffers from the limitation that indeterminate FNA specimens usually contain small numbers of atypical cells, and these cells are often mixed with many inflammatory cells, benign follicular cells, and stromal cells. Therefore, harvesting the cells of interest is the key step in molecular analysis of the FNA specimen.

Preoperative *RET/PTC1* rearrangement analysis in *BRAF* and *RAS* wild-type indeterminate thyroid nodules would permit the formulation of an unambiguous surgical plan, while foregoing the need for other less-specific diagnostic tests like repeat FNA and intraoperative frozen section evaluation. We have previously reported the value of the preoperative *BRAF* and *RAS* mutation analysis in diagnosing PTC in routine air-dried FNA specimens [18,49–51]. In our institution, we recommend surgery for *BRAF* or *RAS*-positive thyroid nodules with preoperative cytological diagnosis of AUS/FLUS and FN/SFN categories, and have been able to detect considerable numbers of PTCs in cytologically-indeterminate nodules [50]. Considering that 88% of the PTCs harbor either a *BRAF* or a *RAS* mutation (Thyroid, 2017, Epub ahead of time), we hypothesized that detection of *RET/PTC* rearrangements on *BRAF* and *RAS* mutation wild-type FNA specimens of the indeterminate thyroid nodules will improve the diagnostic yield of PTC. An algorithmic approach is cost-effective and efficient—especially in *BRAF* mutation-prevalent populations.

In this study, we investigated the clinical feasibility of preoperative *RET/PTC1* rearrangement analysis as an ancillary diagnostic tool in routine air-dried FNA samples. We also evaluated the *RET/PTC1* rearrangement status for 76 *BRAF* and *RAS* wild-type classical PTC cases.

2. Results

2.1. Detection of the RET/PTC1 Rearrangement in a Fresh TPC1 Cell Line

The C_t value was increased when the cell numbers used for analysis were decreased and showed an inverse correlation (Table 1 and Figure 1). *RET/PTC1* rearrangement was detectable after 35 PCR cycles when 100 TPC1 cells were used.

Table 1. C_t values and cell counts of *RET/PTC1* rearrangement analysis by RT-PCR using fresh cultured TPC1 cells.

Cell Number	RET/PTC1 (C_t)	GAPDH (C_t)
1000	30.3	23.1
500	31.2	24.2
250	34.1	27.1
100	35.6	29.4
50	36.7	30.6

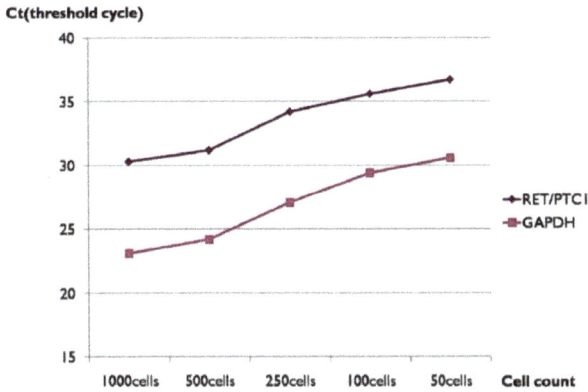

Figure 1. Threshold cycle (C_t) values and cell counts of *RET/PTC1* rearrangement analysis by real-time reverse transcription-polymerase chain reaction (RT-PCR) using fresh cultured TPC1 cells. *GAPDH*: glyceraldehyde-3-phosphate dehydrogenase.

2.2. Detection of the RET/PTC1 Rearrangement in Routine Air-Dried TPC1 Cell Line

When *RET/PTC1* rearrangement was analyzed using various numbers of smeared, alcohol-fixed, and Papanicolaou-stained PTC1 cells, the cell number and the threshold cycle (C_t) value also showed an inverse correlation (Table 2 and Figure 2). The expression of *RET/PTC1* rearrangement was detectable after 33 PCR cycles when 100 cells were used.

Table 2. C_t values and cell counts of *RET/PTC1* rearrangement analysis by RT-PCR using RNA extracted from routine air-dried and Papanicolaou-stained TPC1 cells.

Cell Number	RET/PTC1 (C_t)	GAPDH (C_t)
1000	32.3	29.4
500	33.9	31.1
250	34.2	32.8
100	33.3	31.1
50	34.6	33.3

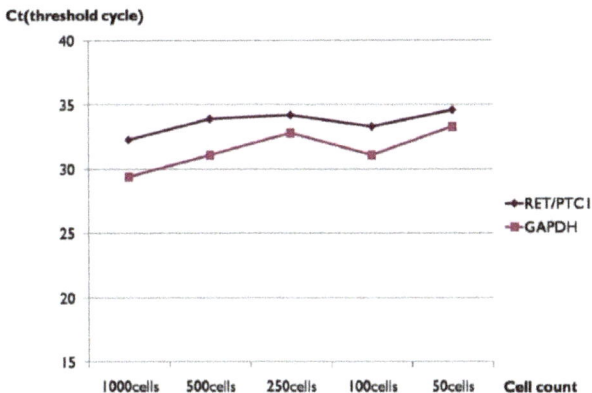

Figure 2. C_t values and cell counts of *RET/PTC1* rearrangement analysis by RT-PCR using RNA extracted from routine air-dried and Papanicolaou-stained TPC1 cells.

2.3. Correlation of the RET/PTC1 Rearrangement between Routine Air-Dried FNA and Paired FFPE PTC Tissue Specimens

When *RET/PTC1* rearrangement was analyzed using PTC cells from archival air-dried FNA slides aspirated from the patients proven to have a histopathological diagnosis of PTC, *RET/PTC1* rearrangement was detected in all 6 cases even though the C_t values of the archival specimen were higher than those of formalin-fixed paraffin-embedded (FFPE) PTC tissue specimens (Table 3). Four cases lacking *RET/PTC1* rearrangement in tissue specimen also failed to reveal rearrangement in FNA samples. These results confirmed that *RET/PTC1* rearrangement analysis by RT-PCR can be applied in preoperative FNA samples as an ancillary diagnostic tool.

Table 3. Correlation of *RET/PTC1* rearrangement status between routine air-dried fine-needle aspiration (FNA) and paired formalin-fixed paraffin-embedded (FFPE) specimens.

	C_t of FFPE Specimen			C_t of FNA Specimen	
Case	*RET/PTC1*	*GAPDH*	Case	*RET/PTC1*	*GAPDH*
1	26.15	27.40	1	35.05	35.86
2	26.15	29.23	2	36.41	37.54
3	26.32	28.96	3	37.03	34.79
4	40	34.67	4	>50	36.74
5	24.64	28.84	5	37.66	35.86
6	31.71	29.92	6	34.26	35.47
7	24.71	26.98	7	36.62	36.78
8	>50.00	28.64	8	>50.00	31.69
9	>50.00	27.55	9	>50.00	30.34
10	>50.00	29.31	10	>50.00	30.01

2.4. Detection of the RET/PTC1 Rearrangement in Resected BRAF and RAS Wild-Type PTC Cases Using FFPE Tissue Specimen

Of 600 surgically resected FFPE specimens histologically diagnosed as PTC, classical type, 518 had *BRAF* mutations and 6 had *RAS* mutations. Among 76 *BRAF* and *RAS* wild-type PTCs, 29 (38.2%) cases turned out to have *RET/PTC1* rearrangement. Considering that alteration of *BRAF*, *RAS*, and *RET* genes are mutually exclusive, 29 (4.8%) of 600 classical PTC cases harbored *RET/PTC1* rearrangement.

2.5. Comparative Analysis of RT-PCR with the NanoString nCounter Gene Expression Assay for Detecting RET/PTC1 Rearrangement in FFPE PTC Tissue Specimen

Twenty-six cases showed correlation on both methods (5 positives and 21 negatives), whereas five cases showed discrepancy between the two methods (three cases positive for Nanostring but not for RT-PCR, two cases positive for RT-PCR but not for Nanostring). Two different analysis methods showed moderate agreement with a κ value of 0.56 ($p = 0.002$).

3. Discussion

The value of molecular markers on preoperative FNA specimens has been described in various thyroid nodules [18,47,49–53]. *RET/PTC* rearrangements are commonly found in adult sporadic PTCs with a marked variable prevalence in different studies owing to geographic variability or different sensitivity of the detection methods [17,29–37]. The reported prevalence rates of *RET/PTC* rearrangements varied largely among studies. While geographical factors and radiation exposure can partially account for this wide range of prevalence, the methodology applied appears to be the most important factor to explain this variability. Searching for *RET/PTC* rearrangements by a less sensitive method may have the drawback of leaving some PTCs undiagnosed, but has the advantage of reducing false positive findings. Indeed, while sporadic cells harboring *RET/PTC* rearrangements can be present in benign nodules, its clonal occurrence is exclusive to PTC. Hence, the less sensitive

RT-PCR seems to be more suitable for diagnostic purposes. *RET/PTC* rearrangements analysis on thyroid tumor has not been extensively performed in Korea, given the prevalence of the *BRAF* V600E mutation in PTC in Korea.

In this report, we assessed the clinical usability of preoperative *RET/PTC1* rearrangement analysis as an ancillary diagnostic tool for *BRAF* and *RAS* wild-type indeterminate thyroid nodules, and explored the *RET/PTC1* rearrangement status in a large number of PTC cases. These explorations have never been done in Korea, to our knowledge. We routinely use atypical follicular cells marked by the cytopathologists and dissected from routine air-dried FNA samples to increase the sensitivity. Since clinical FNA samples contain limited numbers of cells to perform several steps required for deciding optimum number of cells for successful analysis and cutoff values, we performed same analysis using fresh TPC1 cells which are equivalent to the fresh FNA samples in step 1 and air-dried Papanicolaou-stained TPC1 cells equivalent to the archival FNA slides in step 2.

The C_t values in Table 2 tend to decrease when the cell numbers were increased; however, both C_t values of the *RET/PTC1* and *GAPDH* expression using 250 cells are greater than those using 100 cells. Since C_t values of the housekeeping gene expression also showed the same phenomenon, we assumed that a considerable amount of RNA in 250-cell groups might have been deteriorated. At any rate, we found that *RET/PTC1* expression could be measured when 50–100 air-dried Papanicolaou-stained TPC1 cells were used.

When we compared the C_t values of fresh and air-dried Papanicolaou-stained TPC1 cells according to the cell numbers, the C_t values were slightly decreased when air-dried Papanicolaou-stained cells were used. We assumed that this finding might have resulted from the imprecise cell count in step 2. Fresh cells were counted using a hemocytometer, whereas air-dried and Papanicolaou-stained cells were counted on a slide using a square micrometer under the microscope. The expression of *RET/PTC1* rearrangement detectable after 33 PCR cycles when routine air-dried 100 TPC1 cells were used suggests that *RET/PTC1* expression could be detected in routine air-dried FNA samples containing 100 cells.

When *RET/PTC1* rearrangement status from ten FNA and paired FFPE samples were compared, the results showed complete agreement. The higher C_t value of FNA samples compared to the matched FFPE samples could be attributed to the much smaller numbers of cells in FNA samples. The other reason might be the different RNA extraction method used for the two different samples.

Since *BRAF* mutations, *RAS* mutations, and *RET/PTC* rearrangements are mutually exclusive, we analyzed the *RET/PTC1* rearrangement status on both *BRAF* (V600E and K601E) and *RAS* (*NRAS* codons 12, 13, 61; *HRAS* codons 12, 13, 61; *KRAS* codons 12, 13, 61) wild-type PTC cases to save cost and effort. The main limitation of our experiment is that we only performed *RET/PTC1* rearrangement, even though the prevalence for *RET/PTC3* arrangement in previous Korean report was 0%. Another reason for analyzing *RET/PTC1* is that we were able to secure a cell line harboring only the *RET/PTC1* rearrangement.

Among 76 surgically resected both *BRAF* and *RAS* wild-type FFPE specimens histopathologically diagnosed as PTC, classical type, 29 (38.2%) cases turned out to have *RET/PTC1* rearrangement; this means that *RET/PTC1* rearrangement was detected in 29 (4.8%) of 600 classical-type PTCs. Two previous studies reported *REP/PTC* rearrangements in Korea. One study failed to identify any *RET/PTC1, 2, 3* rearrangements in 24 cases of PTC by RT-PCR [31]. The other study detected 2 (6.5%) *RET/PTC1, 2* (6.5%) *RET/PTC2*, and no (0%) *RET/PTC3* rearrangements in 31 PTCs by RT-PCR [38]. Both studies used fresh frozen tumor tissue. The slight discrepancy could be explained by the difference of the sample size. The slightly lower prevalence of *RET/PTC1* rearrangement compared to the second study might also be attributed to the poor RNA preservation in the FFPE specimens.

The NanoString nCounter Gene Expression Assay is a robust and highly reproducible method for detecting the expression of up to 800 genes in a single reaction with high sensitivity and linearity across a broad range of expression levels. The methodology serves to bridge the gap between genome-wide (microarrays) and targeted (real-time quantitative PCR) expression profiling. The nCounter assay is based on direct digital detection of mRNA molecules of interest using target-specific, color-coded probe pairs. It does not require the conversion of mRNA to cDNA by reverse transcription or the amplification of the resulting cDNA by PCR. The expression level of a gene is measured by counting the number of times the color-coded barcode for that gene is detected, and the barcode counts are then tabulated [54]. Comparative analysis of RT-PCR with the Nanostring method for detecting *RET/PTC1* rearrangement in FFPE PTC tissue showed moderate agreement with a *k* value of 0.56 (*p* = 0.002). There is a discrepancy between these two methods (three cases positive for Nanostring but not for RT-PCR, two cases positive for RT-PCR but not for Nanostring). The discrepancy might be attributed to the different RNA extraction methods and cut-off values of each method. In three cases with discrepancy, the results were near to the cut-off value. Another reason might be attributed to the difference of tumor portion used in two different methods. Since we did not initially plan to compare RT-PCR with Nanostring, we made the tumor sections only for RT-PCR analysis. Therefore, the tumor portions which were used for Nanostring might have been slightly different from the initial tumor portion. Next generation sequencing (NGS) is being used to study genetic alterations in institutions worldwide. However, it may be a long time until NGS becomes a routine part of thyroid cancer practice in Korea, since only NGS panels relevant for the therapeutic modality have been approved by the Korean government. Furthermore, only large institutions like university hospitals can adopt NGS in practice. Therefore, algorithmic approach of *BRAF* mutation analysis followed by *RAS* mutation and *RET/PTC1* rearrangement may be of more practical help to refine FNA diagnosis of indeterminate thyroid nodules.

We confirmed the technical applicability of *RET/PTC1* rearrangement analysis using routine air-dried FNA samples as an ancillary diagnostic tool through several steps of the experiment. The presence of *RET/TPC1* rearrangement in a significant proportion (38.2%) of the patients with *BRAF* and *RAS* wild-type PTCs can be used to diagnose and manage patients with *BRAF* and *RAS* wild-type indeterminate thyroid nodules. Since the *BRAF* V600E mutation, *NRAS* codon 61 mutation, and *RET/PTC1* rearrangement comprise more than 90%, 75%, and 50% of the *BRAF* mutations, *RAS* mutations, and *RET/PTC* rearrangements [18,50], an algorithmic approach of *BRAF* V600E mutation analysis followed by *NRAS* 61 mutation and *RET/PTC1* rearrangement analysis would cost-effectively and efficiently overcome a diagnostic limitation of the thyroid FNA by triaging considerable numbers of PTCs in cytologically indeterminate nodules.

4. Materials and Methods

4.1. Total RNA Extraction and First-Strand Synthesis

Total RNA from fresh and fixed TPC1 cells (derived from human thyroid papillary carcinoma, classic type and harboring *RET/PTC1* rearrangement) was extracted using MasterPure Complete DNA and RNA Purification Kit (Epicentre, Madison, WI, USA). Total RNA from formalin-fixed paraffin-embedded (FFPE) specimen was extracted using a High Pure FFPE RNA isolation kit (Roche Diagnostics, Mannheim, Germany). First-strand synthesis was performed on 2 μg of total RNA using a Tetro cDNA synthesis kit (Bioline, London, UK). Tetro reverse transcriptase with diethyl pyrocarbonate water and cDNA reverse transcribed product from the TPC1 cells were used as negative and positive controls, respectively.

4.2. RT-PCR

Amplification was performed by RT-PCR using a LightCycler 480 Instrument (Roche Diagnostics), and measurement was performed using LightCycler quantification software version 1.5 (Roche Diagnostics). The RT-PCR reaction mixture was prepared in a Light Cycler® 480 Multiwell Plate 96 containing 0.5 μM of each primer set (*RET/PTC1* and *glyceraldehyde-3-phosphate isomerase, GAPDH*), 0.25 μM of the probes, 2X of LightCycler 480 Probes Master (Roche Diagnostics), and 1–2 μg (1 μg for the cell and 2 μg for the FFPE tissue) of cDNA template in a final reaction volume of 20 μL (Table 4).

Table 4. Primers and probes sequences for RT-PCR.

RET/PTC1	Primers and Probes Sequences
Forward primer (5′–3′)	CGC GAC CTG CGC AAA
Reverse primer (5′–3′)	CAA GTT CTT CCG AGG GAA TTC C
TaqMan Probe (5′–3′)	FAM-CCA GCG TTA CCA TCG AGG ATC AA AGT-BHQ1
GAPDH	
Forward primer (5′–3′)	GTT CGA CAG TCA GCC GCA TC
Reverse primer (5′–3′)	GGA ATT TGC CAT GGG TGG A
TaqMan Probe (5′–3′)	FAM-ACC AGG CGC CCA ATA CGA CCA A-BHQ1

4.3. NanoString nCounter Gene Expression Assay

Tumor portion on the hematoxylin and eosin-stained FFPE tissue slides was marked by the pathologist, and total RNA was isolated from two to three FFPE tissue sections (10 μm thick) using an miRNeasy FFPE Kit (Qiagen, Hilden, Germany) according to the manufacturer's instructions. The probe sets were custom designed and synthesized by NanoString Technologies (Seattle, WA, USA), and nCounter assays were performed according to the manufacturer's protocol. Briefly, 500 ng of total RNA was hybridized to nCounter probe sets for 16 hours at 65 °C. Samples were then processed using an automated nCounter Sample Prep Station (NanoString Technologies, Inc., Seattle, WA, USA). Cartridges containing immobilized and aligned reporter complexes were subsequently imaged on an nCounter Digital Analyzer (NanoString Technologies, Inc.). Reporter counts were collected using the NanoString's nSolver analysis software version 1, normalized, and analyzed. A total of eight expression probes were designed, four (5′-1 to 5′-4) proximal and four distal (3′-1 to 3′-4) to most commonly-known junction sites for *RET* fusions. An imbalance between 5′ and 3′ probe signals was indicative of the presence of a *RET* fusion transcript. We used a cutoff of three-fold for 3′/5′ ratio. Therefore, a case was considered positive for rearrangement if 3′/5′ imbalance was three-fold or more.

We used Cohen's κ coefficient to measure agreement between RT-PCR and Nanostring method.

4.4. Detection of the RET/PTC1 Rearrangement in Fresh TPC1 Cell Line

Total RNA was extracted by Master Pure Complete DNA and RNA Purification Kit (Epicentre) using a fresh cell colony formed from 1000 cultured TPC1 cells (provided by Nagataki, Nakasaki University, Japan). RT-PCR was performed and the minimum number of cycles (C_t value) needed to detect the expression of *RET/PTC1* rearrangement and *GAPDH* was determined. Similarly, the number of the cells was reduced to 500, 250, 100, and 50, and RT-PCR was performed to evaluate C_t values according to cell count. The whole procedure was performed in triplicate after TPC1 cells were harvested

4.5. Detection of RET/PTC1 Rearrangement in Routine Air-Dried TPC1 Cell Line

To make a condition identical to that in the routine air-dried FNA preparation, cultured TPC1 cells were smeared on a slide and fixed with 95% ethanol according to the routine FNA preparation in our cytology laboratory. The fixed cells were stained by the routine Papanicolaou procedure. After the coverslips were removed from the smeared slides, the atypical cells of interest were dissected with a 26-gauge needle under the light microscope. Approximately 50, 100, 250, 500, and 1000 cells were dissected using a square micrometer under the microscope. A needle tip was carefully submerged in a tube containing extraction buffer supplied by MasterPure Complete DNA and RNA Purification Kit (Epicentre), and total RNA was extracted. RT-PCR was performed, and C_t values for the expression of *RET/PTC1* rearrangement and *GAPDH* were evaluated using 50, 100, 250, 500, and 1000 air-dried and alcohol fixed TPC1 cells, respectively. The whole procedure was performed in triplicate after TPC1 cells were harvested.

4.6. Correlation of RET/PTC1 Rearrangement between Routine Air-Dried FNA and Paired FFPE PTC Tissue Specimens

PTC cells from the archival FNA slides from the Department of Pathology, Konkuk University Medical Center were used. The slides that were selected were from samples aspirated from ten thyroid nodules with histopathological diagnosis of classical-type PTC. Study approval was obtained from the Institutional Review Board (KUH1210043). After the coverslips were removed from the slides, approximately 100 atypical follicular cells were dissected with a 26-gauge needle under the light microscope, and total RNA was extracted using MasterPure Complete DNA and RNA Purification Kit (Epicentre). RT-PCR was performed, and C_t values of the *RET/PTC1* rearrangement and *GAPDH* expression were evaluated. Tumor portion on the hematoxylin and eosin-stained FFPE tissue slides was marked by the pathologist, and total RNA was isolated from two-to-three FFPE tissue sections (10 μm thick) using High Pure FFPE RNA isolation kit (Roche Diagnostics). RT-PCR was performed and C_t values of the *RET/PTC1* rearrangement and *GAPDH* expression were evaluated. The C_t values defining the analysis as positive is greater than 40 cycles.

4.7. Detection of the RET/PTC1 Rearrangement in Resected BRAF and RAS Wild-Type PTC Cases Using FFPE Tissue Specimen

Archival thyroid neoplasm that had been surgically removed between 2010 and 2014 at Konkuk University Medical Center were blindly re-evaluated according to the 2004 World Health Organization classification of thyroid neoplasm by the two pathologists (Tae Sook Hwang, who is an endocrine pathologist, and Young Sin Ko). In case of a disagreement and to reach a consensus, another endocrine pathologist (Chan-Kwon Jung) independently reviewed the cases. Of the 600 classical PTC cases selected, 518 had *BRAF* mutation and 6 had *RAS* mutation. Finally, 76 *BRAF* and *RAS* wild-type classical PTC cases were selected. Tumor portion on the hematoxylin and eosin-stained FFPE tissue slides was marked by the pathologist, and total RNA was isolated from two-to-three FFPE tissue sections (10 μm thick) using High Pure FFPE RNA isolation kit (Roche Diagnostics). RT-PCR was performed, and C_t values of the *RET/PTC1* rearrangement and *GAPDH* expression were evaluated. The C_t value defining the analysis as positive is greater than 40 cycles.

4.8. Comparison Analysis of RT-PCR with the NanoString nCounter Gene Expression Assay for Detecting RET/PTC1 Rearrangement

RET/PTC1 rearrangement status was also analyzed by the Nanostring method, using 31 cases having sufficient cancer tissue remaining for the comparative analysis.

5. Conclusions

RET/PTC1 rearrangement analysis by RT-PCR using routine air-dried FNA specimen was confirmed to be technically applicable and significant population (38.2%) of the *BRAF* and *RAS* wild type PTCs harbor *RET/PTC1* rearrangement. Preoperative *RET/PTC1* rearrangement analysis in *BRAF* and *RAS* wild type indeterminate thyroid nodules would permit a formulation of unambiguous surgical plan, while foregoing the need for other less specific diagnostic test such as repeat FNA and intraoperative frozen section evaluation. An algorithmic approach is cost-effective and efficient especially in *BRAF* mutation prevalent populations.

Acknowledgments: This paper was supported by Konkuk University. The authors thank Chan-Kwon Jung (Department of Pathology, College of Medicine, The Catholic University, Seoul, Korea) for reviewing thyroid tissue slides.

Author Contributions: Young Sin Ko designed the experiments, conducted the main experiments and prepared the manuscript; Tae Sook Hwang conceived and designed the experiments and revised the manuscript; Ja Yeon Kim also conducted the main experiments and analyzed the data; Seung Eun Lee conducted Nanostring and analyzed the data; Yoon-La Choi, Hye Seung Han, Wan Seop Kim, Suk Kyeong Kim, and Kyoung Sik Park contributed clarifications and guidance on the manuscript. All authors read and approved the manuscript.

Conflicts of Interest: The authors declare no conflict of interest.

Abbreviations

AUS	Atypia of undetermined significance
C_t	Threshold cycle
FFPE	Formalin-fixed paraffin-embedded
FN	Follicular neoplasm;
FNA	Fine needle aspiration
FLUS	Follicular lesion of undetermined significance
FVPTC	Follicular variant of papillary thyroid carcinoma
FTC	Follicular thyroid carcinoma
NGS	Next generation sequencing
RT-PCR	Real-time reverse transcription-polymerase chain reaction
SFN	Suspicious for follicular neoplasm
PTC	Papillary thyroid carcinoma
TPC1	Thyroid papillary carcinoma 1

References

1. Vander, J.B.; Gaston, E.A.; Dawber, T.R. The significance of nontoxic thyroid nodules. Final report of a 15-year study of the incidence of thyroid malignancy. *Ann. Intern. Med.* **1968**, *69*, 537–540. [CrossRef] [PubMed]
2. Tunbridge, W.M.; Evered, D.C.; Hall, R.; Appleton, D.; Brewis, M.; Clark, F.; Evans, J.G.; Young, E.; Bird, T.; Smith, P.A. The spectrum of thyroid disease in a community: The whickham survey. *Clin. Endocrinol.* **1977**, *7*, 481–493. [CrossRef]
3. Tan, G.H.; Gharib, H. Thyroid incidentalomas: Management approaches to nonpalpable nodules discovered incidentally on thyroid imaging. *Ann. Intern. Med.* **1997**, *126*, 226–231. [CrossRef] [PubMed]
4. Guth, S.; Theune, U.; Aberle, J.; Galach, A.; Bamberger, C.M. Very high prevalence of thyroid nodules detected by high frequency (13 MHz) ultrasound examination. *Eur. J. Clin. Investig.* **2009**, *39*, 699–706. [CrossRef] [PubMed]
5. Hegedus, L. Clinical practice. The thyroid nodule. *N. Engl. J. Med.* **2004**, *351*, 1764–1771. [CrossRef] [PubMed]
6. Mandel, S.J. A 64-year-old woman with a thyroid nodule. *JAMA* **2004**, *292*, 2632–2642. [CrossRef] [PubMed]
7. Sherman, S.I. Thyroid carcinoma. *Lancet* **2003**, *361*, 501–511. [CrossRef]
8. Siegel, R.; Ma, J.; Zou, Z.; Jemal, A. Cancer statistics, 2014. *CA Cancer J. Clin.* **2014**, *64*, 9–29. [CrossRef] [PubMed]
9. Davies, L.; Welch, H.G. Current thyroid cancer trends in the United States. *JAMA Otolaryngol. Head Neck Surg.* **2014**, *140*, 317–322. [CrossRef] [PubMed]
10. *National Cancer Registration and Statistics in Korea 2015*; Korea Central Cancer Registry: Goyang, Korea, 2015.
11. Gharib, H.; Goellner, J.R. Fine-needle aspiration biopsy of the thyroid: An appraisal. *Ann. Intern. Med.* **1993**, *118*, 282–289. [CrossRef] [PubMed]
12. Udelsman, R.; Chen, H. The current management of thyroid cancer. *Adv. Surg.* **1999**, *33*, 1–27. [PubMed]
13. Haugen, B.R.; Alexander, E.K.; Bible, K.C.; Doherty, G.M.; Mandel, S.J.; Nikiforov, Y.E.; Pacini, F.; Randolph, G.W.; Sawka, A.M.; Schlumberger, M.; et al. 2015 American thyroid association management guidelines for adult patients with thyroid nodules and differentiated thyroid cancer: The American thyroid association guidelines task force on thyroid nodules and differentiated thyroid cancer. *Thyroid* **2016**, *26*, 1–133. [CrossRef] [PubMed]
14. Goutas, N.; Vlachodimitropoulos, D.; Bouka, M.; Lazaris, A.C.; Nasioulas, G.; Gazouli, M. BRAF and K-RAS mutation in a Greek papillary and medullary thyroid carcinoma cohort. *Anticancer Res.* **2008**, *28*, 305–308. [PubMed]
15. Kim, S.K.; Song, K.H.; Lim, S.D.; Lim, Y.C.; Yoo, Y.B.; Kim, J.S.; Hwang, T.S. Clinical and pathological features and the BRAF^{V600E} mutation in patients with papillary thyroid carcinoma with and without concurrent hashimoto thyroiditis. *Thyroid* **2009**, *19*, 137–141. [CrossRef] [PubMed]

16. Davies, L.; Welch, H.G. Increasing incidence of thyroid cancer in the United States, 1973–2002. *JAMA* **2006**, *295*, 2164–2167. [CrossRef] [PubMed]
17. Jung, C.K.; Little, M.P.; Lubin, J.H.; Brenner, A.V.; Wells, S.A., Jr.; Sigurdson, A.J.; Nikiforov, Y.E. The increase in thyroid cancer incidence during the last four decades is accompanied by a high frequency of BRAF mutations and a sharp increase in RAS mutations. *J. Clin. Endocrinol. Metab.* **2014**, *99*, E276–E285. [CrossRef] [PubMed]
18. Park, J.Y.; Kim, W.Y.; Hwang, T.S.; Lee, S.S.; Kim, H.; Han, H.S.; Lim, S.D.; Kim, W.S.; Yoo, Y.B.; Park, K.S. BRAF and RAS mutations in follicular variants of papillary thyroid carcinoma. *Endocr. Pathol.* **2013**, *24*, 69–76. [CrossRef] [PubMed]
19. Lee, S.R.; Jung, C.K.; Kim, T.E.; Bae, J.S.; Jung, S.L.; Choi, Y.J.; Kang, C.S. Molecular genotyping of follicular variant of papillary thyroid carcinoma correlates with diagnostic category of fine-needle aspiration cytology: Values of RAS mutation testing. *Thyroid* **2013**, *23*, 1416–1422. [CrossRef] [PubMed]
20. Bongarzone, I.; Butti, M.G.; Coronelli, S.; Borrello, M.G.; Santoro, M.; Mondellini, P.; Pilotti, S.; Fusco, A.; Della Porta, G.; Pierotti, M.A. Frequent activation of ret protooncogene by fusion with a new activating gene in papillary thyroid carcinomas. *Cancer Res.* **1994**, *54*, 2979–2985. [PubMed]
21. Bongarzone, I.; Monzini, N.; Borrello, M.G.; Carcano, C.; Ferraresi, G.; Arighi, E.; Mondellini, P.; Della Porta, G.; Pierotti, M.A. Molecular characterization of a thyroid tumor-specific transforming sequence formed by the fusion of ret tyrosine kinase and the regulatory subunit RI α of cyclic AMP-dependent protein kinase A. *Mol. Cell. Biol.* **1993**, *13*, 358–366. [CrossRef] [PubMed]
22. Grieco, M.; Santoro, M.; Berlingieri, M.T.; Melillo, R.M.; Donghi, R.; Bongarzone, I.; Pierotti, M.A.; Della Porta, G.; Fusco, A.; Vecchio, G. PTC is a novel rearranged form of the RET proto-oncogene and is frequently detected in vivo in human thyroid papillary carcinomas. *Cell* **1990**, *60*, 557–563. [CrossRef]
23. Jhiang, S.M.; Smanik, P.A.; Mazzaferri, E.L. Development of a single-step duplex RT-PCR detecting different forms of RET activation, and identification of the third form of in vivo ret activation in human papillary thyroid carcinoma. *Cancer Lett.* **1994**, *78*, 69–76. [CrossRef]
24. Santoro, M.; Dathan, N.A.; Berlingieri, M.T.; Bongarzone, I.; Paulin, C.; Grieco, M.; Pierotti, M.A.; Vecchio, G.; Fusco, A. Molecular characterization of RET/PTC3; a novel rearranged version of the retproto-oncogene in a human thyroid papillary carcinoma. *Oncogene* **1994**, *9*, 509–516. [PubMed]
25. Fenton, C.L.; Lukes, Y.; Nicholson, D.; Dinauer, C.A.; Francis, G.L.; Tuttle, R.M. The RET/PTC mutations are common in sporadic papillary thyroid carcinoma of children and young adults. *J. Clin. Endocrinol. Metab.* **2000**, *85*, 1170–1175. [CrossRef] [PubMed]
26. Rabes, H.M.; Demidchik, E.P.; Sidorow, J.D.; Lengfelder, E.; Beimfohr, C.; Hoelzel, D.; Klugbauer, S. Pattern of radiation-induced ret and NTRK1 rearrangements in 191 post-chernobyl papillary thyroid carcinomas: Biological, phenotypic, and clinical implications. *Clin. Cancer Res.* **2000**, *6*, 1093–1103. [PubMed]
27. Unger, K.; Zitzelsberger, H.; Salvatore, G.; Santoro, M.; Bogdanova, T.; Braselmann, H.; Kastner, P.; Zurnadzhy, L.; Tronko, N.; Hutzler, P.; et al. Heterogeneity in the distribution of RET/PTC rearrangements within individual post-chernobyl papillary thyroid carcinomas. *J. Clin. Endocrinol. Metab.* **2004**, *89*, 4272–4279. [CrossRef] [PubMed]
28. Zhu, Z.; Ciampi, R.; Nikiforova, M.N.; Gandhi, M.; Nikiforov, Y.E. Prevalence of RET/PTC rearrangements in thyroid papillary carcinomas: Effects of the detection methods and genetic heterogeneity. *J. Clin. Endocrinol. Metab.* **2006**, *91*, 3603–3610. [CrossRef] [PubMed]
29. Namba, H.; Yamashita, S.; Pei, H.C.; Ishikawa, N.; Villadolid, M.C.; Tominaga, T.; Kimura, H.; Tsuruta, M.; Yokoyama, N.; Izumi, M.; et al. Lack of *PTC* gene (RET proto-oncogene rearrangement) in human thyroid tumors. *Endocrinol. Jpn.* **1991**, *38*, 627–632. [CrossRef] [PubMed]
30. Nikiforov, Y.E.; Rowland, J.M.; Bove, K.E.; Monforte-Munoz, H.; Fagin, J.A. Distinct pattern of RET oncogene rearrangements in morphological variants of radiation-induced and sporadic thyroid papillary carcinomas in children. *Cancer Res.* **1997**, *57*, 1690–1694. [PubMed]
31. Park, K.Y.; Koh, J.M.; Kim, Y.I.; Park, H.J.; Gong, G.; Hong, S.J.; Ahn, I.M. Prevalences of GS α, ras, p53 mutations and RET/PTC rearrangement in differentiated thyroid tumours in a Korean population. *Clin. Endocrinol.* **1998**, *49*, 317–323. [CrossRef]
32. Lee, C.H.; Hsu, L.S.; Chi, C.W.; Chen, G.D.; Yang, A.H.; Chen, J.Y. High frequency of rearrangement of the ret protooncogene (RET/PTC) in chinese papillary thyroid carcinomas. *J. Clin. Endocrinol. Metab.* **1998**, *83*, 1629–1632. [CrossRef] [PubMed]

33. Rhoden, K.J.; Johnson, C.; Brandao, G.; Howe, J.G.; Smith, B.R.; Tallini, G. Real-time quantitative RT-PCR identifies distinct C-RET, RET/PTC1 and RET/PTC3 expression patterns in papillary thyroid carcinoma. *Lab. Investig.* **2004**, *84*, 1557–1570. [CrossRef] [PubMed]

34. Di Cristofaro, J.; Vasko, V.; Savchenko, V.; Cherenko, S.; Larin, A.; Ringel, M.D.; Saji, M.; Marcy, M.; Henry, J.F.; Carayon, P.; et al. Ret/PTC1 and RET/PTC3 in thyroid tumors from chernobyl liquidators: Comparison with sporadic tumors from ukrainian and french patients. *Endocr. Relat. Cancer* **2005**, *12*, 173–183. [CrossRef] [PubMed]

35. Mayr, B.; Potter, E.; Goretzki, P.; Ruschoff, J.; Dietmaier, W.; Hoang-Vu, C.; Dralle, H.; Brabant, G. Expression of RET/PTC1, -2, -3, -Δ3 and -4 in German papillary thyroid carcinoma. *Br. J. Cancer* **1998**, *77*, 903–906. [CrossRef] [PubMed]

36. Nikiforov, Y.E. RET/PTC rearrangement in thyroid tumors. *Endocr. Pathol.* **2002**, *13*, 3–16. [CrossRef] [PubMed]

37. Tallini, G.; Asa, S.L. Ret oncogene activation in papillary thyroid carcinoma. *Adv. Anat. Pathol.* **2001**, *8*, 345–354. [CrossRef] [PubMed]

38. Chung, J.H.; Hahm, J.R.; Min, Y.K.; Lee, M.S.; Lee, M.K.; Kim, K.W.; Nam, S.J.; Yang, J.H.; Ree, H.J. Detection of RET/PTC oncogene rearrangements in Korean papillary thyroid carcinomas. *Thyroid* **1999**, *9*, 1237–1243. [CrossRef] [PubMed]

39. Motomura, T.; Nikiforov, Y.E.; Namba, H.; Ashizawa, K.; Nagataki, S.; Yamashita, S.; Fagin, J.A. Ret rearrangements in Japanese pediatric and adult papillary thyroid cancers. *Thyroid* **1998**, *8*, 485–489. [CrossRef] [PubMed]

40. Guerra, A.; Sapio, M.R.; Marotta, V.; Campanile, E.; Moretti, M.I.; Deandrea, M.; Motta, M.; Limone, P.P.; Fenzi, G.; Rossi, G.; et al. Prevalence of RET/PTC rearrangement in benign and malignant thyroid nodules and its clinical application. *Endocr. J.* **2011**, *58*, 31–38. [CrossRef] [PubMed]

41. Marotta, V.; Guerra, A.; Sapio, M.R.; Campanile, E.; Motta, M.; Fenzi, G.; Rossi, G.; Vitale, M. Growing thyroid nodules with benign histology and RET rearrangement. *Endocr. J.* **2010**, *57*, 1081–1087. [CrossRef] [PubMed]

42. Sapio, M.R.; Guerra, A.; Marotta, V.; Campanile, E.; Formisano, R.; Deandrea, M.; Motta, M.; Limone, P.P.; Fenzi, G.; Rossi, G.; et al. High growth rate of benign thyroid nodules bearing RET/PTC rearrangements. *J. Clin. Endocrinol. Metab.* **2011**, *96*, E916–E919. [CrossRef] [PubMed]

43. Eszlinger, M.; Krogdahl, A.; Munz, S.; Rehfeld, C.; Precht Jensen, E.M.; Ferraz, C.; Bosenberg, E.; Drieschner, N.; Scholz, M.; Hegedus, L.; et al. Impact of molecular screening for point mutations and rearrangements in routine air-dried fine-needle aspiration samples of thyroid nodules. *Thyroid* **2014**, *24*, 305–313. [CrossRef] [PubMed]

44. Ferraz, C.; Rehfeld, C.; Krogdahl, A.; Precht Jensen, E.M.; Bosenberg, E.; Narz, F.; Hegedus, L.; Paschke, R.; Eszlinger, M. Detection of PAX8/PPARG and RET/PTC rearrangements is feasible in routine air-dried fine needle aspiration smears. *Thyroid* **2012**, *22*, 1025–1030. [CrossRef] [PubMed]

45. Cheung, C.C.; Carydis, B.; Ezzat, S.; Bedard, Y.C.; Asa, S.L. Analysis of RET/PTC gene rearrangements refines the fine needle aspiration diagnosis of thyroid cancer. *J. Clin. Endocrinol. Metab.* **2001**, *86*, 2187–2190. [CrossRef] [PubMed]

46. Musholt, T.J.; Fottner, C.; Weber, M.M.; Eichhorn, W.; Pohlenz, J.; Musholt, P.B.; Springer, E.; Schad, A. Detection of papillary thyroid carcinoma by analysis of BRAF and RET/PTC1 mutations in fine-needle aspiration biopsies of thyroid nodules. *World J. Surg.* **2010**, *34*, 2595–2603. [CrossRef] [PubMed]

47. Nikiforov, Y.E.; Steward, D.L.; Robinson-Smith, T.M.; Haugen, B.R.; Klopper, J.P.; Zhu, Z.; Fagin, J.A.; Falciglia, M.; Weber, K.; Nikiforova, M.N. Molecular testing for mutations in improving the fine-needle aspiration diagnosis of thyroid nodules. *J. Clin. Endocrinol. Metab.* **2009**, *94*, 2092–2098. [CrossRef] [PubMed]

48. Salvatore, G.; Giannini, R.; Faviana, P.; Caleo, A.; Migliaccio, I.; Fagin, J.A.; Nikiforov, Y.E.; Troncone, G.; Palombini, L.; Basolo, F.; et al. Analysis of BRAF point mutation and RET/PTC rearrangement refines the fine-needle aspiration diagnosis of papillary thyroid carcinoma. *J. Clin. Endocrinol. Metab.* **2004**, *89*, 5175–5180. [CrossRef] [PubMed]

49. An, J.H.; Song, K.H.; Kim, S.K.; Park, K.S.; Yoo, Y.B.; Yang, J.H.; Hwang, T.S.; Kim, D.L. Ras mutations in indeterminate thyroid nodules are predictive of the follicular variant of papillary thyroid carcinoma. *Clin. Endocrinol.* **2015**, *82*, 760–766. [CrossRef] [PubMed]

50. Hwang, T.S.; Kim, W.Y.; Han, H.S.; Lim, S.D.; Kim, W.S.; Yoo, Y.B.; Park, K.S.; Oh, S.Y.; Kim, S.K.; Yang, J.H. Preoperative RAS mutational analysis is of great value in predicting follicular variant of papillary thyroid carcinoma. *BioMed Res. Int.* **2015**, *2015*, 697068. [CrossRef] [PubMed]

51. Kim, S.K.; Kim, D.L.; Han, H.S.; Kim, W.S.; Kim, S.J.; Moon, W.J.; Oh, S.Y.; Hwang, T.S. Pyrosequencing analysis for detection of a BRAF^{V600E} mutation in an fnab specimen of thyroid nodules. *Diagn. Mol. Pathol.* **2008**, *17*, 118–125. [CrossRef] [PubMed]

52. Cantara, S.; Capezzone, M.; Marchisotta, S.; Capuano, S.; Busonero, G.; Toti, P.; Di Santo, A.; Caruso, G.; Carli, A.F.; Brilli, L.; et al. Impact of proto-oncogene mutation detection in cytological specimens from thyroid nodules improves the diagnostic accuracy of cytology. *J. Clin. Endocrinol. Metab.* **2010**, *95*, 1365–1369. [CrossRef] [PubMed]

53. Ohori, N.P.; Nikiforova, M.N.; Schoedel, K.E.; LeBeau, S.O.; Hodak, S.P.; Seethala, R.R.; Carty, S.E.; Ogilvie, J.B.; Yip, L.; Nikiforov, Y.E. Contribution of molecular testing to thyroid fine-needle aspiration cytology of "follicular lesion of undetermined significance/atypia of undetermined significance". *Cancer Cytopathol.* **2010**, *118*, 17–23. [CrossRef] [PubMed]

54. Kulkarni, M.M. Digital multiplexed gene expression analysis using the nanostring ncounter system. *Curr. Protoc. Mol. Biol.* **2011**. [CrossRef]

International Journal of
Molecular Sciences

MDPI

Article

Integrity and Quantity of Total Cell-Free DNA in the Diagnosis of Thyroid Cancer: Correlation with Cytological Classification

Francesca Salvianti [1], Corinna Giuliani [2], Luisa Petrone [2], Irene Mancini [1], Vania Vezzosi [3], Cinzia Pupilli [4] and Pamela Pinzani [1,*]

[1]   Department of Experimental and Clinical Biomedical Sciences, Molecular and Clinical Biochemistry Unit, Careggi Hospital and University of Florence, 50134 Florence, Italy; francesca.salvianti@unifi.it (F.S.); irene.mancini@unifi.it (I.M.)
[2]   Department of Clinical, Experimental and Biomedical Sciences, Endocrinology Unit, Careggi Hospital and University of Florence, 50134 Florence, Italy; corinna.giuliani@tiscali.it (C.G.); luisa.petrone@aouc.unifi.it (L.P.)
[3]   Division of Pathological Anatomy, University of Florence, 50121 Florence, Italy; vvezzosi@unifi.it
[4]   Endocrinology Unit, Santa Maria Nuova Hospital, Florence, Azienda USL Toscana Centro, 50122 Florence, Italy; cinzia.pupilli@uslcentro.toscana.it
*   Correspondence: p.pinzani@dfc.unifi.it; Tel.: +39-055-275-8233

Received: 18 May 2017; Accepted: 22 June 2017; Published: 24 June 2017

Abstract: Cell-free DNA (cfDNA) quantity and quality in plasma has been investigated as a non-invasive biomarker in cancer. Previous studies have demonstrated increased cfDNA amount and length in different types of cancer with respect to healthy controls. The present study aims to test the hypothesis that the presence of longer DNA strands circulating in plasma can be considered a biomarker for tumor presence in thyroid cancer. We adopted a quantitative real-time PCR (qPCR) approach based on the quantification of two amplicons of different length (67 and 180 bp respectively) to evaluate the integrity index 180/67. Cell-free DNA quantity and integrity were higher in patients affected by nodular thyroid diseases than in healthy controls. Importantly, cfDNA integrity index was higher in patients with cytological diagnosis of thyroid carcinoma (Thy4/Thy5) than in subjects with benign nodules (Thy2). Therefore, cfDNA integrity index 180/67 is a suitable parameter for monitoring cfDNA fragmentation in thyroid cancer patients and a promising circulating biomarker in the diagnosis of thyroid nodules.

Keywords: cell-free DNA; integrity index; plasma; qPCR; papillary thyroid carcinoma

---

1. Introduction

Cancer-derived DNA in blood represents a promising biomarker for cancer diagnosis. Previous studies have demonstrated an increase of cell-free circulating DNA in different types of cancer [1] in comparison to the general population.

Even if it is well known that DNA concentration in plasma is elevated in cancer patients [1] and can be influenced by tumor characteristics [2], the hypotheses on its origin are still controversial and details on the mechanism of release are not completely disclosed [3]. Circulating free-DNA is released from apoptotic or necrotic cells, reflecting a differential DNA origin. Necrosis is common in solid malignant cancers and generates a spectrum of DNA fragments of different size, due to random digestion by DNases. In contrast, cell death in normal blood nucleated cells occurs mostly via apoptosis that generates small and uniform DNA fragments. Support for this hypothesis has been reported by several papers [2,4–12] and confirmed in recent studies demonstrating increased DNA length in plasma from patients with breast [13,14], prostate [15], colorectal [16–18] and lung cancer [19].

In addition, total cell-free DNA (cfDNA) concentration may also be altered in patients with various benign diseases such as trauma, stroke, burns, sepsis, and autoimmune diseases, thus limiting its value for diagnosis of cancer [20]. For this reason, the simple cell-free DNA quantitative analysis cannot provide the expected clinical specificity.

To this purpose, the search of qualitative alterations of DNA, such as mutations, loss of heterozygosity (LOH), microsatellite instability and epigenetic changes, were shown to improve the cancer specificity [2]. Tumor biomarkers identified in plasma of cancer patients may have a high diagnostic and prognostic value, however, the detection of these alterations is limited by the frequency of their occurrence in each tumor type and their tumor-specificity so that the development of different assays can be necessary when dealing with tumors with different mutational signatures.

Our attention is focused on fragmentation of plasma DNA. We extensively studied the characteristics of cfDNA in plasma of melanoma patients, evidencing that total quantity and integrity index could provide useful information to discriminate the tumor affected population from healthy individuals [9]. The test is based on the hypothesis that DNA fragments in plasma are longer than those of healthy individuals on account of an inefficient nuclease activity.

This manuscript studies patients affected by differentiated papillary thyroid carcinoma to test the hypothesis that the presence of longer DNA strands circulating in plasma can be considered a biomarker for tumor presence also in the case of thyroid cancer. We adopted a qPCR approach based on the quantification of two amplicons of different length (67 and 180 bp respectively) to evaluate the integrity index 180/67 whose performance had been previously evaluated in a case study composed of melanoma patients. Control subjects and subjects affected by benign pathologies have been considered as the references. Our aim is to investigate the ability of this parameter to provide diagnostic information related to the cytomorphological classification of this tumor performed on fine needle aspirates (FNA) of the nodular lesion.

2. Results

2.1. Plasma DNA Concentration and Integrity in Basal Blood Samples

Total cfDNA was quantified by the qPCR assay targeting the 67 bp amplicon on the *APP* gene (see Methods section). Quantitative values of cfDNA concentration in plasma are reported in Table 1 and graphically represented in Figure 1a. Patients affected by nodular thyroid diseases (respectively Thy2, Thy3 and Thy4/Thy5 cytology) showed higher levels of total cfDNA than healthy individuals ($p < 0.001$).

Table 1. Quantitative values of cell-free DNA (cfDNA) markers in the case study.

		(cfDNA) (ng/mL pl)	App 180 (ng/mL pl)	Integrity Index 180/67
Healthy $n = 49$	Median	5.12	2.42	0.56
	Range	0.99–26.71	0.40–12.68	0.08–1.81
Thy2 $n = 25$	Median	11.88	9.03	0.67
	Range	5.10–296.52	4.55–261.04	0.22–1.24
Thy3 $n = 44$	Median	12.41	9.02	0.83
	Range	2.26–128.44	0.20–199.07	0.01–2.21
Thy4 + Thy5 $n = 28$	Median	11.47	10.15	1.02
	Range	1.31–62.60	2.19–82.65	0.22–2.02

Analogously, cfDNA quantity according to the qPCR assay targeting the 180 bp amplicon on the *APP* gene (see Methods section) was constantly higher in each patient's cytology group (Thy2, Thy3 and Thy4/Thy5) than healthy controls with a *p* value lower than 0.001. The results are reported in Table 1 and Figure 1b.

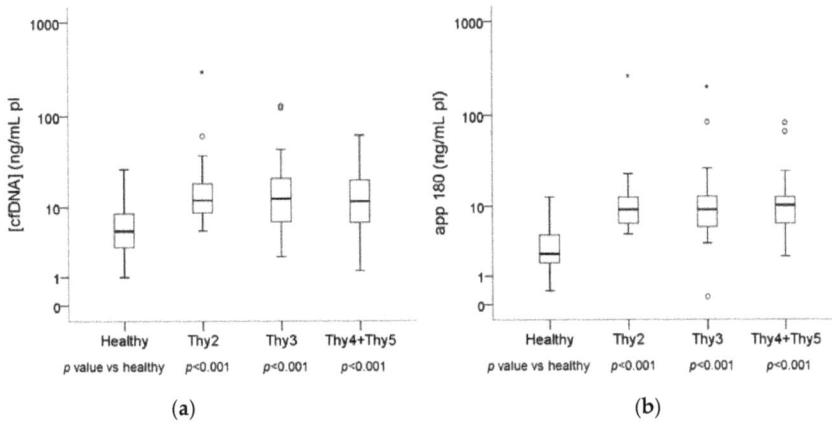

Figure 1. Box plots reflecting the distribution in cases (Thy2, Thy3 and Thy4/Thy5) and controls (healthy subjects) of total cfDNA quantity (**a**), and cfDNA quantity according to a qPCR assay targeting a 180 bp amplicon on the *APP* gene (**b**). Each box indicates the 25th and 75th percentiles. The horizontal line inside the box indicates the median, and the whiskers indicate the extreme measured values. Dots and stars represent outliers.

Cell-free DNA integrity assessed by means of the integrity index 180/67 (see Methods section) was significantly higher in Thy2 ($p = 0.01$), Thy3 ($p = 0.002$) and Thy4/Thy5 ($p < 0.001$) patients than control subjects.

Subjects affected by benign thyroid nodules (Thy2) showed an integrity index significantly lower ($p = 0.013$) than that found in patients with cytological diagnosis of thyroid cancer (Thy4/Thy5).

The values of integrity index for each patient's group are reported in Table 1 and Figure 2.

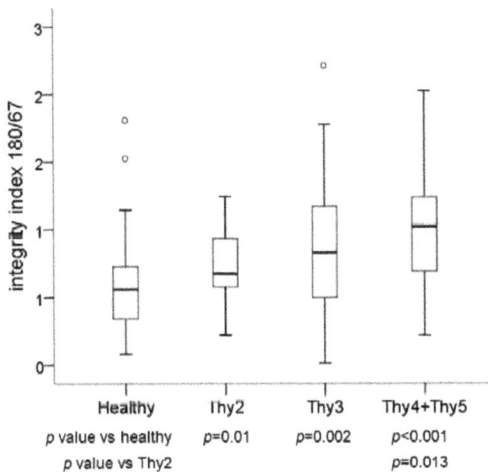

Figure 2. Cell-free DNA integrity in the categories of the case study: healthy subjects, Thy2, Thy3 and Thy4/Thy5. Box plots indicate the 25th and 75th percentiles. The horizontal line inside the box indicates the median, and the whiskers indicate the extreme measured values. Dots represent outliers.

We did not find any significant difference in cfDNA quantity and integrity stratifying patients on the basis of sex and age.

2.2. cfDNA Fragments

The absolute concentration of cfDNA fragments with length ranging from 67 to 180 bp, calculated by subtracting the absolute concentration of the longer amplicon from that of the shorter one, was significantly lower in Thy4/Thy5 patients (median = −0.12, range −20.05–28.62) than subjects with Thy2 nodules (median = 3.40, range −2.24–47.94, p = 0.010) and healthy controls (median = 2.22, range −1.98–14.16, p = 0.013).

Alternatively, the percentage of cfDNA fragments between 67 and 180 bp was calculated by subtracting the absolute concentration of the longer amplicon to that of the shorter one and normalizing for total cfDNA quantity (assessed by the shorter amplicon).

Healthy controls presented about 44% of these fragments, with a significant difference with respect to Thy2 (33%, p = 0.010), Thy3 (17%, p = 0.002) and Thy4/Thy5 (−2%, p < 0.001) patients.

Thy2 subjects showed a higher percentage of fragments between 67 and 180 bp than Thy4/Thy5 patients (p = 0.013).

2.3. Plasma DNA Integrity in Post-Surgery Blood Samples

The specificity of the assay was demonstrated considering the variation of plasma DNA concentration and integrity from the pre-surgery value to that found after treatment, few months later. Specifically, for a group of 17 patients, an additional blood draw was taken 3–6 months after surgery and radioactive iodine treatment when appropriate.

While no statistical differences were evidenced, independently from the amplicon length, between pre-surgery total cell-free DNA concentration and that found at the follow-up time, the after surgery sample showed a lower integrity index (median = 0.59, range 0.36–1.67) than that taken before surgery (median = 0.87, range 0.32–1.30, p = 0.035, Figure 3).

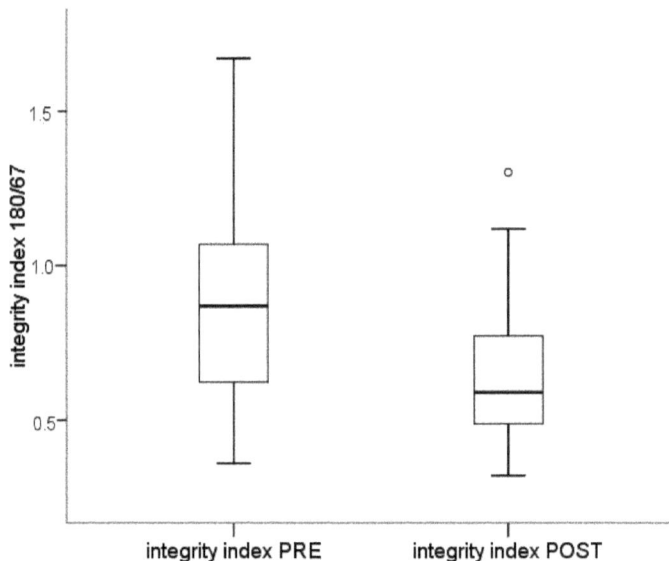

Figure 3. Integrity index in samples before (PRE) and after (POST) surgery. Box plots indicate the 25th and 75th percentiles. The horizontal line inside the box indicates the median, and the whiskers indicate the extreme measured values. Dots represent outliers.

After surgery, thyroid cancer patients showed a higher percentage of small (67–180 bp) DNA fragments (median = 41.02 range −30.28–67.87) in comparison to the pre-surgery condition (median = 12.99 range −67.18–64.00, $p = 0.035$).

2.4. ROC (Receiver Operating Characteristic) Curve Analysis

The predictive capability (i.e., diagnostic performance) of cfDNA quantity and integrity in thyroid cancer was investigated by means of the area under the ROC curve by comparing healthy subjects with patients with cytological diagnosis of thyroid carcinoma (Thy4/Thy5).

All the three markers showed a good predictive capability with an area under the ROC curve (AUC) of 0.765 ($p < 0.001$), 0.982 ($p < 0.001$) and 0.796 ($p < 0.001$) for cfDNA quantity by 67 bp amplicon, cfDNA quantity by 180 bp amplicon and integrity index, respectively (Figure 4).

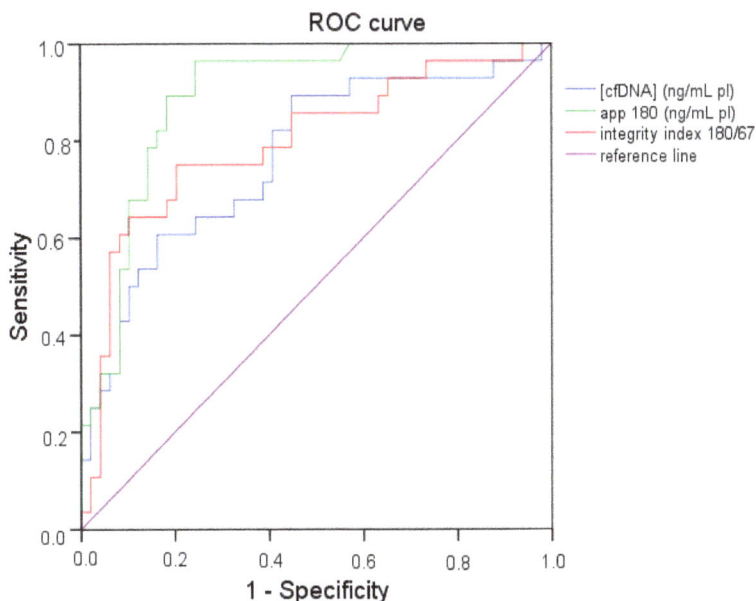

Figure 4. Receiver Operating Characteristic (ROC) curve of cfDNA quantity assessed by two qPCR assays targeting respectively a 67 and a 180 bp amplicon on the *APP* gene and cfDNA integrity index 180/67 in patients with cytological diagnosis of thyroid carcinoma (Thy4/Thy5) and control subjects.

ROC curve analysis was also used to investigate the diagnostic performance of the markers under study by comparing Thy2 subjects with Thy4/Thy5 patients. Among the three considered parameters, only cfDNA integrity showed a good AUC (0.699) with a significant p value ($p = 0.013$).

3. Discussion

Assays performed on tumor surrogate samples are attractive mainly because of limited invasiveness of sample collection. CfDNA determination may represent an affordable way to impact on a potential diagnostic, prognostic and monitoring tool in oncology.

Notwithstanding a presumed lack of specificity of the simple estimation of the quantity and quality of total cfDNA, our previous results in melanoma patients show that, by jointly considering a panel of four biomarkers including two tumor-specific ones, the highest predictive capability was given by total cfDNA followed by integrity index 180/67 [21].

Thus, the aim of the present work was to study cfDNA quantity and the qualitative parameter integrity index 180/67 in thyroid cancer patients; to date, the latter aspect has not been investigated.

The study focused on the quantitative determination of circulating DNA by means of two qPCR assays differing in the amplicon length (67 and 180 bp respectively).

It is generally accepted that the 180 bp-fragment reflects apoptosis, which is the prevalent mechanism of cell death in normal cells, while necrosis, producing much longer DNA fragments, seems to occur more frequently in tumor cells [22,23].

A statistically significant increase could be evidenced in nodular goiter patients when compared to healthy subjects for both the absolute measurements of DNA concentration and the deriving integrity index, similarly to already published results for different types of cancer. In fact, notwithstanding a great heterogeneity in the pre-analytical and analytical steps, most of the papers based on qPCR approaches to measure DNA fragmentation in plasma report an increase of integrity index in tumor patients in comparison to the healthy population [6,7,10,13–19].

Besides qPCR, other approaches such as electrophoresis, electron and atomic force microscopy and, more recently, massive parallel sequencing have been adopted for the determination of cfDNA fragment size [5], providing insights into different aspects of cfDNA. However, among the above mentioned approaches qPCR represents a fast, reliable and cheap method to investigate cfDNA integrity.

The most relevant result in our study is the significant difference in cfDNA integrity index between subjects with benign nodules (Thy2) and patients with cytological diagnosis of thyroid carcinoma (Thy4/Thy5). This finding supports cfDNA integrity as a promising biomarker in the diagnosis of thyroid nodules.

Another important finding is the significant reduction of the considered integrity index evidenced 3–6 months after surgery. This result could be an indication of successful removal of the tumor, analogously to what is reported by Gang et al. [24], and demonstrates the relationship between cfDNA integrity index and tumor presence.

By evaluating the percentage of fragments in the range of length 67–180 bp, we evidenced differences between the cytological categories of the case study (Thy2, Thy3 and Thy4/Thy5) and control subjects, and a higher percentage of fragments in subjects with Thy2 nodules with respect to Thy4/Thy5 patients. Analogously, post-surgical thyroid cancer patients show a higher percentage of small DNA fragments (67–180 bp) in comparison to the pre-surgery condition. The same fragment distribution could be evidenced in our previous study on melanoma patients [9] with the smaller fragment being more abundant in control subjects with respect to melanoma and after tumor removal. These analogous results confirm that total cfDNA analysis (quantity and quality) can be considered a tumor-independent marker.

In conclusion, cfDNA integrity index 180/67 turned out to be a suitable parameter for monitoring cfDNA fragmentation in thyroid cancer patients. These data support the hypothesis that a panel of circulating biomolecular markers can be a non-invasive, valuable tool in the diagnosis of differentiated thyroid cancer and are in line with our previous findings on the detection of BRAFV600E mutation in plasma from patients with this disease [25]. As a perspective in thyroid patients, further studies using DNA integrity index to monitor patients' outcome and the effect of therapy are advisable. Moreover, based on our previous experience in melanoma, a multimarker approach, taking into account different biomarkers related to both cfDNA quantity and quality [21], as well as markers of tumor origin, should be pursued.

4. Materials and Methods

4.1. Patients

The presence of cfDNA circulating in plasma and its integrity features were evaluated in 97 patients (71 females and 26 males, age range 18–90 years, median 56 years) admitted to the Endocrinology Unit of Careggi Teaching Hospital because of nodular goiter between 2011 and 2015.

Patients were submitted to US-FNA according to the adopted guidelines [26] and the cytological diagnoses were made in accordance with the five diagnostic groups of the British Thyroid Association, 2007: Thy 1—non-diagnostic; Thy 2—non-neoplastic, Thy 3—follicular lesions, Thy 4—suspicious of malignancy, Thy 5—diagnostic of malignancy [27]. When more than 1 nodule was sampled by US-FNA in multinodular goiters, the worst cytology was considered.

Our case study is composed as follows: Thy2 (*n* = 25), Thy3 (*n* = 44), Thy4 (*n* = 24) and Thy5 (*n* = 4).

Seventeen patients affected by PTC were submitted to a second blood draw 3–6 months after surgery (mean 4.4 months, range 2.7–6.4 months). In this period of time, 12 patients also underwent radioactive iodine (RAI) treatment according to the adopted guidelines [28,29].

Healthy subjects were used as control populations, (*n* = 49, 26 females and 23 males, age range 25–89 years, median 53 years).

The research protocol was conducted in accordance with the guidelines in the Declaration of Helsinki and approved by the local review board "Comitato Etico Regione Toscana Sezione Area Vasta Centro" (code: CEAVC BIO 16.026, 2017); all the patients signed an informed consent.

4.2. DNA Extraction

Peripheral blood (5 mL) was collected in an ethylenediaminetetraacetic acid (EDTA) tube, transported within one hour to the laboratory and centrifuged twice at 4 °C for 10 min (1600 and 14,000 rcf). Plasma aliquots were stored at −80 °C before use. DNA was extracted from 2 mL of plasma, using the QIAsymphony Circulating DNA Kit (Qiagen, Hilden, Germany).

4.3. Plasma DNA Integrity Index 180/67 by qPCR

The quantity and integrity of the cell-free DNA circulating in plasma was evaluated by a quantitative real-time PCR (qPCR) targeting the human *APP* (Amyloid Precursor protein, chr. 21q21.2) gene (accession NM_000484). The assays were designed in a way that the forward primer and the probe were the same for all amplicons, whereas the reverse primer varied (see ref. [9] for sequences and qPCR protocol). The lengths of the amplicons selected for this study were 67 and 180 bps respectively.

Absolute quantification of the shorter amplicon (67 bp) on the gene *APP* was performed in plasma samples to accurately measure the amount of free circulating DNA per mL plasma, using primers and probe previously reported [30]. This assay was assumed to be able to measure the total amount of circulating plasma DNA, including fragments down to 67 bp of length. Quantification of DNA concentration was obtained by interpolation on an external reference curve ranging from 10 to 10^5 pg/reaction of genomic DNA. (A DNA preparation obtained by Sigma-Aldrich was employed as the standard).

The ratio between the absolute concentration of the longer amplicon (180 bp) and the shorter one (67 bp) defined the integrity index 180/67, which was used to assess the fragmentation of cfDNA. Higher integrity index values indicate that all the cfDNA molecules are at least 180 bp in length in the *APP* gene. Lower integrity indexes mean that cfDNA contains fragments below 180 bp in the same target sequence.

4.4. Statistical Analysis

Statistical analysis was carried out using the SPSS statistics software package version 24 (IBM, Armonk, NY, USA). Quantitative results were evaluated by Mann–Whitney and Wilcoxon signed-rank test. *p* values lower than 0.05 were considered statistically significant. The predictive capability (i.e., diagnostic performance) of each biomarker was investigated by means of the area under the ROC (Receiver-Operating Characteristics) curve (AUC). The ROC curve measures the accuracy of biomarkers when their expression is detected on a continuous scale, displaying the relationship between sensitivity (true-positive rate, *y*-axes) and 1-specificity (false-positive rate, *x*-axes) across all possible threshold values of the considered biomarker. A useful way to summarize the overall diagnostic accuracy of the biomarker is the area under the ROC curve (AUC), the value of which is expected to be 0.5 in the absence of predictive capability, whereas it tends to be 1.00 in the case of high predictive capacity [31].

Acknowledgments: Italian Ministry of Health grant RF-2011-02352294 (Project title: "Circulating cell-free biomarkers in the diagnosis and follow up of differentiated thyroid cancer").

Author Contributions: Pamela Pinzani conceived and designed the experiments; Francesca Salvianti and Irene Mancini performed the experiments; Francesca Salvianti and Pamela Pinzani analyzed the data; Corinna Giuliani, Luisa Petrone and Cinzia Pupilli enrolled the patients and managed clinical data; Vania Vezzosi is the anatomo-pathologist responsible for citomorphological analysis of fine needle aspirates; Pamela Pinzani and Francesca Salvianti wrote the paper; Cinzia Pupilli is the PI of the project granted by the Italian Ministry of Health; Cinzia Pupilli revised the manuscript.

Conflicts of Interest: The authors declare no conflict of interest.

Abbreviations

qPCR   quantitative real-time PCR
cfDNA   cell-free DNA

References

1. Fleischhacker, M.; Schmidt, B. Circulating nucleic acids (CNAs) and cancer—A survey. *Biochim. Biophys. Acta* **2007**, *1775*, 181–232. [CrossRef] [PubMed]
2. Jung, K.; Fleischhacker, M.; Rabien, A. Cell-free DNA in the blood as a solid tumor biomarker—A critical appraisal of the literature. *Clin. Chim. Acta* **2010**, *411*, 1611–1624. [CrossRef] [PubMed]
3. Van der Vaart, M.; Pretorius, P.J. Circulating DNA— Its origin and fluctuation. *Ann. N. Y. Acad. Sci.* **2008**, *1137*, 18–26. [CrossRef] [PubMed]
4. Pinzani, P.; Salvianti, F.; Pazzagli, M.; Orlando, C. Circulating nucleic acids in cancer and pregnancy. *Methods* **2010**, *50*, 302–307. [CrossRef] [PubMed]
5. Jiang, P.; Lo, Y.M. The long and short of circulating cell-free DNA and the Ins and Outs of molecular diagnostics. *Trends Genet.* **2016**, *32*, 360–371. [CrossRef] [PubMed]
6. Umetani, N.; Giuliano, A.E.; Hiramatsu, S.H.; Amersi, F.; Nakagawa, T.; Martino, S.; Hoon, D.S. Prediction of breast tumor progression by integrity of free circulating DNA in serum. *J. Clin. Oncol.* **2006**, *24*, 4270–4276. [CrossRef] [PubMed]
7. Umetani, N.; Kim, J.; Hiramatsu, S.; Reber, H.A.; Hines, O.J.; Bilchik, A.J.; Hoon, D.S. Increased integrity of free circulating DNA in sera of patients with colorectal or periampullary cancer: Direct quantitative PCR for ALU repeats. *Clin. Chem.* **2006**, *52*, 1062–1069. [CrossRef] [PubMed]
8. Schmidt, B.; Weickmann, S.; Witt, C.; Fleischhacker, M. Integrity of cell-free plasma DNA in patients with lung cancer and nonmalignant lung disease. *Ann. N. Y. Acad. Sci.* **2008**, *1137*, 207–213. [CrossRef] [PubMed]
9. Pinzani, P.; Salvianti, F.; Zaccara, S.; Massi, D.; De Giorgi, V.; Pazzagli, M.; Orlando, C. Circulating cell-free DNA in plasma of melanoma patients: Qualitative and quantitative considerations. *Clin. Chim. Acta* **2011**, *412*, 2141–2145. [CrossRef] [PubMed]
10. Wang, B.G.; Huang, H.Y.; Chen, Y.C.; Bristow, R.E.; Kassauei, K.; Cheng, C.C.; Roden, R.; Sokoll, L.J.; Chan, D.W.; Shih, I. Increased plasma DNA integrity in cancer patients. *Cancer Res.* **2003**, *63*, 3966–3968. [PubMed]
11. Jiang, W.W.; Zahurak, M.; Goldenberg, D.; Milman, Y.; Park, H.L.; Westra, W.H.; Koch, W.; Sidransky, D.; Califano, J. Increased plasma DNA integrity index in head and neck cancer patients. *Int. J. Cancer* **2006**, *119*, 2673–2676. [CrossRef] [PubMed]
12. Hanley, R.; Rieger-Christ, K.M.; Canes, D.; Emara, N.R.; Shuber, A.P.; Boynton, K.A.; Libertino, J.A.; Summerhayes, I.C. DNA integrity assay: A plasma-based screening tool for the detection of prostate cancer. *Clin. Cancer Res.* **2006**, *12*, 4569–4574. [CrossRef] [PubMed]
13. Iqbal, S.; Vishnubhatla, S.; Raina, V.; Sharma, S.; Gogia, A.; Deo, S.S.; Mathur, S.; Shukla, N.K. Circulating cell-free DNA and its integrity as a prognostic marker for breast cancer. *Springerplus* **2015**, *4*, 265. [CrossRef] [PubMed]

14. Kamel, A.M.; Teama, S.; Fawzy, A.; El Deftar, M. Plasma DNA integrity index as a potential molecular diagnostic marker for breast cancer. *Tumour Biol.* **2016**, *37*, 7565–7572. [CrossRef] [PubMed]
15. Fawzy, A.; Sweify, K.M.; El-Fayoumy, H.M.; Nofal, N. Quantitative analysis of plasma cell-free DNA and its DNA integrity in patients with metastatic prostate cancer using ALU sequence. *J. Egypt Natl. Canc. Inst.* **2016**, *28*, 235–242. [CrossRef] [PubMed]
16. Leszinski, G.; Lehner, J.; Gezer, U.; Holdenrieder, S. Increased DNA integrity in colorectal cancer. *In Vivo* **2014**, *28*, 299–303. [PubMed]
17. El-Gayar, D.; El-Abd, N.; Hassan, N.; Ali, R. Increased Free Circulating DNA Integrity Index as a Serum Biomarker in Patients with Colorectal Carcinoma. *Asian Pac. J. Cancer Prev.* **2016**, *17*, 939–944. [CrossRef] [PubMed]
18. Bedin, C.; Enzo, M.V.; Del Bianco, P.; Pucciarelli, S.; Nitti, D.; Agostini, M. Diagnostic and prognostic role of cell-free DNA testing for colorectal cancer patients. *Int. J. Cancer* **2017**, *140*, 1888–1898. [CrossRef] [PubMed]
19. Szpechcinski, A.; Rudzinski, P.; Kupis, W.; Langfort, R.; Orlowski, T.; Chorostowska-Wynimko, J. Plasma cell-free DNA levels and integrity in patients with chest radiological findings: NSCLC versus benign lung nodules. *Cancer Lett.* **2016**, *374*, 202–207. [CrossRef] [PubMed]
20. Holdenrieder, S.; Burges, A.; Reich, O.; Spelsberg, F.W.; Stieber, P. DNA integrity in plasma and serum of patients with malignant and benign diseases. *Ann. N. Y. Acad Sci.* **2008**, *1137*, 162–170. [CrossRef] [PubMed]
21. Salvianti, F.; Pinzani, P.; Verderio, P.; Ciniselli, C.M.; Massi, D.; De Giorgi, V.; Grazzini, M.; Pazzagli, M.; Orlando, C. Multiparametric analysis of cell-free DNA in melanoma patients. *PLoS ONE* **2012**, *7*, e49843. [CrossRef] [PubMed]
22. Jahr, S.; Hentze, H.; Englisch, S.; Hardt, D.; Fackelmayer, F.O.; Hesch, R.D.; Knippers, R. DNA fragments in the blood plasma of cancer patients quantitations and evidence for their origin from apoptotic and necrotic cells. *Cancer Res.* **2001**, *61*, 1659–1665. [PubMed]
23. Suzuki, N.; Kamataki, A.; Yamaki, J.; Homma, Y. Characterization of circulating DNA in healthy human plasma. *Clin. Chim. Acta* **2008**, *387*, 55. [CrossRef] [PubMed]
24. Gang, F.; Guorong, L.; An, Z.; Anne, G.P.; Christian, G.; Jacques, T. Prediction of clear cell renal cell carcinoma by integrity of cell-free DNA in serum. *Urology* **2010**, *75*, 262–265. [CrossRef] [PubMed]
25. Pupilli, C.; Pinzani, P.; Salvianti, F.; Fibbi, B.; Rossi, M.; Petrone, L.; Perigli, G.; de Feo, M.L.; Vezzosi, V.; Pazzagli, M.; et al. Circulating BRAFV600E in the diagnosis and follow-up of differentiated papillary thyroid carcinoma. *J. Clin. Endocrinol. Metab.* **2013**, *98*, 3359–3365. [CrossRef] [PubMed]
26. Gharib, H.; Papini, E.; Paschke, R.; Duick, D.S.; Valcavi, R.; Hegedüs, L.; Vitti, P.; AACE/AME/ETA Task Force on Thyroid Nodules. American Association of Clinical Endocrinologists, Associazione Medici Endocrinologi, and European Thyroid Association Medical Guidelines for Clinical Practice for the Diagnosis and Management of Thyroid Nodules. *Endocr. Pract.* **2010**, *16* (Suppl. 1), 1–43. [CrossRef] [PubMed]
27. British Thyroid Association; Royal College of Physicians. *Diagnostic Categories. Guidelines for the Management of Thyroid Cancer*, 2nd ed.; Lavenham Press: Suffolk, UK, 2007; p. 10.
28. Cooper, D.S.; Doherty, G.M.; Haugen, B.R.; Kloos, R.T.; Lee, S.L.; Mandel, S.J.; Mazzaferri, E.L.; McIver, B.; Pacini, F.; Schlumberger, M.; et al. Revised American Thyroid Association management guidelines for patients with thyroid nodules and differentiated thyroid cancer. *Thyroid* **2009**, *19*, 1167–1214. [CrossRef] [PubMed]
29. Pacini, F.; Castagna, M.G.; Brilli, L.; Pentheroudakis, G.; ESMO Guidelines Working Group. Thyroid cancer: ESMO Clinical Practice Guidelines for diagnosis, treatment and follow-up. *Ann. Oncol.* **2012**, *23* (Suppl. 7), VII110–VII119. [CrossRef] [PubMed]
30. Lehmann, U.; Glöckner, S.; Kleeberger, W.; von Wasielewski, H.F.; Kreipe, H. Detection of gene amplification in archival breast cancer specimens by laser-assisted microdissection and quantitative real-time polymerase chain reaction. *Am. J. Pathol.* **2000**, *156*, 1855–1864. [CrossRef]
31. Hanley, J.A.; McNeil, B.J. The meaning and use of the area under a receiver operating characteristic (ROC) curve. *Radiology* **1982**, *143*, 29–36. [CrossRef] [PubMed]

International Journal of
Molecular Sciences

MDPI

Article

Gene Expression (mRNA) Markers for Differentiating between Malignant and Benign Follicular Thyroid Tumours

Bartosz Wojtas [1,2,†], Aleksandra Pfeifer [1,3,†], Malgorzata Oczko-Wojciechowska [1], Jolanta Krajewska [1], Agnieszka Czarniecka [4], Aleksandra Kukulska [1], Markus Eszlinger [5], Thomas Musholt [6], Tomasz Stokowy [1,3,7], Michal Swierniak [1,8], Ewa Stobiecka [9], Ewa Chmielik [9], Dagmara Rusinek [1], Tomasz Tyszkiewicz [1], Monika Halczok [1], Steffen Hauptmann [10], Dariusz Lange [9], Michal Jarzab [11], Ralf Paschke [12] and Barbara Jarzab [1,*]

1   Department of Nuclear Medicine and Endocrine Oncology, Maria Sklodowska-Curie Institute—Oncology Center, Gliwice Branch, Wybrzeze Armii Krajowej 15, 44-101 Gliwice, Poland; Bartosz.Wojtas@io.gliwice.pl (B.W.); Aleksandra.Pfeifer@io.gliwice.pl (A.P.); Malgorzata.Oczko-Wojciechowska@io.gliwice.pl (M.O.-W.); Jolanta.Krajewska@io.gliwice.pl (J.K.); Aleksandra.Kukulska@io.gliwice.pl (A.K.); tomasz.stokowy@k2.uib.no (T.S.); michal.swierniak@wum.edu.pl (M.S.); Dagmara.Rusinek@io.gliwice.pl (D.R.); Tomasz.Tyszkiewicz@io.gliwice.pl (T.T.); Monika.Kowal@io.gliwice.pl (M.H.)
2   Laboratory of Molecular Neurobiology, Neurobiology Center, Nencki Institute of Experimental Biology, Pasteura 3, 02-093 Warsaw, Poland
3   Faculty of Automatic Control, Electronics and Computer Science, Silesian University of Technology, Akademicka 2A, 44-100 Gliwice, Poland
4   The Oncologic and Reconstructive Surgery Clinic, Maria Sklodowska-Curie Institute—Oncology Center, Gliwice Branch, Wybrzeze Armii Krajowej 15, 44-101 Gliwice, Poland; Agnieszka.Czarniecka@io.gliwice.pl
5   Department of Oncology & Arnie Charbonneau Cancer Institute, Cumming School of Medicine, University of Calgary, Calgary, AB T2N 4N1, Canada; markus.eszlinger1@ucalgary.ca
6   Department of General, Visceral, and Transplantation Surgery, University Medical Center of the Johannes Gutenberg University, D55099 Mainz, Germany; musholt@uni-mainz.de
7   Department of Clinical Science, University of Bergen, 5020 Bergen, Norway
8   Genomic Medicine, Department of General, Transplant, and Liver Surgery, Medical University of Warsaw, Zwirki i Wigury 61, 02-093 Warsaw, Poland
9   Tumor Pathology Department, Maria Sklodowska-Curie Institute—Oncology Center, Gliwice Branch, Wybrzeze Armii Krajowej 15, 44-101 Gliwice, Poland; Ewa.Stobiecka@io.gliwice.pl (E.S.); Ewa.Chmielik@io.gliwice.pl (E.C.); dlange693@gmail.com (D.L.)
10 Department of Pathology, Martin Luther University Halle-Wittenberg, 06108 Halle (Saale), Germany; steffen.hauptmann@patho-ao.de
11 III Department of Radiotherapy and Chemotherapy, Maria Sklodowska-Curie Institute—Oncology Center, Gliwice Branch, Wybrzeze Armii Krajowej 15, 44-101 Gliwice, Poland; Michal.Jarzab@io.gliwice.pl
12 Division of Endocrinology, Departments of Medicine, Pathology, Biochemistry & Molecular Biology, and Oncology, and Arnie Charbonneau Cancer Institute, Cumming School of Medicine, University of Calgary, Calgary, Alberta T2N 4N1, Canada; ralf.paschke@ucalgary.ca
*   Correspondence: Barbara.Jarzab@io.gliwice.pl; Tel.: +48-32-278-93-01; Fax: +48-32-231-93-10
†   These authors contributed equally to this work.

Academic Editor: Daniela Gabriele Grimm
Received: 3 April 2017; Accepted: 28 May 2017; Published: 2 June 2017

Abstract: Distinguishing between follicular thyroid cancer (FTC) and follicular thyroid adenoma (FTA) constitutes a long-standing diagnostic problem resulting in equivocal histopathological diagnoses. There is therefore a need for additional molecular markers. To identify molecular differences between FTC and FTA, we analyzed the gene expression microarray data of 52 follicular neoplasms. We also performed a meta-analysis involving 14 studies employing high throughput methods (365 follicular neoplasms analyzed). Based on these two analyses, we selected 18 genes differentially expressed between FTA and FTC. We validated them by quantitative real-time polymerase chain reaction (qRT-PCR) in an independent set of 71 follicular neoplasms from formaldehyde-fixed paraffin embedded (FFPE) tissue material. We confirmed differential expression for 7 genes (*CPQ, PLVAP, TFF3, ACVRL1, ZFYVE21, FAM189A2,* and *CLEC3B*). Finally, we created a classifier that distinguished between FTC and FTA with an accuracy of 78%, sensitivity of 76%, and specificity of 80%, based on the expression of 4 genes (*CPQ, PLVAP, TFF3, ACVRL1*). In our study, we have demonstrated that meta-analysis is a valuable method for selecting possible molecular markers. Based on our results, we conclude that there might exist a plausible limit of gene classifier accuracy of approximately 80%, when follicular tumors are discriminated based on formalin-fixed postoperative material.

Keywords: follicular thyroid adenoma; follicular thyroid cancer; gene expression; microarray; meta-analysis

1. Introduction

Follicular neoplasms are the most controversial area in the thyroid pathology. According to World Health Organization (WHO) follicular adenoma is a benign, encapsulated tumor of the thyroid showing follicular cell differentiation [1]. This tumor demonstrates no evidence of capsular or vascular invasion. Follicular carcinoma is a malignant tumor showing evidence of follicular cell differentiation. The distinction between follicular adenoma and carcinoma is based on the presence of capsular and/or vascular invasion. Capsular invasion is defined by tumor penetration through the entire thickness of the capsule [1]. The invading tumor nests should present a connection with main tumor mass. The interpretation of capsular invasion may be sometimes problematic. According to the literature data and our experience there is a group of patients with only partial capsular invasion but presenting metastases of follicular carcinoma [2]. Yamashina analyzed entire circumference of tumor capsules of follicular neoplasms and observed that tumors with only capsular invasion in initial sections also presented vascular invasion on additional slices adjacent to tumor capsule [3]. Therefore it would be advisable to evaluate gene expression of follicular adenomas and follicular carcinomas.

Between 2000 and 2014, numerous studies have investigated the gene expression (mRNA) profile that would differentiate follicular thyroid adenoma (FTA) from follicular thyroid cancer (FTC) to improve the diagnostic process and to find features of follicular thyroid tumours important for malignant potential (Table S1) [4–17]. However, reproducibility of results obtained between mentioned publications was rather low. This could be a consequence of slight molecular differences between FTC and FTA [18,19] or the insufficient sample size used in these studies. Genetic alterations, such as *RAS* gene family somatic mutations or *PAX8/PPARG* translocations, although very promising in initial studies, were not found to be specific for follicular carcinoma, as these genetic alterations occurred in both FTCs and FTAs with similar frequencies [20–22]. These doubts stimulated us to carry on a meta-analysis.

In our study, we also raised the problem of oncocytic tumors. WHO involves oncocytic thyroid carcinoma (OTC) to FTC and respectively oncocytic adenoma to FTA. Oncocytic tumors (Hurthle cell tumors) are believed to have a different gene expression profile [23,24]. Ganly et al. demonstrated on the basis of mutational, transcriptional, and copy number profiles that Hurthle cell carcinoma was a unique thyroid cancer distinct from papillary thyroid cancer (PTC) and FTC [24].

In the present study we decided to base on FTC definition, proposed by the WHO. Nevertheless, we tried to check whether an inclusion of oncocytic follicular carcinoma does not influence on molecular markers selection. OTC is composed predominantly of oncocytic cells. These tumors are associated with a higher frequency of extrathyroidal extension, local recurrence, nodal metastases in more than 30% of cases and occasionally distant lung and bone metastases [1]. Compared with conventional follicular carcinomas, oncocytic follicular carcinomas are more aggressive [1]. Therefore, it may be reasonable to involve oncocytic feature in our analysis.

Most recent thyroid studies have focused on identifying molecular markers supporting pre-operative FNAB examination to exclude malignancy [25,26]. In 2010, Chudova et al. published a study focused on determining the general preoperative distinction between benign and malignant thyroid nodules, which appeared promising and resulted in the establishment of the Afirma classifier [25]. Our approach, used in the present study, is different.

In our study, we utilised two different approaches to select new gene-expression markers for differentiating between FTC and FTA tumours. We performed a two-step analysis: first a statistical testing of a large gene expression microarray dataset of FTC and FTA previously generated in our laboratory [18,27], and next, a meta-analysis of all available datasets, to select the most robustly represented markers [4–17] (Figure 1). Such approach allowed us to select independent genes coming from own dataset and from a meta-analysis. Meta-analysis by combining the results of various studies enabled us to draw common conclusions. The results of both analyses were further validated by quantitative real-time polymerase chain reaction (qRT-PCR) using an independent dataset of follicular tumours.

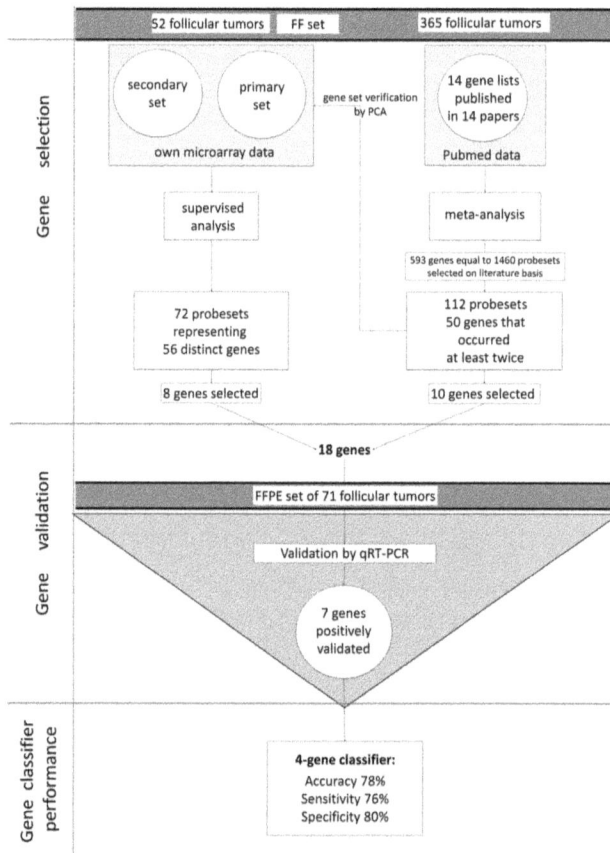

Figure 1. Presentation of a study scheme.

2. Results

2.1. Supervised Analysis of Gene Expression Microarrays

Fresh-frozen (FF) material from 52 tumors (27 FTC, 25 FTA) was used for our gene expression microarray experiment and divided into primary and secondary sets. The primary one was considered as highly reliable dataset and contained all samples that were independently and concordantly diagnosed by two thyroid pathology experts. The secondary set contained samples that were diagnosed by only one expert, equivocal samples diagnosed by two experts and a one sample that was discordantly diagnosed according to malignancy.

To select potential molecular markers useful in the distinction between FTA and FTC, we considered genes that were differentially expressed in the primary and secondary microarray datasets. We compared the lists of genes obtained in the analysis of the primary and the secondary sets and selected only those that were significant in both sets. Our secondary microarray set contained borderline and ambiguous cases, and we established genes as valuable and characteristic when they were also differentially expressed in this set.

There were 72 differentially expressed probe sets (representing 56 distinct genes) and 6 non-annotated probe sets. Eight genes were selected (*ACVRL1, CLEC3B, DIP2B, GABARAPL2, ZFYVE21, LIMK2, ZMYND11,* and *MAFB*) for validation by qRT-PCR (Table 1). Those genes were characterised by low false discovery rate (FDR) value, high fold-change, and from our point of view, they could be biologically interesting. Another selection criterion was that these genes were not previously validated as markers differentiating FTCs from FTAs.

As it has been shown that the oncocytic FTC is a unique thyroid cancer distinct from non-oncocytic FTC [24] we decided to perform an additional analysis. We excluded oncocytic samples from microarray dataset (just for the sake of this particular analysis) and evaluated the significance of eight selected genes in the dataset comprising of non-oncocytic samples only to investigate the differences between FTC and FTA (7 FTC and 11 FTA). All these genes showed significant differential expression between FTC and FTA in this dataset (Table 1).

Table 1. Differentially expressed genes selected based on analysis of our own microarray dataset.

No.	Gene Symbol	Gene Name	Affy ID	Primary Dataset				Primary Dataset—Evaluation of Non-Oncocytic Samples Only
				FDR Corrected p-Value	Mean Expression in FTC	Mean Expression in FTA	Fold-Change	FDR Corrected p-Value
1	ACVRL1	activin A receptor type II-like 1	226950_at	0.07	5.52	7.02	0.35	0.12
2	CLEC3B	C-type lectin domain family 3, member B	205200_at	0.08	7.54	9.52	0.25	0.13
3	GABARAPL2	GABA(A) receptor-associated protein-like 2	209046_s_at	0.08	11.05	11.84	0.58	0.15
4	ZFYVE21	zinc finger, FYVE domain containing 21	219929_s_at	0.07	7.39	8.67	0.41	0.04
5	LIMK2	LIM domain kinase 2	217475_s_at	0.07	4.32	5.84	0.35	0.12
6	ZMYND11	zinc finger, MYND domain containing 11	1554159_a_at	0.10	6.60	8.05	0.37	0.15
7	DIP2B	DIP2 disco-interacting protein 2 homolog B (Drosophila)	224872_at	0.11	8.23	7.40	1.78	0.16
8	MAFB	v-maf musculoaponeurotic fibrosarcoma oncogene homolog B (avian)	222670_s_at	0.08	8.23	9.78	0.34	0.13

The genes were selected for validation from the genes differentially expressed both in primary and secondary microarray set. Values represented in the table are from analysis of the primary microarray data set.

2.2. Meta-Analysis

We included 14 papers in which the difference in gene expression between FTC and FTA was assessed by a high throughput method (expression microarrays, serial analysis of gene expression (SAGE), high-throughput differential screening by serial analysis of gene expression (HDSS), adapter-tagged competitive polymerase chain reaction (ATAC-PCR)) (Table S1). The papers were published during the years 2000–2014 and in total 365 samples (201 FTA and 164 FTC) were analyzed.

All reported genes differentiating FTC and FTA were extracted from these publications. We identified 600 genes reported in at least one publication, while 57 genes were reported in more than one publication. Fifty out of those 57 genes were reported with concordant direction of change (Table 2). Seven genes (*CA4, EGR2, FAM189A2, KCNAB1, CPQ, SLC26A4, TFF3*) were reported in 3 publications. Two of these genes (*CA4,* and *KCNAB1*) were already evaluated by qRT-PCR as described in our previous study [27].

Among the genes selected based on the meta-analysis, ten genes were chosen for qRT-PCR validation. We chose five down-regulated genes that occurred in three papers (*EGR2, FAM189A2, SLC26A4, TFF3, CPQ*), four up-regulated genes that occurred in two papers (*CKS2, GDF15, ASNS, DDIT3*), and one down-regulated gene that occurred in two papers and simultaneously showed significant differences in expression in our primary microarray dataset (*PLVAP*).

Table 2. The results of a meta-analysis of 14 papers, in which differences in gene expression profile between follicular thyroid cancers (FTC) and follicular thyroid adenomas (FTA) were assessed by a high throughput method. Ten genes (highlighted in bold) were selected for our qRT-PCR validation.

No.	Entrez Gene ID	Symbol	Name	Number of Papers	References	Gene Regulation
1	762	*CA4*	carbonic anhydrase IV	3	[5,9,16]	down
2	1959	**EGR2**	**early growth response 2**	**3**	**[5,14,16]**	**down**
3	9413	**FAM189A2**	**family with sequence similarity 189, member A2**	**3**	**[5,9,12]**	**down**
4	7881	*KCNAB1*	potassium voltage-gated channel, shaker-related subfamily, beta member 1	3	[6,9,16] Confirmed by us [27]	down
5	10404	**CPQ**	**carboxypeptidase Q**	**3**	**[9,11,14]**	**down**
6	5172	**SLC26A4**	**solute carrier family 26 (anion exchanger), member 4**	**3**	**[6,14,16]**	**down**
7	7033	**TFF3**	**trefoil factor 3 (intestinal)**	**3**	**[5,6,10]**	**down**
8	185	*AGTR1*	angiotensin II receptor, type 1	2	[13,16]	down
9	822	*CAPG*	capping protein (actin filament), gelsolin-like	2	[14,17]	down
10	1306	*COL15A1*	collagen, type XV, alpha 1	2	[5,13]	down
11	1363	*CPE*	carboxypeptidase E	2	[9,17]	down
12	3491	*CYR61*	cysteine-rich, angiogenic inducer, 61	2	[8,16]	down
13	1733	*DIO1*	deiodinase, iodothyronine, type I	2	[6,12]	down
14	11072	*DUSP14*	dual specificity phosphatase 14	2	[5,16]	down
15	129080	*EMID1*	EMI domain containing 1	2	[5,7]	down
16	953	*ENTPD1*	ectonucleoside triphosphate diphosphohydrolase 1	2	[9,14]	down
17	8857	*FCGBP*	Fc fragment of IgG binding protein	2	[5,17]	down
18	2354	*FOSB*	FBJ murine osteosarcoma viral oncogene homolog B	2	[16,17]	down
19	2697	*GJA1*	gap junction protein, alpha 1, 43 kDa	2	[5,11]	down
20	55830	*GLT8D1*	glycosyltransferase 8 domain containing 1	2	[5,11]	down
21	221395	*GPR116*	G protein-coupled receptor 116	2	[5,9]	down
22	3043	*HBB*	hemoglobin, beta	2	[12,13]	down

Table 2. *Cont.*

No.	Entrez Gene ID	Symbol	Name	Number of Papers	References	Gene Regulation
23	3309	HSPA5	heat shock 70 kDa protein 5 (glucose-regulated protein, 78 kDa)	2	[9,17]	down
24	3400	ID4	inhibitor of DNA binding 4, dominant negative helix-loop-helix protein	2	[5,8]	down
25	3590	IL11RA	interleukin 11 receptor, alpha	2	[5,11]	down
26	9452	ITM2A	integral membrane protein 2A	2	[9,16]	down
27	3708	ITPR1	inositol 1,4,5-trisphosphate receptor, type 1	2	[5,11]	down
28	3725	JUN	jun proto-oncogene	2	[5,16]	down
29	3912	LAMB1	laminin, beta 1	2	[5,11]	down
30	744	MPPED2	metallophosphoesterase domain containing 2	2	[16,17]	down
31	22795	NID2	nidogen 2 (osteonidogen)	2	[5,7]	down
32	3164	NR4A1	nuclear receptor subfamily 4, group A, member 1	2	[12,16]	down
33	22925	PLA2R1	phospholipase A2 receptor 1, 180 kDa	2	[12,16]	down
34	**83483**	**PLVAP**	**plasmalemma vesicle associated protein**	**2**	**[9,13]**	**down**
35	5583	PRKCH	protein kinase C, eta	2	[9,14]	down
36	23180	RFTN1	raftlin, lipid raft linker 1	2	[5,9]	down
37	8490	RGS5	regulator of G-protein signaling 5	2	[9,13]	down
38	6414	SEPP1	selenoprotein P, plasma, 1	2	[5,14]	down
39	7038	TG	Thyroglobulin	2	[10,17]	down
40	4982	TNFRSF11B	tumor necrosis factor receptor superfamily, member 11b	2	[5,11]	down
41	7173	TPO	thyroid peroxidase	2	[10,17]	down
42	**440**	**ASNS**	**asparagine synthetase (glutamine-hydrolyzing)**	**2**	**[5,9]**	**up**
43	771	CA12	carbonic anhydrase XII	2	[5,12]	up
44	**1164**	**CKS2**	**CDC28 protein kinase regulatory subunit 2**	**2**	**[16,17]**	**up**
45	**1649**	**DDIT3**	**DNA-damage-inducible transcript 3**	**2**	**[5,7]**	**up**
46	2358	FPR2	formyl peptide receptor 2	2	[5,11]	up
47	**9518**	**GDF15**	**growth differentiation factor 15**	**2**	**[9,17]**	**up**
48	2896	GRN	Granulin	2	[4,8]	up
49	3486	IGFBP3	insulin-like growth factor binding protein 3	2	[5,10]	up
50	23089	PEG10	paternally expressed 10	2	[5,11]	up

Table 2 shows the Entrez ID, gene symbol, gene name, number of papers in which a particular gene occurs, references to the papers, regulation direction (up–up-regulated in FTC; down–down-regulated in FTC).

2.3. Principal Component Analysis

We selected 593 genes that occurred at least once in the meta-analysis (excluding seven genes with discordant direction of change). We identified HG-U133 PLUS 2 Affymetrix microarray probe sets for these genes. There were 1460 such probe sets (for some genes there was more than one probe set). Next, we performed PCA of our own microarray samples (combined primary and secondary dataset) based on these 1460 probe sets (Figure 2, upper plot). Similarly, we selected 50 genes that occurred at least twice in investigated papers (excluding the genes with discordant direction of change). We identified HG-U133 PLUS 2 Affymetrix microarray probe sets for these genes. There were 112 such probe sets. We performed PCA based on these 112 probe sets (Figure 2, lower plot). Although gene selection was independent of the microarray dataset, we achieved good discrimination of benign and malignant tumors in both analyses. However, the discrimination was not perfect, because a few FTA samples clustered with the FTC group, and a few FTC samples clustered with the FTA group.

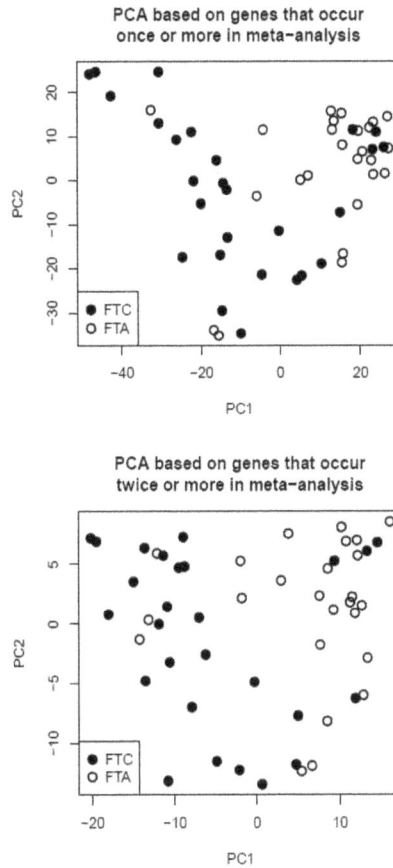

Figure 2. Principal component analysis (PCA) results. PCA plots of samples from our own microarray dataset, based on genes selected in the meta-analysis that occurred in at least one paper (**upper** plot) or at least two papers (**lower** plot).

2.4. qRT-PCR Validation

qRT-PCR was used to validate 18 genes selected based on the analysis of our own microarray dataset and the meta-analysis (Table 3). *GABARAPL2*, *DDIT3*, and *SLC26A4* amplification was not possible in the FFPE samples (probably due to low endogenous expression), and therefore, it was excluded from validation.

Log-transformed expression levels of the remaining 15 genes were analysed using the Student's *t*-test (Table 3). Two FTC samples were extreme outliers (the expression was higher than third quartile (Q3) + 6 × interquartile range (IQR)) in two distinct genes. These samples were excluded from further analysis. Differential expression of *CPQ*, *PLVAP*, *TFF3*, *ACVRL1*, *ZFYVE21*, *FAM189A2*, and *CLEC3B* was confirmed by qRT-PCR contrary to the expression of *ZMYND11*, *LIMK2*, *DIP2B*, *MAFB*, *CKS2*, *ASNS*, *EGR2*, and *GDF15*. All confirmed genes were downregulated in FTC and the direction of change agreed between qRT-PCR data and microarray/meta-analysis data. Boxplots of qRT-PCR results for significantly differentially expressed genes are shown on Figure 3. Based on our results, the following genes that most significantly differentiated between FTC/FTA were selected by a meta-analysis: *CPQ* (*PGCP*), *PLVAP*, and *TFF3*.

Table 3. Comparison of gene expression between FTC (29 samples) and FTA (40 samples) in qRT-PCR dataset (*t*-test and two-way ANOVA calculated *p*-values corrected for multiple tests by FDR method). FDR corrected *p*-values below 0.05 are highlighted in bold.

No.	Gene	Gene Selection	*t*-Test—FDR Corrected *p*-Value	Fold Change (FTC/FTA)	Two-Way ANOVA—FDR Corrected *p*-Value
1	*ACVRL1*	Microarrays	**0.0017**	0.58	**0.0036**
2	*ZFYVE21*	Microarrays	**0.0024**	0.69	**0.0036**
3	*CLEC3B*	Microarrays	**0.027**	0.75	**0.045**
4	*ZMYND11*	Microarrays	0.068	0.81	0.17
5	*LIMK2*	Microarrays	0.093	0.79	0.17
6	*DIP2B*	Microarrays	0.23	0.86	**0.04**
7	*MAFB*	Microarrays	0.44	0.89	0.56
8	*GABARAPL2*	Microarrays	Amplification not possible in FFPE samples		
9	*CPQ*	Meta-analysis	**0.000001**	0.49	**0.0004**
10	*PLVAP*	Meta-analysis	**0.00001**	0.51	**0.0001**
11	*TFF3*	Meta-analysis	**0.0004**	0.48	**0.0036**
12	*FAM189A2*	Meta-analysis	**0.0094**	0.68	**0.016**
13	*GDF15*	Meta-analysis	0.058	1.49	0.99
14	*CKS2*	Meta-analysis	0.69	1.07	0.94
15	*ASNS*	Meta-analysis	0.90	1.02	0.17
16	*EGR2*	Meta-analysis	0.90	0.97	0.89
17	*DDIT3*	Meta-analysis	Amplification not possible in FFPE samples		
18	*SLC26A4*	Meta-analysis	Amplification not possible in FFPE samples		

Two-way analysis of variance (ANOVA) was used in order to adjust for oncocytic feature and the results are shown in column "two-way ANOVA—FDR corrected *p*-value".

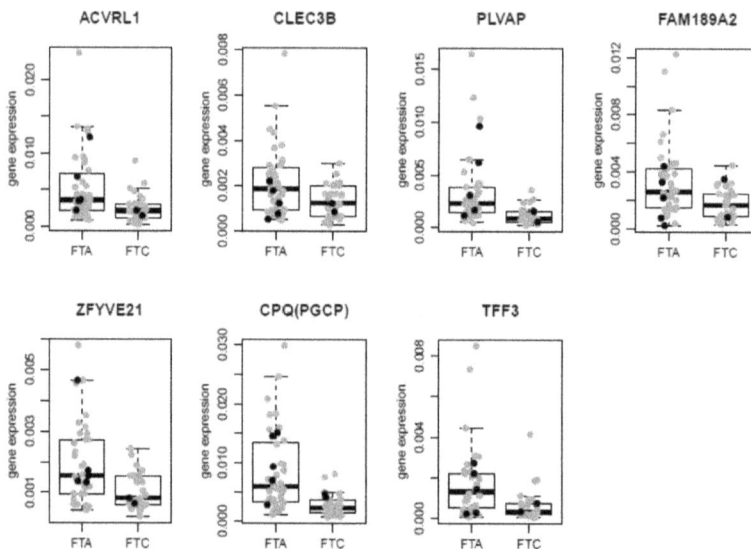

Figure 3. The normalized relative expression levels of positively validated genes in the FFPE dataset of 69 samples. Boxplots superimposed with scatterplots are shown. The line inside each box corresponds to median. Upper and lower edges of boxes correspond to first (Q1) and third (Q3) quartiles, respectively. The whiskers extend to smallest and largest observations within 1.5 times interquartile range (IQR) from the box. Black dots represent *RAS* mutation carrying samples, and grey dots represent samples without *RAS* mutation.

A multivariate ANOVA with two factors: malignancy and oncocytic feature was also performed, in order to evaluate the differential expression between FTC and FTA after adjusting for the effect of oncocytic feature. All seven genes significant in the Student's *t*-test were also significant in this ANOVA analysis (Table 3). Adding the additional variables such as age, gender, and *RAS* mutation status did not substantially modify the ANOVA results.

2.5. Classifier Performance

To evaluate the usefulness of selected genes as diagnostic support, we performed sample classification based on the FFPE dataset. Log-transformation of the gene expression values and a leave-one-out cross-validation of the classifier was performed. In each iteration, the samples were divided into two independent sets: all but one sample were used for significance threshold tuning, gene selection, and classifier training, and the remaining sample was used for testing. Diagonal linear discrimination analysis (DLDA) algorithm was used for the classifier training. After performing all iterations, the classifier's performance was calculated. The accuracy, sensitivity, and specificity were 78% (95% confidence interval (CI): 67–87%), 76% (95% CI: 56–90%), and 80% (95% CI: 64–91%), respectively. The classifier involved 4 genes with *p*-value below 0.0005 in the Student's *t*-test, namely *CPQ*, *PLVAP*, *TFF3*, and *ACVRL1*. When accuracy was calculated for non-oncocytic (45 tumors) and oncocytic (24 tumors) tumors separately it was 84% (95% CI: 71–94%) and 67% (95% CI: 45–84%), respectively.

A receiver operating characteristic (ROC) curve was also created in order to assess the diagnostic efficacy of the classifier (Figure 4). The area under the ROC curve (AUC) equals 0.84.

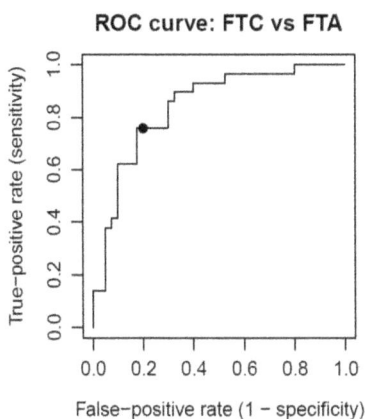

Figure 4. Receiver operating characteristc (ROC) curve analysis for the predictive power of 4-gene classifier, estimated in qRT-PCR dataset. Using a cutoff probability of 50% (marked with black dot), we obtained sensitivity of 76% and specificity of 80%. The calculated area under the ROC curve was 0.84.

2.6. RAS Mutation Status

The presence of the *RAS* gene mutation was investigated in freshly frozen FTC (27) and FTA (25) samples. We identified 3 FTC samples with *NRAS* codon 61 mutation and 1 with *KRAS* codon 61 mutation (in total 14.8%). In the FTA samples, we identified only 1 mutated sample with *NRAS* codon 61 mutation (4%) (Table S2). The frequencies of *RAS* gene mutations in malignant and benign samples did not differ significantly.

The status of the *RAS* gene mutations was also analysed in FFPE specimens, however due to limitations related to sample quantity, 14 samples were not fully profiled (only *NRAS* codon 61 was analysed and mutations were excluded in these samples). Among 31 FFPE FTC specimens, 2 samples with *NRAS* codon 61 mutation and 1 with *HRAS* codon 12 mutation were identified (9.7%) (Table S3). More *RAS* mutations were observed in FFPE FTA samples; however, the difference was not significant. Among 40 FFPE FTA specimens, 3 samples with *NRAS* codon 61 mutations, 1 sample with *KRAS* codon 12 mutation, and 1 with *HRAS* codon 61 mutation were detected (12.5%) (Table S3). However, analysis of the total prevalence of *RAS* mutations in FTC and FTA, regardless of the method used for tissue preservation (FF vs. FFPE) demonstrated that there was no difference in the occurrence of *RAS* mutations between FTC and FTA: 12% and 9.3%, respectively.

3. Discussion

The differential diagnostics between FTC and FTA is still challenging, particularly because in a molecular sense these lesions lie on a continuum, with similar molecular profiles. Perhaps the 2nd or 3rd molecular hit converts adenoma to carcinoma [28,29] In our study, we performed a meta-analysis of markers differentiating FTC and FTA to summarise the results obtained over a 15-year period (2000–2014), and described in multiple papers.

We obtained a list of 50 genes that were significantly differentially expressed in concordant direction in two or more such papers. We selected 10 genes from the meta-analysis and positively validated 4 of them: *CPQ, PLVAP, TFF3,* and *FAM189A2*. While, of the 8 genes selected from our own gene expression microarray dataset, three genes: *ACVRL1, CLEC3B,* and *ZFYVE21*, were positively validated by qRT-PCR (Table 3).

Due to small number of *RAS* mutation positive samples, we were not able to establish its influence on the expression of genes selected for qRT-PCR validation (Figures S1 and S2).

Finally, we created a gene classifier involving 4 genes (*CPQ, PLVAP, TFF3,* and *ACVRL1*) that showed a diagnostic accuracy of 78%, sensitivity of 76%, and specificity of 80% for FTC and FTA differentiation. We are aware that our set of genes requires confirmation by an independent clinical study, similar to the study by Alexander et al. [30], which positively verified the clinical utility of a gene classifier proposed by Chudova et al. [25]. However, there are some important differences between Afirma and our approach. While FNAB-based Afirma classifier, used in a preoperative diagnostics, considered all malignant tumors and differed them from benign ones, our classifier was devoted to discriminate only between FTC and FTA on the basis of postoperative material. We did not consider the results of fine-needle aspiration biopsy (FNAB) at any time during our analyses as well as did not link our results to Bethesda Categories. We hope that our classifier may help in such cases where there is a dilemma in a post-operative diagnostics in FTA/FTC distinction. Thus, our work may not be considered as a kind of confirmation of Afirma results.

Transcription profiling, as a method for selection of gene expression markers for distinguishing follicular neoplasms, has been used for over a decade. However, to date, no powerful molecular markers have been established. Similarly, our previous study did not fully accomplish this goal [27]. Therefore, we decided to strengthen our results by performing a meta-analysis of all available studies related to FTC and FTA differentiation [4–17].

The analysis of genes differentially expressed in FTC and FTA in our own gene expression microarray dataset revealed 56 genes. Genes with higher fold-changes and lower *p* values (Table S4), as well as those related to other types of cancer or tumour aggressiveness were preferably selected for qRT-PCR validation. One of these genes, *ACVRL1* correlated with tumour progression in patients with head and neck cancers [31]; whereas two other genes: *ZFYVE21*, and *CLEC3B* were related to cancer invasiveness [32,33]. Four genes, obtained from the meta-analysis were subsequently positively validated *CPQ, PLVAP, TFF3,* and *FAM189A2*.

Based on the meta-analysis, it appears that building an accurate classifier to differentiate FTCs from FTAs is impossible, even using a large dataset of follicular tumour samples (365 samples in meta-analysis). Therefore, we propose that an accuracy of approximately 80% constitutes a plausible limit of FTC vs. FTA gene classifier performance when analysis is performed in postoperative formalin-fixed material [27].

Possible reason for not satisfying classifier accuracy is that follicular tumours are too similar at the gene expression level. Another hypothetic possible reason is that FTC and FTA classes may have been incorrectly assigned prior to the microarray experiments. Histopathological diagnosis in case of follicular tumours can be influenced by intraobserver variability [34]. To circumvent this, we involved two experienced pathologists in the diagnostic process. It is possible however, that some minimally invasive FTCs did not yet demonstrate any signs of vascular or capsular invasion, and were classified as FTAs.

We assume that FTCs and FTAs are biologically different as they have different clinical outcomes. We are however aware that to date, histopathology constitutes the best option in differential diagnostics of follicular tumours, but a gene-classifier may provide more information in difficult cases. Therefore, we may try to use classifiers ([27], current classifier) to distinguish FTCs and FTAs without histopathological data (unsupervised approach). The results from an unsupervised approach can then be compared to histopathological evaluation, with focus on cases showing discrepancy between the histopathology and classifier data.

It is possible, that we may not reach better classifier performance because of over-simplification that we applied in our analysis. We assumed that both FTC and FTA tumours are internally homogenous, but quite often they are not and they may encompass different zones of differentiation or different histopathological features [35]. Neither FTA nor FTC are completely similar. Considering diversity of biology we cannot expect to cover the whole biological variance with four genes only.

In the present study we decided not to include PTC, because it demonstrated its own, characteristic gene expression profile [36] and the differences between PTC and FTC were quite intense [37]. We believe that an inclusion of PTC to malignant samples may lead to inadequate conclusions, whereas without PTC the study is much cleaner.

The low number of *RAS*-positive samples did not allow an evaluation of the impact of the *RAS* gene mutations on the gene expression profile. However, *RAS*-positive samples did not cluster differentially compared to samples not carrying mutations based on the unsupervised PCA analysis, which suggests small biological differences (Figures S3 and S4). Interestingly, the prevalence of *RAS* somatic mutations in our own FF FTC dataset was 12%, while other studies show the prevalence of *RAS* mutation at 60% [38]. This result might be attributable to the population in the studied region of Europe. Unfortunately, we were not able to analyse of the *RAS* gene mutations in 14 samples due to limited amount of material.

We are aware that our findings would be more robust if we use a single technique of tissue preservation but to a much larger group and the using of FFPE material for validation had a possible limitation. Performing gene expression on FFPE is very challenging and these results could even improve when using cryopreserved samples instead. However, malignant follicular thyroid neoplasms are rare and we had to base on the available material. We did our best to collect as large group as it was possible. We used qRT-PCR with multiple reference genes, to assure that we can amplify sequences coming from reference genes in our tumor samples. Moreover, the results obtained in our study were validated on the independent set of samples. We believe that our results constitute an essential input into the better understanding of molecular biology of follicular thyroid neoplasms.

4. Materials and Methods

4.1. Material

4.1.1. Clinical Materials for Gene Expression Microarray Analysis Using Our Own Thyroid Samples

Fresh-frozen (FF) material from 52 tumours (27 FTC, 25 FTA) was used for our gene expression microarray experiments. The samples and microarray data have been already used in our previous studies and are reused in the current study [18,27]. Surgical procedures on patients were conducted in Polish and German centres, at the MSC Institute—Oncology Center in Gliwice, University of Leipzig, University of Halle, and Mainz University Hospital. Samples collected in hospitals were subsequently sent to our laboratory in Gliwice for microarray molecular profiling. Because the diagnosis of follicular thyroid tumours may be often equivocal [34], we attempted to obtain the evaluation of each pathology slide by two pathologists. However, we had access to the paraffin slides in only a part of the samples. If the slide was available for us, the sample was evaluated by two highly qualified pathologists. If the slide was not available for us, we based on the primary diagnosis, stated in the origin hospital by a single pathologist.

Next, the clinical material was divided into primary and secondary sets of tumors, depending on the concordance in histopathological diagnosis. The primary set contained all samples that were independently and concordantly diagnosed by two thyroid pathology experts (Dariusz Lange, Gliwice, and Steffen Hauptmann, Halle (Saale)). The secondary set contained samples that were diagnosed by only one expert, equivocal samples diagnosed by two experts and one sample that was discordantly diagnosed according to malignancy. A description of the material and the frequency of oncocytic tumors is shown in Table 4 (detailed description is given in Table S2).

The study was approved by the local ethics committees (Bioethics Committee of MSC Institute—Oncology Center in Gliwice; approvals: DK/ZMN-493-1-10/09, 20 November 2002 and KB/492-17/11, 9 February 2011), and informed consent was obtained from all patients.

Table 4. Fresh-frozen material used for microarray analysis.

Set	Histotype	Samples	% of Men	Median Age (Years)	Frequency of Oncocytic Tumours	Concordance of Pathologic Diagnosis by 2 Experts
Primary set	FTC	13	38.5%	66	46.2%	100%
	FTA	13	0%	42	15.4%	100%
Secondary set	FTC	14	21.4%	69	7.1%	28.6%
	FTA	12	25%	49.5	0%	75%
Total	-	52	21.2%	60.5	17.3%	75%

4.1.2. Clinical Materials for Validation Studies

FFPE tissue was used for validation in qRT-PCR experiments. The FFPE tissue consisted of 40 FTA and 31 FTC samples from patients treated in the MSC Institute—Oncology Center in Gliwice. The same set of samples was used in our previous study [27]. Diagnosis of FFPE tumours was based on the independent diagnoses of two pathologists. Material description and frequency of oncocytic tumours is presented in Table 5 (detailed description is given in Table S3 in Supplementary Material). Fresh frozen and FFPE datasets were independent datasets; there was no patient overlap between them.

Table 5. FFPE material used for qRT-PCR validation.

Histopathological Diagnosis	Number of Samples	% of Men	Median Age (Years)	Frequency of Oncocytic Tumours
FTC	31	32.3%	59	61.3%
FTA	40	12.5%	45	15%
Total	71	21.1%	52	35.2%

4.1.3. *RAS* Mutation Screening

All 123 samples of thyroid follicular tumour used for gene expression microarray (52 samples) and qRT-PCR experiments (71 samples) were screened for *RAS* mutations using the Sanger sequencing method with the ABI 3130*xl* Genetic Analyzer. Three *RAS* genes (*H-*, *K-*, *N-RAS*) sequences in commonly mutated codon sites (12, 13, and 61) were analysed. Different primer sets (different size of amplicon) for FF and FFPE samples were used due to sample degradation in FFPE samples (details Table S5).

4.2. Gene Expression Microarray-Based Analysis of Our Own Follicular Tumours

4.2.1. Gene Expression Microarray Experiment

FF materials from 52 follicular thyroid tumours (27 FTC, 25 FTA) were used for microarray analysis. RNA was isolated using the RNeasy Mini kit (Qiagen, Hilden, Germany). The RNA quality was assessed with capillary electrophoresis (Bioanalyzer 2100) and all the samples had the RNA integrity number (RIN) higher than 7. An Affymetrix (Santa Clara, CA, USA) HG-U133 PLUS 2 array experiment was performed as described previously [27].

4.2.2. Gene Expression Microarray Data Preprocessing

All microarray data analyses were performed in an R/Bioconductor environment. The microarray data preprocessing was performed in the same way as described in our previous study [27]. The quality of the microarray data was analysed using arrayMvout 1.12.0 library [39]. The raw data were preprocessed using the GCRMA method [40]. The microarray data discussed in this publication have been deposited in NCBI's Gene Expression Omnibus [41], and are accessible through GEO Series accession number GSE82208 (available online: https://www.ncbi.nlm.nih.gov/geo/query/acc.cgi?acc=GSE82208).

4.2.3. Supervised Analysis of Our Own Gene Expression Microarray Data

The selection of differentially expressed genes was performed independently on the primary and secondary microarray dataset (FF material) (Figure 1), in order to take into account the different levels of diagnosis certainty in the two sets. The following criteria were used for the primary dataset: normalized mean expression of the gene above 4.5, the variance of the gene above the 20th percentile, p-value in Student's t-test below 0.001, a fold-change above 1.5 in either direction of the change. The following, less strict criterion was used for the secondary dataset: p-value in Student's t-test < 0.05.

For genes selected for validation study, an additional analysis was performed in order to assess the significance of difference between FTC and FTA in microarray dataset comprised of non-oncocytic samples only. The genes were considered significant if the unadjusted p-value in the Student's t-test was below 0.005 in primary dataset and below 0.05 in secondary dataset.

In order to adjust p-values for multiple comparisons, false discovery rate (FDR) was estimated by Benjamini and Hochberg procedure [42].

4.3. Meta-Analysis of All Published Papers

The meta-analysis included all 14 papers in which the difference in gene expression between FTC and FTA was assessed by a high throughput method (gene expression microarrays, SAGE, HDSS, ATAC-PCR); which were published during 2000–2014; and found in PubMed, Google Scholar, or by screening the reference lists of selected papers (Table S1). The following criteria were used for the selection of papers: "follicular thyroid carcinoma/cancer/tumour/adenoma AND microarray/gene expression".

The lists of genes that were reported by the authors as differentially expressed between FTC and FTA, were extracted from each paper. Different types of gene identifiers were used in each study, such as gene symbols, gene names, GenBank accession numbers, cDNA sequences, Affymetrix identifiers, RefSeq accession numbers, and UniGene accession numbers. All gene identifiers were converted to EntrezID, the lists of genes were compared, and common genes were extracted. Finally, ten genes among the most frequently occurring ones were chosen for qRT-PCR validation (Figure 1).

Principal Component Analysis of Microarrays Based on the Meta-Analysis Identified Genes

To visually inspect whether the genes selected in the meta-analysis are able to separate FTC and FTA on an independent dataset, Principal Component Analysis (PCA) was conducted. We performed

PCA on our own microarray samples, based on genes that occurred at least once in the meta-analysis (Figure 2 upper plot). We also performed PCA on these samples, based on the genes that occurred at least twice in the meta-analysis (Figure 2 lower plot).

4.4. qRT-PCR Validation

4.4.1. qRT-PCR Experiment

FFPE materials from 71 follicular thyroid tumours (31 FTC, 40 FTA) were used for qRT-PCR analysis. RNA was isolated using the FFPE RNeasy Mini Kit (Qiagen) from 5 slices of paraffin blocks selected by a histopathologist. qRT-PCR was carried out for 18 genes (gene names given in Table S6, primer probe design given in Table S7). This experiment was performed with the 7900HT Fast Real-Time PCR (Life Technologies, Carlsbad, CA, USA) using Universal Probe Library fluorescent probes (Roche, Basel, Switzerland) and the 5'-nuclease assay, starting from 200 ng of total RNA. All experiments were performed twice. Results were normalised using the Pfaffl method [43] and the GeNorm application [44] with a combination of 3 normalisation genes: *EIF3A* (eukaryotic translation initiation factor 3, subunit A), *EIF5* (eukaryotic translation initiation factor 5), and *HADHA* (hydroxyacyl-CoA dehydrogenase/3-ketoacyl-CoA thiolase/enoyl-CoA hydratase (trifunctional protein), alpha subunit). Obtained normalised relative expression levels were further log-transformed (Figure S5).

Differences between FTC and FTA were tested using the Student's *t*-test. In addition, two-way analysis of variance (ANOVA) was used in order to adjust for oncocytic feature. False Discovery Rate (FDR) correction was applied and genes with FDR < 0.05 in both analyses were considered as significant.

4.4.2. Classifier Performance

The classifier was created and validated on the FFPE dataset using CMA package [45] in R/Bioconductor environment. The DLDA was used as a classification algorithm. Student's *t*-test was used for gene selection with significance level threshold tuned over a grid of significance levels. The performance of the classifier was evaluated by the doubly nested leave-one-out cross validation (LOOCV) approach in order to obtain an unbiased estimate of the accuracy [46]. The outer loop was used for estimating the classifier accuracy, and the inner loop was used for optimising the significance level threshold.

The ROC curve was also created to assess the diagnostic efficacy of the classifier (Figure 4). In the outer leave-one-out loop, for each sample, the probability that the sample belongs to the FTC class was calculated, based on DLDA algorithm. Varying the threshold for the probability, the ROC curve was plotted.

5. Conclusions

In our study, we have demonstrated that meta-analysis is a valuable method for selecting possible molecular markers. We showed that genes *CPQ*, *PLVAP*, *TFF3*, *ACVRL1*, *ZFYVE21*, *FAM189A2*, and *CLEC3B* are differentially expressed between FTC and FTA. Furthermore, we propose a 4-gene classifier, which discriminates between benign and malignant follicular neoplasms with the accuracy of 78%. Based on our results, we conclude that there might exist a plausible limit of gene classifier accuracy of approximately 80%, when follicular tumors are discriminated based on postoperative formalin-fixed material.

Supplementary Materials: Supplementary materials can be found at www.mdpi.com/1422-0067/18/6/1184/s1.

Acknowledgments: This study was supported by the Ministry of Science and Higher Education (grant number N N401 072637); Polish National Science Center (decision number DEC-2011/03/N/NZ5/05623); Foundation for Polish Science (MPD Program "Molecular Genomics, Transcriptomics and Bioinformatics in Cancer"); Postgraduate School of Molecular Medicine (fellowships to Bartosz Wojtas and Tomasz Stokowy); National Center for Research and Development project under the program "Prevention practices and treatment of civilisation diseases" STRATEGMED (MILESTONE, project number STRATEGMED2/267398/4/NCBR/2015);

and DFG (grant number ES162/4-1 to Markus Eszlinger). The costs to publish in open access were covered by Maria Sklodowska-Curie Institute—Oncology Center, Gliwice Branch.

Author Contributions: Bartosz Wojtas, Aleksandra Pfeifer, Jolanta Krajewska, Michal Jarzab and Ewa Chmielik contributed to the writing of the manuscript; Michal Jarzab, Markus Eszlinger, Ralf Paschke, and Barbara Jarzab designed and coordinated the study; Agnieszka Czarniecka and Thomas Musholt collected the tissue material; Aleksandra Kukulska and Jolanta Krajewska analyzed patient data; Ewa Stobiecka, Ewa Chmielik, Steffen Hauptmann, and Dariusz Lange performed the histopathological examination of the samples; Bartosz Wojtas, Malgorzata Oczko-Wojciechowska, Dagmara Rusinek, Tomasz Tyszkiewicz and Monika Halczok performed the experiments described in this study; Aleksandra Pfeifer, Tomasz Stokowy and Michal Swierniak performed the bioinformatics analysis.

Conflicts of Interest: The authors declare no conflict of interest.

Abbreviations

ANOVA	Analysis of variance
ATAC-PCR	Adapter-tagged competitive polymerase chain reaction
AUC	Area under the ROC curve
DLDA	Diagonal linear discrimination analysis
FDR	False discovery rate
FF	Fresh-frozen
FFPE	Formaldehyde-fixed paraffin embedded
FNAB	Fine needle aspiration biopsy
FTA	Follicular thyroid adenoma
FTC	Follicular thyroid cancer
HDSS	High-throughput differential screening by serial analysis of gene expression
IQR	Interquartile range
LOOCV	Leave-one-out cross validation
OTC	Oncocytic thyroid carcinoma
PCA	Principal component analysis
Q1	First quartile
Q3	Third quartile
qRT-PCR	Quantitative real-time polymerase chain reaction
RIN	RNA integrity number
ROC	Receiver operating characteristc
SAGE	Serial analysis of gene expression
WHO	World Health Organization

References

1. DeLellis, R.; Lloyd, R.; Heitz, P.; Eng, C. *WHO Pathology and Genetics. Tumours of Endocrine Organs*; IARC Press: Lyon, France, 2004.

2. LiVolsi, V.A.; Baloch, Z.W. Follicular-patterned tumors of the thyroid: The battle of benign vs. malignant vs. so-called uncertain. *Endocr. Pathol.* **2011**, *22*, 184–189. [CrossRef] [PubMed]

3. Yamashina, M. Follicular neoplasms of the thyroid. Total circumferential evaluation of the fibrous capsule. *Am. J. Surg. Pathol.* **1992**, *16*, 392–400. [CrossRef] [PubMed]

4. Takano, T.; Hasegawa, Y.; Matsuzuka, F.; Miyauchi, A.; Yoshida, H.; Higashiyama, T.; Kuma, K.; Amino, N. Gene expression profiles in thyroid carcinomas. *Br. J. Cancer* **2000**, *83*, 1495–1502. [CrossRef] [PubMed]

5. Barden, C.B.; Shister, K.W.; Zhu, B.; Guiter, G.; Greenblatt, D.Y.; Zeiger, M.A.; Fahey, T.J., III. Classification of follicular thyroid tumors by molecular signature: Results of gene profiling. *Clin. Cancer Res.* **2003**, *9*, 1792–1800. [PubMed]

6. Takano, T.; Miyauchi, A.; Yoshida, H.; Kuma, K.; Amino, N. High-throughput differential screening of mRNAs by serial analysis of gene expression: Decreased expression of trefoil factor 3 mRNA in thyroid follicular carcinomas. *Br. J. Cancer* **2004**, *90*, 1600–1605. [CrossRef] [PubMed]

7. Cerutti, J.M.; Delcelo, R.; Amadei, M.J.; Nakabashi, C.; Maciel, R.M.; Peterson, B.; Shoemaker, J.; Riggins, G.J. A preoperative diagnostic test that distinguishes benign from malignant thyroid carcinoma based on gene expression. *J. Clin. Investig.* **2004**, *113*, 1234–1242. [CrossRef] [PubMed]

8. Chevillard, S.; Ugolin, N.; Vielh, P.; Ory, K.; Levalois, C.; Elliott, D.; Clayman, G.L.; El-Naggar, A.K. Gene expression profiling of differentiated thyroid neoplasms: Diagnostic and clinical implications. *Clin. Cancer Res.* **2004**, *10*, 6586–6597. [CrossRef] [PubMed]

9. Weber, F.; Shen, L.; Aldred, M.A.; Morrison, C.D.; Frilling, A.; Saji, M.; Schuppert, F.; Broelsch, C.E.; Ringel, M.D.; Eng, C. Genetic classification of benign and malignant thyroid follicular neoplasia based on a three-gene combination. *J. Clin. Endocrinol. Metab.* **2005**, *90*, 2512–2521. [CrossRef] [PubMed]

10. Taniguchi, K.; Takano, T.; Miyauchi, A.; Koizumi, K.; Ito, Y.; Takamura, Y.; Ishitobi, M.; Miyoshi, Y.; Taguchi, T.; Tamaki, Y.; et al. Differentiation of follicular thyroid adenoma from carcinoma by means of gene expression profiling with adapter-tagged competitive polymerase chain reaction. *Oncology* **2005**, *69*, 428–435. [CrossRef] [PubMed]

11. Lubitz, C.C.; Gallagher, L.A.; Finley, D.J.; Zhu, B.; Fahey, T.J., III. Molecular analysis of minimally invasive follicular carcinomas by gene profiling. *Surgery* **2005**, *138*, 1042–1048. [CrossRef] [PubMed]

12. Fryknas, M.; Wickenberg-Bolin, U.; Goransson, H.; Gustafsson, M.G.; Foukakis, T.; Lee, J.J.; Landegren, U.; Hoog, A.; Larsson, C.; Grimelius, L.; et al. Molecular markers for discrimination of benign and malignant follicular thyroid tumors. *Tumour Biol.* **2006**, *27*, 211–220. [PubMed]

13. Stolf, B.S.; Santos, M.M.; Simao, D.F.; Diaz, J.P.; Cristo, E.B.; Hirata, R., Jr.; Curado, M.P.; Neves, E.J.; Kowalski, L.P.; Carvalho, A.F. Class distinction between follicular adenomas and follicular carcinomas of the thyroid gland on the basis of their signature expression. *Cancer* **2006**, *106*, 1891–1900. [CrossRef] [PubMed]

14. Zhao, J.; Leonard, C.; Gemsenjager, E.; Heitz, P.U.; Moch, H.; Odermatt, B. Differentiation of human follicular thyroid adenomas from carcinomas by gene expression profiling. *Oncol. Rep.* **2008**, *19*, 329–337. [CrossRef]

15. Hinsch, N.; Frank, M.; Doring, C.; Vorlander, C.; Hansmann, M.L. QPRT: A potential marker for follicular thyroid carcinoma including minimal invasive variant; a gene expression, RNA and immunohistochemical study. *BMC Cancer* **2009**, *9*, 93. [CrossRef] [PubMed]

16. Borup, R.; Rossing, M.; Henao, R.; Yamamoto, Y.; Krogdahl, A.; Godballe, C.; Winther, O.; Kiss, K.; Christensen, L.; Hogdall, E.; et al. Molecular signatures of thyroid follicular neoplasia. *Endocr. Relat. Cancer* **2010**, *17*, 691–708. [CrossRef] [PubMed]

17. Williams, M.D.; Zhang, L.; Elliott, D.D.; Perrier, N.D.; Lozano, G.; Clayman, G.L.; El-Naggar, A.K. Differential gene expression profiling of aggressive and nonaggressive follicular carcinomas. *Hum. Pathol.* **2011**, *42*, 1213–1220. [CrossRef] [PubMed]

18. Wojtas, B.; Pfeifer, A.; Jarzab, M.; Czarniecka, A.; Krajewska, J.; Swierniak, M.; Stokowy, T.; Rusinek, D.; Kowal, M.; Zebracka-Gala, J.; et al. Unsupervised analysis of follicular thyroid tumours transcriptome by oligonucleotide microarray gene expression profiling. *Endokrynol. Pol.* **2013**, *64*, 328–334. [CrossRef]

19. Swierniak, M.; Pfeifer, A.; Stokowy, T.; Rusinek, D.; Chekan, M.; Lange, D.; Krajewska, J.; Oczko-Wojciechowska, M.; Czarniecka, A.; Jarzab, M.; et al. Somatic mutation profiling of follicular thyroid cancer by next generation sequencing. *Mol. Cell. Endocrinol.* **2016**, *433*, 130–137. [CrossRef] [PubMed]

20. Cheung, L.; Messina, M.; Gill, A.; Clarkson, A.; Learoyd, D.; Delbridge, L.; Wentworth, J.; Philips, J.; Clifton-Bligh, R.; Robinson, B.G. Detection of the PAX8-PPARγ fusion oncogene in both follicular thyroid carcinomas and adenomas. *J. Clin. Endocrinol. Metab.* **2003**, *88*, 354–357. [CrossRef] [PubMed]

21. Sahin, M.; Allard, B.L.; Yates, M.; Powell, J.G.; Wang, X.L.; Hay, I.D.; Zhao, Y.; Goellner, J.R.; Sebo, T.J.; Grebe, S.K.; et al. PPARγ staining as a surrogate for PAX8/PPARγ fusion oncogene expression in follicular neoplasms: Clinicopathological correlation and histopathological diagnostic value. *J. Clin. Endocrinol. Metab.* **2005**, *90*, 463–468. [CrossRef] [PubMed]

22. Kloos, R.T.; Reynolds, J.D.; Walsh, P.S.; Wilde, J.I.; Tom, E.Y.; Pagan, M.; Barbacioru, C.; Chudova, D.I.; Wong, M.; Friedman, L.; et al. Does addition of BRAF V600E mutation testing modify sensitivity or specificity of the Afirma Gene Expression Classifier in cytologically indeterminate thyroid nodules? *J. Clin. Endocrinol. Metab.* **2013**, *98*, 761–768. [CrossRef] [PubMed]

23. Baris, O.; Savagner, F.; Nasser, V.; Loriod, B.; Granjeaud, S.; Guyetant, S.; Franc, B.; Rodien, P.; Rohmer, V.; Bertucci, F.; et al. Transcriptional profiling reveals coordinated up-regulation of oxidative metabolism genes in thyroid oncocytic tumors. *J. Clin. Endocrinol. Metab.* **2004**, *89*, 994–1005. [CrossRef] [PubMed]

24. Ganly, I.; Ricarte Filho, J.; Eng, S.; Ghossein, R.; Morris, L.G.; Liang, Y.; Socci, N.; Kannan, K.; Mo, Q.; Fagin, J.A.; et al. Genomic dissection of Hurthle cell carcinoma reveals a unique class of thyroid malignancy. *J. Clin. Endocrinol. Metab.* **2013**, *98*, 962–972. [CrossRef] [PubMed]

25. Chudova, D.; Wilde, J.I.; Wang, E.T.; Wang, H.; Rabbee, N.; Egidio, C.M.; Reynolds, J.; Tom, E.; Pagan, M.; Rigl, C.T.; et al. Molecular classification of thyroid nodules using high-dimensionality genomic data. *J. Clin. Endocrinol. Metab.* **2010**, *95*, 5296–5304. [CrossRef] [PubMed]

26. Keutgen, X.M.; Filicori, F.; Crowley, M.J.; Wang, Y.; Scognamiglio, T.; Hoda, R.; Buitrago, D.; Cooper, D.; Zeiger, M.A.; Zarnegar, R.; et al. A panel of four miRNAs accurately differentiates malignant from benign indeterminate thyroid lesions on fine needle aspiration. *Clin. Cancer Res.* **2012**, *18*, 2032–2038. [CrossRef] [PubMed]

27. Pfeifer, A.; Wojtas, B.; Oczko-Wojciechowska, M.; Kukulska, A.; Czarniecka, A.; Eszlinger, M.; Musholt, T.; Stokowy, T.; Swierniak, M.; Stobiecka, E.; et al. Molecular differential diagnosis of follicular thyroid carcinoma and adenoma based on gene expression profiling by using formalin-fixed paraffin-embedded tissues. *BMC Med. Genom.* **2013**, *6*, 38. [CrossRef] [PubMed]

28. Evans, H.L.; Vassilopoulou-Sellin, R. Follicular and Hurthle cell carcinomas of the thyroid: A comparative study. *Am. J. Surg. Pathol.* **1998**, *22*, 1512–1520. [CrossRef] [PubMed]

29. Parameswaran, R.; Brooks, S.; Sadler, G.P. Molecular pathogenesis of follicular cell derived thyroid cancers. *Int. J. Surg.* **2010**, *8*, 186–193. [CrossRef] [PubMed]

30. Alexander, E.K.; Kennedy, G.C.; Baloch, Z.W.; Cibas, E.S.; Chudova, D.; Diggans, J.; Friedman, L.; Kloos, R.T.; LiVolsi, V.A.; Mandel, S.J.; et al. Preoperative diagnosis of benign thyroid nodules with indeterminate cytology. *N. Engl. J. Med.* **2012**, *367*, 705–715. [CrossRef] [PubMed]

31. Chien, C.Y.; Chuang, H.C.; Chen, C.H.; Fang, F.M.; Chen, W.C.; Huang, C.C.; Huang, H.Y. The expression of activin receptor-like kinase 1 among patients with head and neck cancer. *Otolaryngol. Head Neck Surg.* **2013**, *148*, 965–973. [CrossRef] [PubMed]

32. Arvanitis, D.L.; Kamper, E.F.; Kopeikina, L.; Stavridou, A.; Sgantzos, M.N.; Kallioras, V.; Athanasiou, E.; Kanavaros, P. Tetranectin expression in gastric adenocarcinomas. *Histol. Histopathol.* **2002**, *17*, 471–475. [PubMed]

33. Hoshino, D.; Nagano, M.; Saitoh, A.; Koshikawa, N.; Suzuki, T.; Seiki, M. The phosphoinositide-binding protein ZF21 regulates ECM degradation by invadopodia. *PLoS ONE* **2013**, *8*, e50825. [CrossRef] [PubMed]

34. Franc, B.; de la Salmonière, P.; Lange, F.; Hoang, C.; Louvel, A.; de Roquancourt, A.; Vilde, F.; Hejblum, G.; Chevret, S.; Chastang, C. Interobserver and intraobserver reproducibility in the histopathology of follicular thyroid carcinoma. *Hum. Pathol.* **2003**, *34*, 1092–1100. [CrossRef]

35. Da, S.L.; James, D.; Simpson, P.T.; Walker, D.; Vargas, A.C.; Jayanthan, J.; Lakhani, S.R.; McNicol, A.M. Tumor heterogeneity in a follicular carcinoma of thyroid: A study by comparative genomic hybridization. *Endocr. Pathol.* **2011**, *22*, 103–107.

36. Jarzab, B.; Wiench, M.; Fujarewicz, K.; Simek, K.; Jarzab, M.; Oczko-Wojciechowska, M.; Wloch, J.; Czarniecka, A.; Chmielik, E.; Lange, D.; et al. Gene expression profile of papillary thyroid cancer: Sources of variability and diagnostic implications. *Cancer Res.* **2005**, *65*, 1587–1597. [CrossRef] [PubMed]

37. Aldred, M.A.; Huang, Y.; Liyanarachchi, S.; Pellegata, N.S.; Gimm, O.; Jhiang, S.; Davuluri, R.V.; de la Chapelle, A.; Eng, C. Papillary and follicular thyroid carcinomas show distinctly different microarray expression profiles and can be distinguished by a minimum of five genes. *J. Clin. Oncol.* **2004**, *22*, 3531–3539. [CrossRef] [PubMed]

38. Rodrigues, H.G.; de Pontes, A.A.; Adan, L.F. Use of molecular markers in samples obtained from preoperative aspiration of thyroid. *Endocr. J.* **2012**, *59*, 417–424. [CrossRef] [PubMed]

39. Asare, A.L.; Gao, Z.; Carey, V.J.; Wang, R.; Seyfert-Margolis, V. Power enhancement via multivariate outlier testing with gene expression arrays. *Bioinformatics* **2009**, *25*, 48–53. [CrossRef] [PubMed]

40. Wu, Z.J.; Irizarry, R.A.; Gentleman, R.; Martinez-Murillo, F.; Spencer, F. A model-based background adjustment for oligonucleotide expression arrays. *J. Am. Stat. Assoc.* **2004**, *99*, 909–917. [CrossRef]

41. Edgar, R.; Domrachev, M.; Lash, A.E. Gene expression omnibus: NCBI gene expression and hybridization array data repository. *Nucleic Acids Res.* **2002**, *30*, 207–210. [CrossRef] [PubMed]

42. Benjamini, Y.; Hochberg, Y. Controlling the false discovery rate: A practical and powerful approach to multiple testing. *J. R. Stat. Soc. Ser. B* **1995**, *57*, 289–300.

43. Pfaffl, M.W. A new mathematical model for relative quantification in real-time RT-PCR. *Nucleic Acids Res.* **2001**, *29*, e45. [CrossRef] [PubMed]

44. Vandesompele, J.; de Preter, K.; Pattyn, F.; Poppe, B.; van Roy, N.; de Paepe, A.; Speleman, F. Accurate normalization of real-time quantitative RT-PCR data by geometric averaging of multiple internal control genes. *Genome Biol.* **2002**, *3*, 1–11. [CrossRef]
45. Slawski, M.; Daumer, M.; Boulesteix, A.L. CMA: A comprehensive bioconductor package for supervised classification with high dimensional data. *BMC Bioinform.* **2008**, *9*, 439. [CrossRef] [PubMed]
46. Varma, S.; Simon, R. Bias in error estimation when using cross-validation for model selection. *BMC Bioinform.* **2006**, *7*. [CrossRef] [PubMed]

International Journal of
Molecular Sciences

MDPI

Article

Histogram Analysis of Diffusion Weighted Imaging at 3T is Useful for Prediction of Lymphatic Metastatic Spread, Proliferative Activity, and Cellularity in Thyroid Cancer

Stefan Schob [1,*], Hans Jonas Meyer [2], Julia Dieckow [3], Bhogal Pervinder [4], Nikolaos Pazaitis [5], Anne Kathrin Höhn [6], Nikita Garnov [2], Diana Horvath-Rizea [4], Karl-Titus Hoffmann [1] and Alexey Surov [2]

[1] Department for Neuroradiology, University Hospital Leipzig, Leipzig 04103, Germany; karl-titus.hoffmann@medizin.uni-leipzig.de
[2] Department for Diagnostic and Interventional Radiology, University Hospital Leipzig, Leipzig 04103, Germany; jonas90.meyer@web.de (H.J.M.); nikita@garnov.de (N.G.); alexey.surov@medizin.uni-leipzig.de (A.S.)
[3] Department for Ophthalmology, University Hospital Leipzig, Leipzig 04103, Germany; julia@dieckow.de
[4] Department for Diagnostic and Interventional Neuroradiology, Katharinenhospital Stuttgart, Stuttgart 70174, Germany; bhogalweb@aol.com (B.P.); dihorvath@freenet.de (D.H.-R.)
[5] Institute for Pathology, University Hospital Halle-Wittenberg, Martin-Luther-University Halle-Wittenberg, Halle 06112, Germany; nikolaos.pazaitis@uk-halle.de
[6] Institute for Pathology, University Hospital Leipzig, Leipzig 04103, Germany; annekathrin.hoehn@medizin.uni-leipzig.de
* Correspondence: stefan.schob@medizin.uni-leipzig.de; Tel.: +49-341-971-6984; Fax: +49-341-971-7490

Academic Editor: Daniela Gabriele Grimm
Received: 9 March 2017; Accepted: 10 April 2017; Published: 12 April 2017

Abstract: Pre-surgical diffusion weighted imaging (DWI) is increasingly important in the context of thyroid cancer for identification of the optimal treatment strategy. It has exemplarily been shown that DWI at 3T can distinguish undifferentiated from well-differentiated thyroid carcinoma, which has decisive implications for the magnitude of surgery. This study used DWI histogram analysis of whole tumor apparent diffusion coefficient (ADC) maps. The primary aim was to discriminate thyroid carcinomas which had already gained the capacity to metastasize lymphatically from those not yet being able to spread via the lymphatic system. The secondary aim was to reflect prognostically important tumor-biological features like cellularity and proliferative activity with ADC histogram analysis. Fifteen patients with follicular-cell derived thyroid cancer were enrolled. Lymph node status, extent of infiltration of surrounding tissue, and Ki-67 and p53 expression were assessed in these patients. DWI was obtained in a 3T system using b values of 0, 400, and 800 s/mm^2. Whole tumor ADC volumes were analyzed using a histogram-based approach. Several ADC parameters showed significant correlations with immunohistopathological parameters. Most importantly, ADC histogram skewness and ADC histogram kurtosis were able to differentiate between nodal negative and nodal positive thyroid carcinoma. Conclusions: histogram analysis of whole ADC tumor volumes has the potential to provide valuable information on tumor biology in thyroid carcinoma. However, further studies are warranted.

Keywords: thyroid carcinoma; diffusion weighted imaging; lymphatic metastatic spread; ADC histogram analysis; histopathologic features; Ki-67; p53

1. Introduction

The incidence of thyroid cancer, being the most abundant endocrine malignancy, is rapidly increasing [1]. The vast majority of thyroid neoplasms is follicular cell-derived and subsumed under the umbrella categories of papillary thyroid cancer, follicular thyroid cancer, poorly differentiated thyroid cancer, and anaplastic thyroid cancer [2]. Although the overall five-year survival rates of thyroid cancer are 94% in women and 85% in men [3], certain entities of the disease are perpetually associated with poor outcomes (for example the tall cell variant of papillary thyroid cancer and undifferentiated thyroid cancer [1]). Some of the differentiated entities—most of all papillary thyroid cancer variants—frequently metastasize locally via the lymphatic system [4], and resultant local recurrence is not an uncommon scenario [5], leading to significant morbidity.

A variety of therapeutic options is available for thyroid cancer [6], but surgery still remains the predominant treatment [7]. Radical surgery is the most important form of therapy for undifferentiated thyroid cancer [8], and surgical treatment of significant nodal disease in well differentiated thyroid cancer is widely accepted to be associated with improved outcomes in terms of survival and recurrence rates [9]. Nonetheless, extensive surgery in this specific context carries a high risk of therapy-related morbidity like phrenic nerve palsy, brachial plexus palsy, cranial nerve injury, chyle leak, and pneumothorax [10].

Considering the broad spectrum of aggressiveness in thyroid cancer and the resulting necessity for customized treatment, employing presurgical imaging is of great importance, as it allows the thyroid surgeon to identify disease subtypes being associated with increased mortality and morbidity such as metastasizing and undifferentiated thyroid cancer.

Diffusion-weighted magnetic resonance imaging (DWI) has the potential to reveal tumor architectural details like cellular density and proliferative activity in different malignant entities [11,12]. Using a standard echo-planar imaging (EPI) technique, DWI has the capability to differentiate between malignant and benign thyroid nodules [13]. Furthermore, DWI can distinguish manifestations of papillary thyroid cancer with extra-glandular growth from those confined to the thyroid [14]. Using a RESOLVE sequence (which is less prone to susceptibility and motion-induced phase artifacts, has less T2* blurring and provides higher resolution than standard EPI DWI, [15]) in a 3T scanner, DWI even has the capability to distinguish between differentiated and undifferentiated subtypes of thyroid carcinoma [16].

However, in the clinical setting, obtained DWI data is commonly analyzed using a two-dimensional region of interest in the slice of the apparent diffusion coefficient (ADC) map representing the maximum diameter of the tumor. This approach does not account for the regularly encountered heterogeneity of whole tumors and certainly does not reflect the complex micro-architectural properties of malignantly transformed tissue.

An enhanced approach using every voxel of the tumor to compute a histogram of intensity levels could help to further increase prediction of histological features of tumors by magnetic resonance imaging (MRI) [17]. This way, the magnitude of tumor heterogeneity probably is revealed in a fashion superior to the commonly used two-dimensional method [17].

To the best of the authors' knowledge, only one study used ADC histogram analysis in thyroid cancer to differentiate benign from malignant nodules and furthermore reveal extra-thyroidal growth of papillary thyroid cancer [18]. So far, no studies demonstrated predictability of lymph node involvement by ADC histogram analysis of the primary tumor. Therefore, the primary aim of this study was to investigate the potential of ADC histogram analysis (including percentiles, entropy, skewness, and kurtosis) on data obtained with RESOLVE DWI to distinguish between nodal-negative and nodal-positive thyroid cancer. The discriminability of metastatic from non-metastatic thyroid cancer is of great clinical importance. Hence, this study investigated a promising translational approach that might have the potential to significantly increase the value of clinical-oncological imaging. The secondary aim was to correlate ADC histogram parameters with expression of important prognostic markers like p53 and Ki-67. Last, it aimed to compare our findings with the results of previous studies, which investigated the potential of DWI to predict histopathological features in thyroid cancer.

2. Results

2.1. Diffusion Weighted Imaging and Immunohistopathology of Thyroid Carcinoma

For reasons of clarity and comprehensibility, results of MRI and histopathology were organized in tables. Figure 1 shows MRI findings of a patient with follicular thyroid carcinoma, presenting as heterogeneous enlargement of the right thyroid lobe. The corresponding immunohistological images are shown in Figure 2. The calculated DWI parameters of all investigated thyroid carcinomas are summarized in Table 1 and the corresponding histopathological data is given in Table 2.

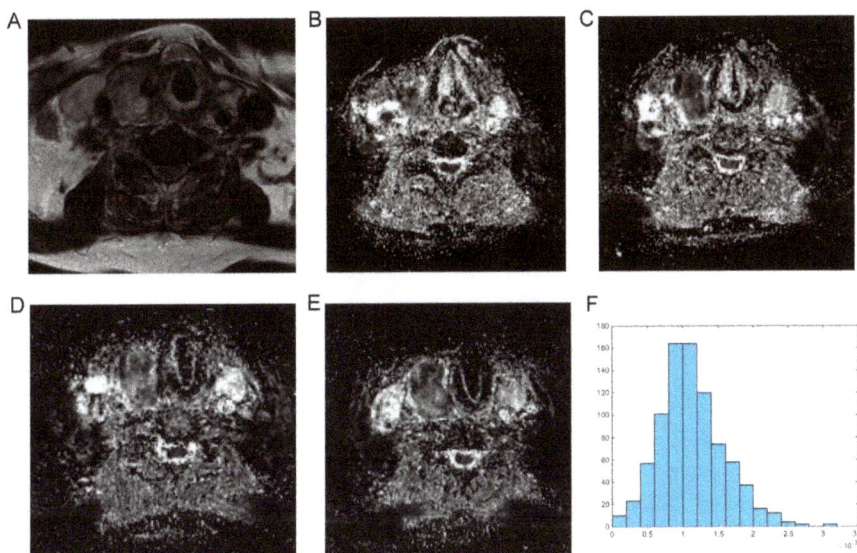

Figure 1. Imaging findings in a patient with follicular thyroid carcinoma. (**A**) Magnetic resonance imaging (T2w axial section) showing a massive inhomogenous enlargement of the right thyroid lobe; (**B–E**) represent the apparent diffusion coefficient (ADC) maps of the tumor; (**F**) is the ADC histogram of the whole lesion. The calculated ADC parameters ($\times 10^{-5}$ mm$^2 \cdot$s^{-1}) are as follows: ADC$_{min}$ = 18.2; ADC$_{mean}$ = 113.3; ADC$_{max}$ = 315.0, mode = 114.4, ADC$_{median}$ = 108.1, P10 = 58.2, P25 = 83.2, P75 = 138.7, and P90 = 176.6. Histogram based parameters are as follows: skewness = 0.59, kurtosis = 3.88, and entropy = 3.21. The z-axis in Figure 1F gives the voxel count.

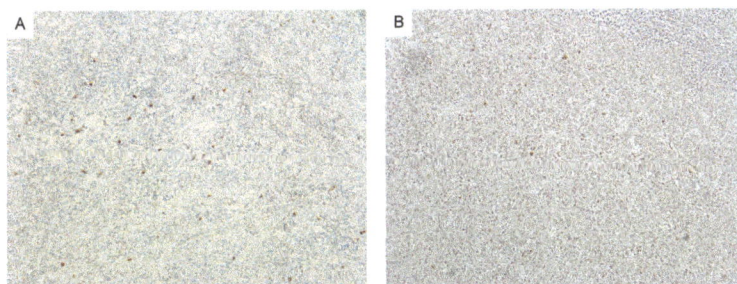

Figure 2. Immunohistochemistry of follicular thyroid carcinoma. (**A**) Shows Ki-67 staining (cell count: 1407, Ki-67 immunoreactiviy: 11%) and (**B**) shows p53 staining (cell count: 1811, p53 immunoreactivity: 36%) of the tumor displayed in Figure 1.

Table 1. Diffusion weighted imaging and related histogram parameters of thyroid carcinoma based on $n = 15$ patients.

DWI Related Parameters	Median	Range	Minimum–Maximum
ADC_{mean}, $\times 10^{-5}$ mm$^2 \cdot$s^{-1}	124.30	90	73–163
ADC_{min}, $\times 10^{-5}$ mm$^2 \cdot$s^{-1}	14.90	53	0.2–53
ADC_{max}, $\times 10^{-5}$ mm$^2 \cdot$s^{-1}	250.70	179	147–325
P10 ADC, $\times 10^{-5}$ mm$^2 \cdot$s^{-1}	72.10	85	30–114
P25 ADC, $\times 10^{-5}$ mm$^2 \cdot$s^{-1}	91.90	84	52–136
P75 ADC, $\times 10^{-5}$ mm$^2 \cdot$s^{-1}	140.40	99	93–192
P90 ADC, $\times 10^{-5}$ mm$^2 \cdot$s^{-1}	172.82	116	97–213
Median ADC, $\times 10^{-5}$ mm$^2 \cdot$s^{-1}	118.00	94	71–165
Mode ADC, $\times 10^{-5}$ mm$^2 \cdot$s^{-1}	101.40	88	53–141
Kurtosis	3.64	1.90	2.89–4.79
Skewness	0.30	1.79	−0.97–0.81
Entropy	3.27	1.98	2.75–4.72

Table 2. Estimated immunohistopathological parameters of thyroid carcinoma ($n = 15$).

Parameters	Median	Range	Minimum–Maximum
Cell count, n	1407	1808	439–2247
Ki 67, %	32.0	90	9–99
p53, %	4.0	94	0–94
Total nuclear area, μm^2	71,735	148,620	14,649–163,269
Average nuclear area, μm^2	53.0	61	33–94

2.2. Correlation Analysis

Table 3 displays results of the correlation analysis between immunohistopathological parameters and ADC fractions as well as histogram related parameters. Correlation analysis identified the following, significant correlations: ADC_{mean} with p53 ($r = 0.548$, $p = 0.034$), ADC_{max} with Ki67 ($r = -0.646$, $p = 0.009$) and p53 ($r = 0.645$, $p = 0.009$), ADCp75 with p53 ($r = 0.537$, $p = 0.025$), ADCp90 with Ki67 ($r = -0.568$, $p = 0.027$) and p53 ($r = 0.588$, $p = 0.021$), ADC_{median} with p53 ($r = 0.556$, $p = 0.032$), ADC_{modus} with p53 ($r = 0.534$, $p = 0.040$), and kurtosis with cell count ($r = -0.571$, $p = 0.026$). Figure 3 summarizes the significant correlations graphically and displays them as dot plots.

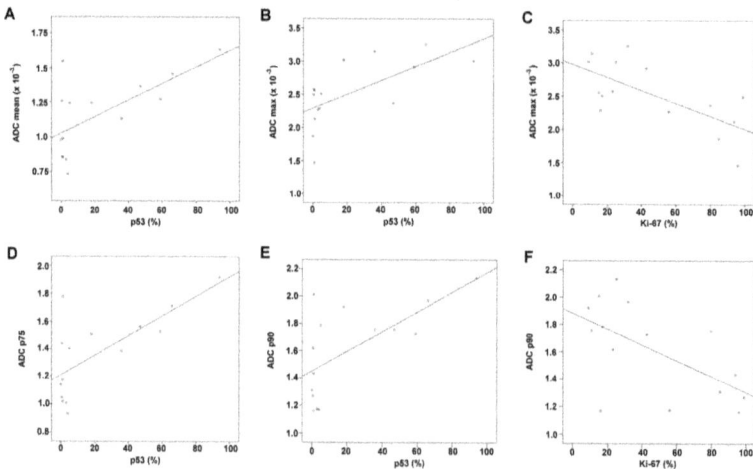

Figure 3. *Cont.*

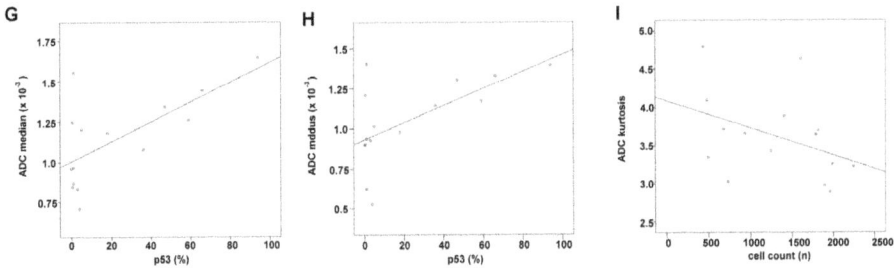

Figure 3. Graphic summary of the significant correlations between imaging and immunohistological findings. R^2-values for the plots shown in Figure 3 are as follows; (**A**) ADC_{mean} & p53: $r^2 = 0.438$; (**B**) ADC_{max} & p53: $r^2 = 0.425$; (**C**) ADC_{max} & Ki-67: $r^2 = 0.464$; (**D**) $ADCp75$ & p53: $r^2 = 0.499$; (**E**) $ADCp90$ & p53: $r^2 = 0.431$; (**F**) $ADCp90$ & Ki-67: $r^2 = 0.360$; (**G**) ADC_{median} & p53: $r^2 = 0.440$; (**H**) ADC_{modus} & p53: $r^2 = 0.377$; (**I**) $ADC_{kurtosis}$ & cell count: $r^2 = 0.160$.

Table 3. Results of Spearman's rank order correlation analysis between DWI and immunohistological parameters ($n = 15$).

ADC Parameters and Histogram Values	Cell Count	p53	Ki-67	Total Nuclear Area	Average Nuclear Area
ADC_{mean}, $\times 10^{-3}$ mm$^2 \cdot$s^{-1}	$r = 0.429$ $p = 0.111$	$r = 0.548$ $p = 0.034$	$r = -0.325$ $p = 0.237$	$r = 0.389$ $p = 0.152$	$r = 0.034$ $p = 0.904$
ADC_{min}, $\times 10^{-3}$ mm$^2 \cdot$s^{-1}	$r = 0.256$ $p = 0.358$	$r = 0.244$ $p = 0.381$	$r = -0.241$ $p = 0.386$	$r = 0.163$ $p = 0.562$	$r = -0.208$ $p = 0.456$
ADC_{max}, $\times 10^{-3}$ mm^2 s^{-1}	$r = 0.372$ $p = 0.173$	$r = 0.645$ $p = 0.009$	$r = -0.646$ $p = 0.009$	$r = 0.461$ $p = 0.084$	$r = 0.155$ $p = 0.580$
ADC p10, $\times 10^{-3}$ mm$^2 \cdot$s^{-1}	$r = 0.361$ $p = 0.187$	$r = 0.409$ $p = 0.130$	$r = 0.289$ $p = 0.296$	$r = 0.275$ $p = 0.321$	$r = -0.079$ $p = 0.781$
ADC p25, $\times 10^{-3}$ mm$^2 \cdot$s^{-1}	$r = 0.375$ $p = 0.168$	$r = 0.509$ $p = 0.053$	$r = 0.361$ $p = 0.187$	$r = 0.311$ $p = 0.260$	$r = -0.064$ $p = 0.820$
ADC p75, $\times 10^{-3}$ mm$^2 \cdot$s^{-1}	$r = 0.450$ $p = 0.092$	$r = 0.537$ $p = 0.025$	$r = -0.343$ $p = 0.211$	$r = 0.411$ $p = 0.128$	$r = 0.055$ $p = 0.845$
ADC p90, $\times 10^{-3}$ mm$^2 \cdot$s^{-1}	$r = 0.289$ $p = 0.296$	$r = 0.588$ $p = 0.021$	$r = -0.568$ $p = 0.027$	$r = 0.300$ $p = 0.277$	$r = 0.075$ $p = 0.790$
Median ADC, $\times 10^{-3}$ mm$^2 \cdot$s^{-1}	$r = 0.414$ $p = 0.125$	$r = 0.556$ $p = 0.032$	$r = -0.314$ $p = 0.254$	$r = 0.361$ $p = 0.187$	$r = -0.020$ $p = 0.945$
Mode ADC, $\times 10^{-3}$ mm$^2 \cdot$s^{-1}	$r = 0.496$ $p = 0.060$	$r = 0.534$ $p = 0.040$	$r = -0.357$ $p = 0.191$	$r = 0.432$ $p = 0.108$	$r = -0.149$ $p = 0.682$
Kurtosis	$r = -0.571$ $p = 0.026$	$r = -0.262$ $p = 0.346$	$r = -0.314$ $p = 0.254$	$r = -0.411$ $p = 0.128$	$r = -0.182$ $p = 0.516$
Skewness	$r = -0.229$ $p = 0.413$	$r = -0.004$ $p = 0.990$	$r = -0.389$ $p = 0.152$	$r = 0.011$ $p = 0.970$	$r = 0.186$ $p = 0.507$
Entropy	$r = 0.243$ $p = 0.383$	$r = -0.240$ $p = 0.389$	$r = 0.289$ $p = 0.296$	$r = 0.225$ $p = 0.420$	$r = 0.316$ $p = 0.251$

2.3. Group Comparisons

Histogram analysis derived ADC values are compared between the nodal negative and the nodal positive group in Figure 4. Levene's Test revealed homoscedasticity for the nodal-negative and the nodal-positive group only regarding $ADC_{skewness}$ ($p = 0.015$). For all remaining ADC derived histogram parameters, Levene's Test showed heterogeneity of variance when comparing the nodal-negative and the nodal-positive group. Hence, group comparisons were performed using unpaired t-test for $ADC_{skewness}$ and Mann-Whitney-U Test for all remaining parameters. The corresponding p-values are given in Table 4. Statistically significant differences were only identified for skewness ($p = 0.031$) and kurtosis ($p = 0.028$). No other significant differences or trends were delineable when comparing thyroid carcinoma patients with restricted vs. advanced infiltration pattern (results not presented).

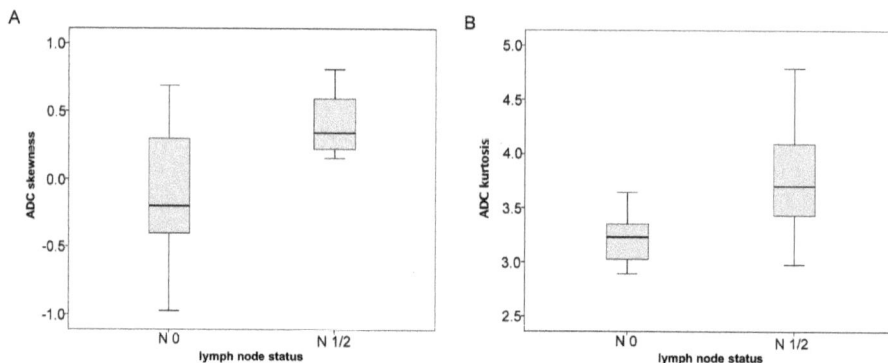

Figure 4. Graphically summarizes the differences in histogram parameters between nodal negative and nodal positive patients with thyroid carcinoma. (**A**) Shows significantly increased ADC histogram skewness in noda-positive compared to nodal-negative patients; (**B**) demonstrates significantly increased values of ADC histogram kurtosis in nodal-positive compared to nodal negative thyroid carcinomas.

Table 4. Group comparison of ADC and histogram parameters of thyroid carcinomas with (N1/2, $n = 10$ patients) and without lymphatic metastatic dissemination (N0, $n = 5$ patients).

ADC Parameters and Histogram Values	N0 Mean ± SD		N1/2 Mean ± SD		Group Comparison: p-Values
ADC_{mean}, $\times 10^{-5}$ mm$^2 \cdot$s^{-1}	125.25	34.1	111.41	25.00	0.513
ADC_{min}, $\times 10^{-5}$ mm$^2 \cdot$s^{-1}	28.26	17.30	14.02	16.90	0.075
ADC_{max}, $\times 10^{-5}$ mm$^2 \cdot$s^{-1}	238.44	69.40	259.43	38.50	0.768
P10 ADC, $\times 10^{-5}$ mm$^2 \cdot$s^{-1}	82.15	26.17	69.14	23.50	0.371
P25 ADC, $\times 10^{-5}$ mm$^2 \cdot$s^{-1}	102.25	30.00	89.19	23.30	0.440
P75 ADC, $\times 10^{-5}$ mm$^2 \cdot$s^{-1}	147.26	39.14	131.75	26.43	0.440
P90 ADC, $\times 10^{-5}$ mm$^2 \cdot$s^{-1}	170.69	44.15	156.55	28.50	0.440
Median ADC, $\times 10^{-5}$ mm$^2 \cdot$s^{-1}	124.14	34.86	109.19	25.50	0.513
Mode ADC, $\times 10^{-5}$ mm$^2 \cdot$s^{-1}	112.32	25.56	101.39	27.50	0.594
Kurtosis	3.23	0.29	3.81	0.57	0.028
Skewness	−0.12	0.64	0.41	0.21	0.031
Entropy	3.56	0.66	3.5	0.71	0.768

3. Discussion

This study aimed to investigate the potential of 3T RESOLVE DWI using an ADC histogram analysis approach to distinguish between limited and advanced thyroid cancer with reference to the status of lymphatic metastatic dissemination. To the author's best knowledge, this work is the first to show differences in ADC histogram parameters between nodal-positive and nodal-negative thyroid cancer.

In detail, skewness and kurtosis of the ADC histograms were significantly increased in nodal-positive compared to nodal-negative thyroid cancer. This finding corresponds to previous studies in other malignant tumors, exemplarily clear cell renal cell carcinoma, and rectal cancer, which revealed that increased skewness of ADC histograms is associated with a more advanced disease stage [19,20]. Furthermore, an increase in ADC histogram skewness was observed in patients suffering from recurrent high grade glioma who showed disease progress under anti-proliferative chemotherapy, indicating ongoing proliferation of glioma cells within the tumor [21]. The association between changes in ADC values and altered cellularity in tumors is a well-known phenomenon [22]. Considering this, the findings of the aforementioned studies and our results we hypothesize that the process of lymphatic metastatic spread of thyroid cancer is linked to profound changes in the tissue microarchitecture, related to proliferation of distinct tumor cell clusters and subsequent migration via the lymphatic system, which finds its reflection in corresponding changes of the ADC histogram.

Additionally, this study found significant correlations between ADC histogram analysis derived values of thyroid cancer and corresponding immune-reactivity for p53. p53 has great importance as tumor suppressor and controls cell fate via induction of apoptosis, cell cycle arrest and senescence [23]. Under normal conditions, p53 remains undetectable for its rapid proteasomic degradation [23]. In thyroid cancer, p53 has been used as prognostic marker being associated with favorable outcome [24,25]. ADC mean, ADC max, ADC median, ADC modus, ADC p75 and ADC p90 correlated significantly with p53 expression. In general, increased ADC values of tumors have been shown to be associated with good therapeutic responses [26]. It was thereupon concluded that increased ADC values of thyroid cancer—in consent with previously published work—indicate a favorable prognosis. Furthermore, a clear inverse correlation of ADC max and ADC p90 with Ki-67 expression was identified. Ki-67 is a nuclear protein strictly associated with cell division and widely used in the clinical routine to assess proliferative activity [27].

Increased proliferation of cells, as indicated by increased expression of Ki-67, consecutively decreases the corresponding extracellular space in a given volume of tissue and thereupon reduces water diffusibility, which is reflected by decreased ADC values [22]. Thus—in accordance to other malignancies [11,28,29]—decreased ADC values are associated with an increased proliferation rate within thyroid cancer tissue.

This study furthermore identified a significant inverse correlation between cell count and kurtosis. Only few studies investigated the potential of ADC kurtosis to reflect histological properties, for example Chandarana and colleagues were able to differentiate clear cell from papillary subtype of renal cell cancer by means of ADC kurtosis [30]. It is therefore concluded that ADC histogram kurtosis provides additional insight in tumor-architectural details, but further studies are necessary to validate this finding in order to further elaborate the significance of this parameter. Conventionally, ADC_{mean} and ADC_{min} were used to investigate histopathological features like cellularity of tumors in vivo [22]. However, classical ADC parameters like ADC_{mean} and ADC_{min} are strongly scanner-dependent and cannot be used to compare patients investigated in different MRI devices without normalization. In contrary, histogram parameters estimate characteristics of the ADC distribution, which is not scanner-dependent like the absolute ADC values. Therefore, ADC derived histogram parameters (skewness, entropy, kurtosis) might be superior when investigating histopathological features in vivo using more than one MRI scanner in a singular study.

This study suffers from few limitations. The major limitation is the small number of patients included in this study. Furthermore, this study did not include all clinically relevant subtypes of thyroid cancer, exemplarily medullary thyroid carcinomas were not investigated. Therefore, future works including greater cohorts with different histopathological subtypes have to confirm these findings and further elucidate the relationship between histopathological findings and ADC alterations. Also, ADC histogram analysis was performed by a single, experienced reader. The suitability of histogram analysis for the clinical routine necessitates assessment of inter-reader and intra-reader variability including readers with different levels of experience. A future work needs to investigate these phenomena in a larger cohort.

ADC histogram analysis can provide more detailed information on diffusion characteristics of tumors than commonly obtained ADC parameters. For example, a previously published study demonstrated that common ADC parameters (mean, max, and min) did not reflect histopathological features like cellularity and proliferative activity in thyroid carcinoma [16]. In contrast, this study demonstrated that certain ADC histogram parameters reflect distinct histopathological features very well. Although it has proven to be a very sensitive tool for detection of microstructural changes, the specificity of ADC histogram parameters for the underlying histological changes is unclear. Characteristic changes of ADC histogram parameters in different tumor entities might be related to very different histological changes. Therefore, the significance of ADC histogram analysis should be investigated in a tumor-specific manner.

4. Materials and Methods

This retrospective study was approved (No. 2014-99) by the local research ethics committee of the Martin-Luther-University Halle-Wittenberg.

4.1. Patients

The radiological database for thyroid carcinoma was reviewed. In total, 20 patients were identified, but only 15 patients with histopathologically confirmed thyroid carcinoma had received proper DWI (using the RESOLVE sequence) and were therefore enrolled in our study.

The patient group was comprised of one male and 14 female patients. The mean age was 67 years (with a standard deviation of 12.9 years). The distribution of histopathological subtypes was as follows; follicular thyroid carcinoma: $n = 4$, papillary thyroid carcinoma: $n = 5$, anaplastic thyroid carcinoma: $n = 6$. Five patients were diagnosed with nodal negative thyroid cancer, and 10 patients had pathologically confirmed lymph node metastases. One patient was diagnosed with distant metastatic disease (pulmonary and pleural manifestation). Infiltration pattern ranged from restriction to the thyroid gland to advanced infiltration including infiltration of the trachea, esophagus, and internal jugular vein. An overview of demographic, clinical and pathological information is given in Table 5.

Table 5. Demographic and pathological data of the investigated thyroid carcinoma patients.

Case	Age	Gender	Histological Subtype	Infiltration Pattern	M Stage	N Stage
1	91	female	anaplastic	trachea	0	1
2	60	female	papillary	trachea	0	1
3	73	male	papillary	trachea, esophagus	0	1
4	68	female	papillary	trachea, esophagus internal jugular vein	0	0
5	73	female	papillary	trachea	0	1
6	67	female	anaplastic	Trachea internal jugular vein	1	?
7	73	female	anaplastic	trachea, esophagus	0	0
8	41	female	follicular	trachea	0	1
9	72	female	anaplastic	none	0	1
10	59	female	anaplastic	trachea	0	1
11	83	female	papillary	trachea	0	0
12	77	female	follicular	trachea	0	1
13	52	female	anaplastic	trachea	0	0
14	51	female	follicular	trachea	0	0
15	66	female	anaplastic	trachea	0	1

4.2. MRI

MRI of the neck was performed for all patients using a 3T device (Magnetom Skyra, Siemens, Erlangen, Germany). The imaging protocol included the following sequences:

1. axial T2 weighted (T2w) turbo spin echo (TSE) sequence (TR/TE: 4000/69, flip angle: 150°, slice thickness: 4 mm, acquisition matrix: 200 × 222, field of view: 100 mm);
2. axial T1 weighted (T1w) turbo spin echo (TSE) sequences (TR/TE: 765/9.5, flip angle: 150°, slice thickness: 5 mm, acquisition matrix: 200 × 222, field of view: 100 mm) before and after intravenous application of contrast medium (gadopentate dimeglumine, Magnevist®, Bayer Schering Pharma, Leverkusen, Germany);

3. axial DWI (readout-segmented, multi-shot EPI sequence; TR/TE: 5400/69, flip angle 180°, slice thickness: 4 mm, acquisition matrix: 200 × 222, field of view: 100 mm) with b values of 0, 400 and 800 s/mm². ADC maps were generated automatically by the implemented software package and analyzed as described previously [28].

All images were available in digital form and were analyzed by an experienced radiologist without knowledge of the histopathological diagnosis on a PACS workstation (Centricity PACS, GE Medical Systems, Milwaukee, WI, USA). Figure 1 shows a representative axial T2 weighted image of follicular thyroid carcinoma and corresponding axial ADC images of the whole tumor, which were used for histogram analysis (also displayed in Figure 1).

4.3. Histogram Analysis of ADC Values

DWI data was transferred in DICOM format and processed offline with a custom-made Matlab-based application (The Mathworks, Natick, MA, USA) on a standard windows operated system. The ADC maps were displayed within a graphical user interface (GUI) that enables the reader to scroll through the slices and draw a volume of interest (VOI) at the tumor's boundary. The VOI was created by manually drawing regions of interest (ROIs) along the margin of the tumor using all slices displaying the tumor (whole lesion measure). All measures were performed by one author (AS). The ROIs were modified in the GUI and saved (in Matlab-specific format) for later processing. After setting the ROIs, the following parameters were calculated and given in a spreadsheet format: ROI volume (cm³), mean (ADC_{mean}), maximum (ADC_{max}), minimum (ADC_{min}), median (ADC_{median}), modus (ADC_{modus}), and the following percentils: 10th (ADCp10), 25th (ADCp25), 75th (ADCp75), and 90th (ADCp90). Additionally, histogram-based characteristics of the VOI—kurtosis, skewness, and entropy—were computed. All calculations were performed using in-build Matlab functions.

4.4. Histopathology and Immunohistochemistry

All thyroid carcinomas were surgically resected and histopathologically analysed. In every case, the proliferation index was estimated on Ki-67 antigen stained specimens using MIB-1 monoclonal antibody (DakoCytomation, Glostrup, Denmark) as reported previously [31]. Furthermore, p53 index was estimated using monoclonal antibody p57, clone DO-7 (DakoCytomation). Two high power fields (0.16 mm² per field, ×400) were analysed. The area with the highest number of positive nuclei was selected. Figure 2 exemplarily shows Ki-67 and p53 immunostaining of a follicular thyroid carcinoma. Additionally, cellular density was calculated for each tumor as average cell count per five high power fields (×400). Furthermore, average nuclear area and total nuclear area were estimated using ImageJ package 1.48v (National Institute of Health, Bethesda, MD, USA) as described previously [11]. All histopathological sections were analysed using a research microscope Jenalumar equipped with a Diagnostic instruments camera 4.2 (Zeiss, Jena, Germany).

4.5. Statistical Analysis

Statistical analysis was performed using IBM SPSS 23™ (SPSS Inc., Chicago, IL, USA). Collected data was first evaluated by means of descriptive statistics. Correlative analysis was then performed using Spearman's correlation coefficient in order to analyze associations between histogram analysis derived values of ADC and (immuno-) histopathological parameters. Subsequently, Levene's Test for homogeneity of variance was performed to assess the equality of variances of ADC derived histogram parameters between different groups of thyroid carcinoma patients in order to identify the suitable test for group comparisons. In case of homoscedasticity, unpaired t test was performed to compare values among different (e.g., the metastatic and the non-metastatic) groups. In case of heteroscedasticity, Mann-Whitney-*U* test was performed to compare values among the different groups. Group comparisons were performed for nodal negative vs. nodal positive patients and patients with restricted (thyroid gland and trachea) vs. advanced (trachea, esophagus, jugular vein) infiltration pattern. Since only one patient with distant metastatic disease was included, a sufficient group comparison between M0 and M1 patients could not be performed. p-Values ≤ 0.05 were considered as statistically significant.

5. Conclusions

This exploratory study revealed significant differences in ADC histogram skewness and kurtosis comparing nodal negative and nodal positive thyroid cancer. Significant correlations between different ADC parameters were identified with p53, Ki-67, and cell count, substantiating the potential of ADC as an important prognostic imaging biomarker. This information certainly has the potential to aid thyroid surgeons in identifying the optimal treatment strategy for patients with thyroid cancer. Further studies investigating a greater cohort of patients are necessary to confirm these findings.

Acknowledgments: We acknowledge funding by the German Research Foundation (DFG) and University Leipzig within the program of open access publishing.

Author Contributions: Stefan Schob and Alexey Surov conceived and designed the experiments and wrote the paper; Nikolaos Pazaitis and Anne Kathrin Höhn performed the immunohistopathological experiments; Diana Horvath-Rizea and Bhogal Pervinder analyzed the data; Nikita Garnov contributed the histogram analysis tool; Hans Jonas Meyer digitalized the immunohistological slides and performed image analysis; Karl-Titus Hoffmann performed MRI; Julia Dieckow wrote the paper.

Conflicts of Interest: The authors declare no conflict of interest.

Abbreviations

MDPI	Multidisciplinary Digital Publishing Institute
DOAJ	Directory of Open Access Journals
TLA	Three Letter Acronym
LD	linear Dichroism

References

1. Katoh, H.; Yamashita, K.; Enomoto, T.; Watanabe, M. Classification and general considerations of thyroid cancer. *Ann. Clin. Pathol.* **2015**, *3*, 1–9.
2. Dralle, H.; Machens, A.; Basa, J.; Fatourechi, V.; Franceschi, S.; Hay, I.D.; Nikiforov, Y.E.; Pacini, F.; Pasieka, J.L.; Sherman, S.I. Follicular cell-derived thyroid cancer. *Nat. Rev. Dis. Prim.* **2015**, *1*, 15077. [CrossRef] [PubMed]
3. Paschke, R.; Lincke, T.; Müller, S.P.; Kreissl, M.C.; Dralle, H.; Fassnacht, M. The treatment of well-differentiated thyroid carcinoma. *Dtsch. Arztebl. Int.* **2015**, *112*, 452–458. [PubMed]
4. Nixon, I.J.; Shaha, A.R. Management of regional nodes in thyroid cancer. *Oral Oncol.* **2013**, *49*, 671–675. [CrossRef] [PubMed]
5. Shaha, A.R. Recurrent differentiated thyroid cancer. *Endocr. Pract.* **2012**, *18*, 600–603. [CrossRef] [PubMed]
6. Ferrari, S.M.; Fallahi, P.; Politti, U.; Materazzi, G.; Baldini, E.; Ulisse, S.; Miccoli, P.; Antonelli, A. Molecular targeted therapies of aggressive thyroid cancer. *Front Endocrinol.* **2015**, *6*, 176. [CrossRef] [PubMed]
7. Cabanillas, M.E.; Dadu, R.; Hu, M.I.; Lu, C.; Gunn, G.B.; Grubbs, E.G.; Lai, S.Y.; Williams, M.D. Thyroid gland malignancies. *Hematol. Oncol. Clin. N. Am.* **2015**, *29*, 1123–1143. [CrossRef] [PubMed]
8. Wendler, J.; Kroiss, M.; Gast, K.; Kreissl, M.C.; Allelein, S.; Lichtenauer, U.; Blaser, R.; Spitzweg, C.; Fassnacht, M.; Schott, M.; et al. Clinical presentation, treatment and outcome of anaplastic thyroid carcinoma: Results of a multicenter study in Germany. *Eur. J. Endocrinol.* **2016**, *175*, 521–529. [CrossRef] [PubMed]
9. Asimakopoulos, P.; Nixon, I.J.; Shaha, A.R. Differentiated and medullary thyroid cancer: Surgical management of cervical lymph nodes. *Clin. Oncol.* **2017**, *29*, 283–289. [CrossRef] [PubMed]
10. Mizrachi, A.; Shaha, A.R. Lymph node dissection for differentiated thyroid cancer. *Mol. Imaging Radionucl. Ther.* **2016**, *26*, 10–15. [CrossRef] [PubMed]
11. Schob, S.; Meyer, J.; Gawlitza, M.; Frydrychowicz, C.; Müller, W.; Preuss, M.; Bure, L.; Quäschling, U.; Hoffmann, K.-T.; Surov, A. Diffusion-weighted MRI reflects proliferative activity in primary CNS lymphoma. *PLoS ONE* **2016**, *11*, e0161386. [CrossRef] [PubMed]

12. Surov, A.; Stumpp, P.; Meyer, H.J.; Gawlitza, M.; Höhn, A.-K.; Boehm, A.; Sabri, O.; Kahn, T.; Purz, S. Simultaneous ^{18}F-FDG-PET/MRI: Associations between diffusion, glucose metabolism and histopathological parameters in patients with head and neck squamous cell carcinoma. *Oral Oncol.* **2016**, *58*, 14–20. [CrossRef] [PubMed]

13. Khizer, A.T.; Raza, S.; Slehria, A.-U.-R. Diffusion-weighted MR imaging and ADC mapping in differentiating benign from malignant thyroid nodules. *J. Coll. Physicians Surg. Pak.* **2015**, *25*, 785–788. [PubMed]

14. Lu, Y.; Moreira, A.L.; Hatzoglou, V.; Stambuk, H.E.; Gonen, M.; Mazaheri, Y.; Deasy, J.O.; Shaha, A.R.; Tuttle, R.M.; Shukla-Dave, A. Using diffusion-weighted MRI to predict aggressive histological features in papillary thyroid carcinoma: A novel tool for pre-operative risk stratification in thyroid cancer. *Thyroid* **2015**, *25*, 672–680. [CrossRef] [PubMed]

15. Porter, D.A.; Heidemann, R.M. High resolution diffusion-weighted imaging using readout-segmented echo-planar imaging, parallel imaging and a two-dimensional navigator-based reacquisition. *Magn. Reson. Med.* **2009**, *62*, 468–475. [CrossRef] [PubMed]

16. Schob, S.; Voigt, P.; Bure, L.; Meyer, H.J.; Wickenhauser, C.; Behrmann, C.; Höhn, A.; Kachel, P.; Dralle, H.; Hoffmann, K.-T.; Surov, A. Diffusion-weighted imaging using a readout-segmented, multishot EPI sequence at 3T distinguishes between morphologically differentiated and undifferentiated subtypes of thyroid carcinoma—A preliminary study. *Transl. Oncol.* **2016**, *9*, 403–410. [CrossRef] [PubMed]

17. Just, N. Improving tumour heterogeneity MRI assessment with histograms. *Br. J. Cancer* **2014**, *111*, 2205–2213. [CrossRef] [PubMed]

18. Hao, Y.; Pan, C.; Chen, W.; Li, T.; Zhu, W.; Qi, J. Differentiation between malignant and benign thyroid nodules and stratification of papillary thyroid cancer with aggressive histological features: Whole-lesion diffusion-weighted imaging histogram analysis. *J. Magn. Reson. Imaging* **2016**, *44*, 1546–1555. [CrossRef] [PubMed]

19. Kierans, A.S.; Rusinek, H.; Lee, A.; Shaikh, M.B.; Triolo, M.; Huang, W.C.; Chandarana, H. Textural differences in apparent diffusion coefficient between low- and high-stage clear cell renal cell carcinoma. *Am. J. Roentgenol.* **2014**, *203*, W637–W644. [CrossRef] [PubMed]

20. Liu, L.; Liu, Y.; Xu, L.; Li, Z.; Lv, H.; Dong, N.; Li, W.; Yang, Z.; Wang, Z.; Jin, E. Application of texture analysis based on apparent diffusion coefficient maps in discriminating different stages of rectal cancer. *J. Magn. Reson. Imaging* **2016**. [CrossRef] [PubMed]

21. Nowosielski, M.; Recheis, W.; Goebel, G.; Güler, O.; Tinkhauser, G.; Kostron, H.; Schocke, M.; Gotwald, T.; Stockhammer, G.; Hutterer, M. ADC histograms predict response to anti-angiogenic therapy in patients with recurrent high-grade glioma. *Neuroradiology* **2011**, *53*, 291–302. [CrossRef] [PubMed]

22. Chen, L.; Liu, M.; Bao, J.; Xia, Y.; Zhang, J.; Zhang, L.; Huang, X.; Wang, J. The correlation between apparent diffusion coefficient and tumor cellularity in patients: A meta-analysis. *PLoS ONE* **2013**, *8*, e79008. [CrossRef] [PubMed]

23. Wang, Z.; Sun, Y. Targeting p53 for novel anticancer therapy. *Transl. Oncol.* **2010**, *3*, 1–12. [CrossRef] [PubMed]

24. Godballe, C.; Asschenfeldt, P.; Jørgensen, K.F.; Bastholt, L.; Clausen, P.P.; Hansen, T.P.; Hansen, O.; Bentzen, S.M. Prognostic factors in papillary and follicular thyroid carcinomas: P53 expression is a significant indicator of prognosis. *Laryngoscope* **1998**, *108*, 243–249. [CrossRef] [PubMed]

25. Bachmann, K.; Pawliska, D.; Kaifi, J.; Schurr, P.; Zörb, J.; Mann, O.; Kahl, H.J.; Izbicki, J.R.; Strate, T. P53 is an independent prognostic factor for survival in thyroid cancer. *Anticancer Res.* **2006**, *27*, 3993–3997.

26. Padhani, A.R.; Liu, G.; Mu-Koh, D.; Chenevert, T.L.; Thoeny, H.C.; Takahara, T.; Dzik-Jurasz, A.; Ross, B.D.; Van Cauteren, M.; Collins, D.; et al. Diffusion-weighted magnetic resonance imaging as a cancer biomarker: Consensus and recommendations. *Neoplasia* **2009**, *11*, 102–125. [CrossRef] [PubMed]

27. Schlüter, C.; Duchrow, M.; Wohlenberg, C. The cell proliferation-associated antigen of antibody Ki-67: A very large, ubiquitous nuclear protein with numerous repeated elements, representing a new kind of cell cycle-maintaining proteins. *J. Cell Biol.* **1993**, *123*, 1–10. [CrossRef]

28. Surov, A.; Caysa, H.; Wienke, A.; Spielmann, R.P.; Fiedler, E. Correlation between different ADC fractions, cell count, Ki-67, total nucleic areas and average nucleic areas in meningothelial meningiomas. *Anticancer Res.* **2015**, *35*, 6841–6846. [PubMed]

29. Chen, L.; Zhang, J.; Chen, Y.; Wang, W.; Zhou, X.; Yan, X.; Wang, J. Relationship between apparent diffusion coefficient and tumour cellularity in lung cancer. *PLoS ONE* **2014**, *9*, e99865. [CrossRef] [PubMed]

30. Chandarana, H.; Rosenkrantz, A.B.; Mussi, T.C.; Kim, S.; Ahmad, A.A.; Raj, S.D.; McMenamy, J.; Melamed, J.; Babb, J.S.; Kiefer, B.; et al. Histogram analysis of whole-lesion enhancement in differentiating clear cell from papillary subtype of renal cell cancer. *Radiology* **2012**, *265*, 790–798. [CrossRef] [PubMed]
31. Surov, A.; Gottschling, S.; Mawrin, C.; Prell, J.; Spielmann, R.P.; Wienke, A.; Fiedler, E. Diffusion-weighted imaging in meningioma: Prediction of tumor grade and association with histopathological parameters. *Transl. Oncol.* **2015**, *8*, 517–523. [CrossRef] [PubMed]

International Journal of
Molecular Sciences

MDPI

Article

HER2 Analysis in Sporadic Thyroid Cancer of Follicular Cell Origin

Rosaria M. Ruggeri [1,*], Alfredo Campennì [2], Giuseppe Giuffrè [3], Luca Giovanella [4],
Massimiliano Siracusa [2], Angela Simone [3], Giovanni Branca [3], Rosa Scarfì [3],
Francesco Trimarchi [1], Antonio Ieni [3] and Giovanni Tuccari [3]

[1] Department of Clinical and Experimental Medicine, Unit of Endocrinology, University of Messina,
    AOU Policlinico G. Martino, 98125 Messina, Italy; Francesco.Trimarchi@unime.it
[2] Department of Biomedical Sciences and Morphological and Functional Images, Unit of Nuclear Medicine,
    University of Messina, AOU Policlinico G. Martino, 98125 Messina, Italy; acampenni@unime.it (A.C.);
    m.siracusadr@alice.it (M.S.)
[3] Department of Human Pathology in Adult and Developmental Age "Gaetano Barresi",
    Unit of Pathological Anatomy, University of Messina, AOU Policlinico G. Martino, 98125 Messina, Italy;
    giuffre@unime.it (G.G.); asimone@unime.it (A.S.); giobranca81@gmail.com (G.B.); rscarfi@unime.it (R.S.);
    aieni@unime.it (A.I.); tuccari@unime.it (G.T.)
[4] Department of Nuclear Medicine, Thyroid and PET/CT Center, Oncology Institute of Southern Switzerland,
    6500 Bellinzona, Switzerland; luca.giovanella@eoc.ch
*   Correspondence: rmruggeri@unime.it; Tel.: +39-090-221-3840; Fax: +39-090-221-3517

Academic Editor: Daniela Gabriele Grimm
Received: 7 November 2016; Accepted: 30 November 2016; Published: 6 December 2016

Abstract: The Epidermal Growth Factor Receoptor (EGFR) family member human epidermal growth factor receptor 2 (HER2) is overexpressed in many human epithelial malignancies, representing a molecular target for specific anti-neoplastic drugs. Few data are available on HER2 status in differentiated thyroid cancer (DTC). The present study was aimed to investigate HER2 status in sporadic cancers of follicular cell origin to better clarify the role of this receptor in the stratification of thyroid cancer. By immunohistochemistry and fluorescence in-situ hybridization, HER2 expression was investigated in formalin-fixed paraffin-embedded surgical specimens from 90 DTC patients, 45 follicular (FTC) and 45 papillary (PTC) histotypes. No HER2 immunostaining was recorded in background thyroid tissue. By contrast, overall HER2 overexpression was found in 20/45 (44%) FTC and 8/45 (18%) PTC, with a significant difference between the two histotypes ($p = 0.046$). Five of the six patients who developed metastatic disease during a median nine-year follow-up had a HER2-positive tumor. Therefore, we suggest that HER2 expression may represent an additional aid to identify a subset of patients who are characterized by a worse prognosis and are potentially eligible for targeted therapy.

Keywords: sporadic differentiated thyroid cancer; HER2 (Human Epidermal Growth Factor Receptor 2); immunohistochemistry; FISH (fluorescence in situ hybridization)

1. Introduction

The human epidermal growth factor receptor 2 (HER2) is a cell surface receptor belonging to the Epidermal growth factor receptor (EGFR) family of receptors, which includes four distinct, but closely related tyrosine kinase receptors: EGFR, HER2 (HER2/c-neu), HER3, and HER4 [1,2]. HER2 has no known cognate ligand and may become active upon hetero-dimerization with other family members, such as EGFR. Upon activation, EGFR and HER2 undergo dimerization and tyrosine auto-phosphorylation, thus leading to activation of proliferative and anti-apoptotic pathways, principally the MAPK, Akt, and JNK pathways. Through such effects on cell-cycle progression, apoptosis, angiogenesis, and tumor-cell motility, HER2 is implicated in the development and progression of cancer [1,2].

The HER2 gene is frequently amplified and the protein overexpressed in several human epithelial malignancies, including breast, gastric, ovarian, and colon-rectal cancers [3–9]. In such tumors, HER2 amplification/overexpression has been linked to a poor overall outcome and a poorly differentiated phenotype [3–7]. It has also been considered a useful indicator of response to specifically targeted therapies, such as trastuzumab, that inhibit the extracellular domain of HER2 [8,9]. Many studies have been addressed to determine the HER2-positive rate, mainly in breast and gastric carcinomas, utilizing a well codified scoring system [10,11], and HER2 status assessment is currently being used in such cancers to determine patient eligibility for treatment with trastuzumab [8,9,12,13].

Studies on HER2 have also been performed in thyroid cancer cells and tissues [14–26]. Interestingly, a wide variation in HER2 overexpression was reported in such studies, with positivity rates varying from 0% up to 70%, which may largely be attributed to inter-study technical and interpretive variations. Due to these conflicting findings, there is no consensus in the currently available literature regarding the potential prognostic and therapeutic value of this marker in thyroid cancer [16,18,23,24]. More recently, HER2 expression has been linked to the expression of estrogen receptors in thyroid tumor tissue [27] and associated with BRAF[V600E] mutation and a more aggressive phenotype in familial papillary thyroid cancers (PTCs) [28].

In the present study, we investigate HER2 expression status in a surgical series of sporadic differentiated thyroid carcinomas of follicular cell origin to better clarify the role of this receptor in the stratification of thyroid cancer.

2. Results

2.1. Clinical-Pathological Findings

The clinical-pathological features of the 90 differentiated thyroid cancer (DTC) patients (73 F and 17 M, mean age 51.6 ± 12.7 years, median 49 years) are summarized in Table 1. The 45 patients with PTC comprised 34 females and 11 males who ranged in ages from 28 to 71 years (median age, 49 years). Histologically, the papillary carcinomas were as follows: 16 classic variant, 21 follicular variant, 4 Hürthle cell variant, and 4 sclerosing variant. The 45 patients with follicular thyroid cancer (FTC) comprised 39 females and 6 males aged 22–76 years (median age, 55 years). The follicular carcinomas included 34 that were minimally invasive and 11 that were widely invasive.

All patients had undergone total or subtotal thyroidectomy. In 61% of the patients, the tumor was <2 cm, pT1 according to TNM classification [29], and 11% had lymph node metastases (Table 1). No patient exhibited distant metastases at the time of surgery.

All patients were followed up for at least five years after thyroidectomy at our Endocrine Unit (median follow-up duration 8.7 years, range 5–20 years). During follow-up, 6 out of 90 (6.7%) patients (all females, aged 43–76 years, median 45 years) developed metastases: one was affected by PTC in the classic variant pT1b stage, two by PTC follicular variant in the pT2 stage, and three by FTC in the pT3 stage. All PTC metastatic patients except for one (with lung metastases) had lymph-nodes metastases, located in the right lateral neck ($n = 1$), left lateral neck ($n = 1$), anterior central compartment ($n = 3$), and upper mediastinum ($n = 1$). The three FTC patients had lung and skeletal metastases.

Table 1. Clinical and pathological features of the 90 differentiated thyroid cancer (DTC) patients at the time of surgery.

Variables	PTC Cases (*n* = 45)	FTC Cases (*n* = 45)
Age (years, mean ± SD)	50.6 ± 12.3	52.7 ± 13.2
Sex		
Male	11	6
Female	34	39
M:F	1:3	1:6.5
Histological features	Classic variant, *n* = 16 Follicular variant, *n* = 21 Hürtle cell variant, *n* = 4 Sclerosing variant, *n* = 4	Minimally invasive, *n* = 34 Widely invasive, *n* = 11
Primary Tumour pT [29]		
T1	34 (13 T1a and 21 T1b)	21 (9 T1a and 12 T1b)
T2	10	13
T3	1	11
T4	/	/
Node metastasis (NX/N0/N1) [29]		
pNX	12	20
pN0	23	25
pN1	10 (N1a)	/

The symbol / means no case.

2.2. Immunohistochemical (IHC) and Fluorescence In Situ Hybridization (FISH) Results

Twenty-seven specimens (17 PTC and 10 FTC) presented unamplified HER2 status and were therefore scored 0. The remaining 63 cases (28 PTC and 35 FTC) stained for HER2 with a variable intensity ranging from 1+ to 3+ (Table 2). A not negligible number of tumors, mostly papillary histotype (15 cases, 13 PTC and 2 FTC), exhibited a low (1+) and patchy expression of HER2, with a granular or diffuse cytoplasmic distribution of the staining. Such cases with no membranous staining were considered negative.

Table 2. Human epidermal growth factor receptor 2 (HER2) expression in thyroid cancer tissue *.

HER2 Status	FTC (*n* = 45)	PTC (*n* = 45)	*p*
IHC Negative/low (0, 1+)	12 (26.6%)	30 (66.6%)	0.020
Equivocal (2+)	15	7	
Positive (3+)	18	8	
IHC/FISH Positive cases (3+ and amplified)	20 (44.4%)	8 (17.7%)	0.046

* HER2 status has been assessed by using immunohistochemistry (IHC) to detect protein expression. Fluorescence in situ hybridization (FISH) was also performed in cases that tested 2+ at IHC, as specified in the Materials and Methods section to detect gene amplification.

HER2 was clearly overexpressed (3+) at IHC with membranous staining in 26 cases: 18 FTC, of which 6 were widely invasive (see example in Figure 1A), and 8 PTC (2 classic, 5 follicular variant, and 1 Hürthle cell variant) (see example in Figure 1B). Twenty-two tumors (15 FTC and 7 PTC) were scored as 2+ by IHC for HER2. All 2+ cases (Figure 2A) were evaluated by FISH: two FTC revealed HER2 amplification (Figure 2B), while the others were unamplified (Figure 2C). Therefore, the overall rate of HER2-positive cases was 31% (28/90 cases). Specifically, HER2 amplification/overexpression was found in 20/45 (44%) FTC and 8/45 (18%) PTC, with a significant difference between the two histotypes (x^2 = 3.96; p = 0.046) (Figure 3). Normal thyroid parenchyma surrounding the tumor lacked expression of HER2.

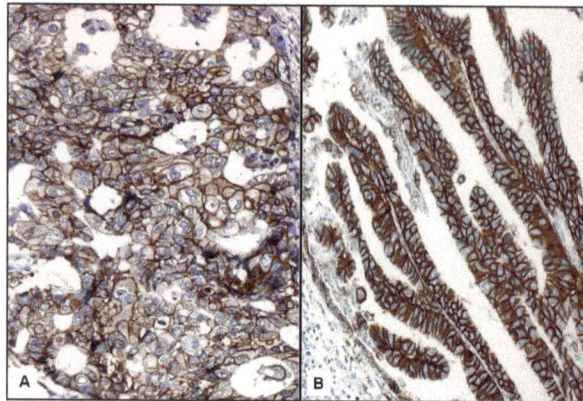

Figure 1. An evident diffuse membranous HER2 immunopositivity was seen in FTC (**A**, ×400) as well as in the classic variant of papillary thyroid cancer (PTC) (**B**, ×400) (Mayer's hemalum counterstain).

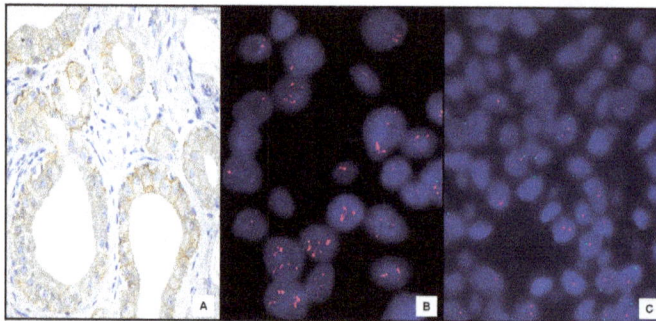

Figure 2. IHC equivocal (2+) follicular thyroid cancer (FTC) case (**A**, ×400) (Mayer's hemalum counterstain) that showed a corresponding HER2 amplification by FISH (**B**, ×660). Another unamplified HER2 FTC case (**C**, ×460).

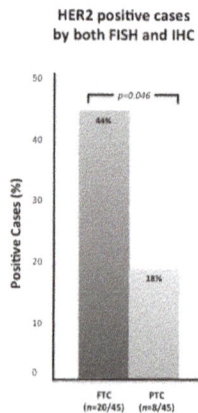

Figure 3. Percentages of FTC and PTC cases which tested positive for HER2 at both FISH and IHC (3+).

No significant correlation was found between HER2 expression and tumor size, as well as lymph node metastases at the time of surgery. However, among the six patients who developed metastatic disease during follow-up, five had a HER2-positive tumor. Three tumors (2 PTC and 1 FTC) stained 3+ at IHC; and the remaining two FTCs stained 2+ at IHC and revealed HER2 amplification at FISH. Thus, 5 out of 28 HER2-positive tumors and 1 out of the remaining 62 HER2-negative tumors developed metastatic disease during the follow-up (x^2 = 8.18; p = 0.004).

None of our cancers was iodine refractory. Concerning iodine uptake, in our cohort of DTC patients, the intensity of 131-radioiodine uptake observed (visual analysis) in thyroid remnant and, mainly, in lymph-node metastases at post-therapy whole body scan (pT-WBS) did not differ in cases overexpressing HER2 compared to negative ones.

3. Discussion

In the present study, we assessed HER2 status in a series of sporadic differentiated thyroid cancers (DTCs) and found that HER2 was overall overexpressed in about one-third of cases; in particular, the expression rate was significantly higher in the follicular (FTC) histotype compared to the papillary (PTC) one.

To date, many studies have evaluated HER2 expression in thyroid cancer with controversial results, largely due to inter-study differences in the size and setting of the examined series and, most of all, to the subjective assessment and lack of uniform methodology [14–26]. Indeed, studies reported in the literature in the past several decades markedly differ by the methodological approach used to assess HER2 status [14–26]. Moreover, the criteria for scoring HER2 expression were different, and, in many studies, cytoplasmic staining patterns were reported as positive immunostaining for HER2 [16–18,22,23,26]. As a consequence, the results of these studies are not comparable and therefore not conclusive.

The present study aimed to evaluate HER2 status using a method as reproducible and as standardized as possible, similar to other cancers. Indeed, a similarly wide variation in HER2 overexpression has been reported in many other tumor types [3–8]. Therefore, as the prognostic and therapeutic relevance of HER2 status has grown, the need to achieve a standardized HER2 assessment method has also arisen in neoplastic sites different from the stomach and breast [11]. In the present study, we utilized the updated ASCO-CAP scoring system reported in breast cancer [10]. Applying such strict criteria to our DTC series, we reported an overall HER2 expression of 31%. In detail, both PTC (18%) and FTC (44%) cases were found to overexpress HER2 in our series, and the expression rate was significantly different between the two histotypes in favor of the follicular one. These data may suggest a prognostic impact of HER2 status in DTC, further supported by the finding that, in nearly all patients who had metastases diagnosed during follow-up, HER2 was overexpressed in the primary tumor. Obviously, such results should be confirmed in a larger series to better determine whether HER2 amplification/overexpression can be considered an additional prognostic aid to identify cases characterized by a more aggressive disease, such as in other epithelial cancers [3,8–10,13].

Other studies have shown the absence of HER2 amplification in normal thyroid tissue, as well as increased expression during malignant progression in DTC [18,19,21,22]. These findings are in agreement with ours, although the importance of a standardized scoring methodology should be emphasized in order to explain the partially contradictory results on DTC. In detail, Sugishita et al. [24] investigated HER2 expression in a surgical series of 69 DTC, including 61 PTC and only 8 FTC, and found that 14 PTC and 2 FTC had a score 3+ at IHC. Amplification of HER2 gene was confirmed via FISH in 10 PTC 3+ plus 4 PTC cases 2+, and in one FTC, with an overall expression rate of 21.7% (23% considering the sole PTC). Therefore, the highest rate of HER2 expression was recorded in PTC. However, there are some limitations in the above-mentioned study [24]. First of all, the authors set an arbitrary cut-off value for the FISH ratio of 1.3 to score their cases because of the low rate of HER2 gene amplification they had found. If they had applied the HER2 amplification criteria for breast and gastric cancer (i.e., FISH ratio > 2.0) strictly to thyroid cancer, as we did, all cases would be judged as negative. Thus, these data are not comparable to ours. Secondly, the series from Sugishita et al. included only

8 FTC, without any morphological evidence of HER2 expression [24]. Successively, another surgical series of 69 DTC has been investigated for HER2 expression [25], but again a low number of FTC was included (11 cases). Although the authors utilized a HER2 scoring methodology equivalent to ours, they reported a very low rate of HER2 overexpression, since no FTC and only four (6.9%) PTCs showed HER-2 overexpression [25]. Therefore, the rate of HER2 expression in PTC was definitely lower than that previously reported elsewhere [24]; in addition, Mdah et al. [25] failed to find HER2 expression in FTC. More recently, Caria and co-workers reported a scattered HER2 expression, restricted to less than 10% of tumor cells, in a few cases of familial PTC [28]. In particular, 5/13 (38.5%) of familial PTC cases showed 5.1%–10% $HER2^+$ cells, while no sporadic PTC cases exceeded the cut-off value [28]. Moreover, when familial PTCs were analyzed via IHC using an anti-c-erbB2 antibody to detect HER2 protein expression, inconsistent results compared to the FISH analysis were obtained, also possibly biased by the age of the available histological sections (7–20 years) [28]. Finally, no data about HER2 expression were recorded about different varieties present in PTC, nor in relation to the metastatic event in PTC [28].

In the present study, we analyzed a large sporadic DTC registry including the same number of PTC and FTC cases. Consequently, our data appear to be statistically more significant in comparison to the reported studies [24,25,28] and reveal a significant HER2 expression related to histotype, greatly favoring FTC. Moreover, its overexpression is more evident in metastatic DTC compared to non-metastatic ones. These findings might have potential practical implications. HER2 may be helpful in identifying a subset of DTC patients characterized by a worse prognosis, but eligible for potential targeted therapies with HER2 inhibitors, such as trastuzumab. This may be relevant in iodine-refractory cancers, since novel molecular targets and therapeutic strategies are currently under investigation for these tumors, whose treatment is still a major challenge [30,31]. Moreover, HER2 expression may be used for prognostic application in the context of other well-accepted clinic-pathological prognostic parameters for DTC (age, gender, pTNM stage, histological subtype), since very few new markers revealed prognostic value *per se* [32,33]. If our observations are confirmed in a larger series, HER2 overexpression may play a role not only in the development and progression of a subset of thyroid carcinomas, but also in their prognostic and therapeutic stratification.

4. Materials and Methods

4.1. Sample Collection

Ninety sporadic differentiated thyroid tumors (DTCs) (45 papillary (PTCs) and 45 follicular (FTCs) thyroid cancers) with available formalin-fixed paraffin-embedded tissue blocks were selected from the files of the Department of Pathology of our University Hospital. The surgical samples were from 90 patients (73 F and 17 M, aged 28–76 years) who were diagnosed and followed up at the Endocrine Unit of our University Hospital over the last twenty years. The clinical records of the 90 patients were reviewed.

Histological classification was performed by two pathologists with experience in thyroid pathology (Giovanni Tuccari, Antonio Ieni and Giovanni Branca), according to the World Health Organization guidelines [34]. Institutional review board approval was obtained.

4.2. Immunohistochemistry

For each case, 5-μm-thick sections from representative tissue blocks of the tumor were obtained. Immunohistochemistry (IHC) was performed twice on each specimen using, firstly, the monoclonal antibody against HER2-pY-1248 (Phosphorylation site specific) (clone PN2A, Dako; w.d. 1:100) and, successively, the Hercep Test (Dako, Glostrup, Denmark), with an automated procedure (DAKO Autostainer Link48), according to the manufacturer's instructions. An antigen retrieval pre-treatment was performed in 3 cycles in a 0.01 M citrate buffer, pH 6.0, in a microwave oven at 750 W. Staining intensity, the percentage of positive cells, and cellular localization were evaluated both in the tumor and in the adjacent non-neoplastic thyroid tissue.

Since there are no established criteria for thyroid cancer, we adapted the current breast criteria for scoring HER2 in our DTC. Staining intensity and cellular localization in the tumor were evaluated and scored according to the updated ASCO-CAP scoring system for breast cancer [10]. Accordingly, the degree of HER2 staining was scored from 0 to 3+. HER2 positivity was defined as 3+ when strong membranous staining was noted in at least 10% of cells, 2+ when weak to moderate complete membranous staining was evident in 10% of tumor cells, 1+ when a faint or weak and incomplete membrane staining was observed, and 0 when no staining was observed or when staining was present in less than 10% of neoplastic cells.

Immunohistochemical evaluations were carried out twice and blindly by two pathologists (Giovanni Tuccari, Antonio Ieni and Giovanni Branca). In the case of disagreement, cases were jointly discussed using a double-headed microscope until agreement was reached.

4.3. Fluorescence In Situ Hybridization

In cases showing 2+ immunostaining, as determined by IHC with the Hercept test, fluorescence in situ hybridization (FISH) analysis was performed using a HER2 FISH PharmDx™ kit (Dako, Glostrup, Denmark), according to the manufacturer's instructions, to detect amplification of the *HER2* gene, as for breast cancer [35]. Gene amplification was recorded when the ratio HER2/centromeric probe for chromosome 17 (CEP17) signal was ≥ 2.0.

Specimens of breast carcinoma were used as appropriate positive controls for both IHC and FISH analysis (Figure 4A). Negative controls were obtained either by omitting the primary antiserum or by replacing the primary antiserum with normal mouse serum, in a parallel section of the same cases (Figure 4B).

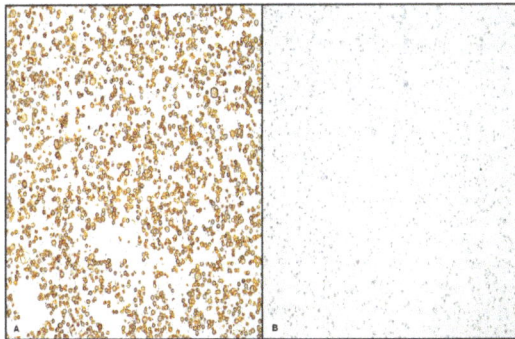

Figure 4. Breast carcinoma tissue section as positive control (A, 200×) and negative control (B, 200×).

4.4. Statistical Analysis

Once tested for normal distribution and variance, data (mean ± standard deviation) were analyzed by the two-tailed Student's *t*-test, a chi-square test with Yates' correction for continuity, and linear regression analysis. The level of statistical significance was always set at $p < 0.05$.

Acknowledgments: We thank Massimo Bongiovanni from the Institute of Pathology, Lausanne University Hospital, (Lausanne, Switzerland) for his useful suggestions.

Author Contributions: Rosaria M. Ruggeri and Giovanni Tuccari conceived of the study, participated in its design and coordination, and drafted the manuscript. Alfredo Campennì and Francesco Trimarchi participated in the design of the study and took care of the clinical approach to patients. Giuseppe Giuffrè and Angela Simone carried out the FISH studies. Angela Simone and Rosi Scarfì carried out the immunohistochemistry analysis. Massimiliano Siracusa was responsible for data collection and database management. Giovanni Branca and Antonio Ieni performed the histopatological and immunohistochemical evaluations of the specimens and data interpretation along with Giovanni Tuccari, and drafted the manuscript. Luca Giovanella participated in writing and critical reviewing of the manuscript. All authors read and approved the final version.

Conflicts of Interest: There is no potential conflict of interest, and the authors have nothing to disclose. This work was not supported by any grant.

References

1. Yarden, Y. The EGFR family and its ligands in human cancer. Signaling mechanisms and therapeutic opportunities. *Eur. J. Cancer* **2001**, *4*, S3–S8. [CrossRef]
2. Salomon, D.S.; Brandt, R.; Ciardiello, F.; Normanno, N. Epidermal growth factor-related peptides and their receptors in human malignancies. *Crit. Rev. Oncol. Hematol.* **1995**, *19*, 183–232. [CrossRef]
3. Ieni, A.; Barresi, V.; Giuffrè, G.; Caruso, R.A.; Lanzafame, S.; Villari, L.; Salomone, E.; Roz, E.; Cabibi, D.; Franco, V.; et al. HER2 status in advanced gastric carcinoma: A retrospective multicentric analysis from Sicily. *Oncol. Lett.* **2013**, *6*, 1591–1594. [PubMed]
4. Ieni, A.; Barresi, V.; Caltabiano, R.; Caleo, A.; Bonetti, L.R.; Lanzafame, S.; Zeppa, P.; Caruso, R.A.; Tuccari, G. Discordance rate of HER2 status in primary gastric carcinomas and synchronous lymph node metastases: A multicenter retrospective analysis. *Int. J. Mol. Sci.* **2014**, *15*, 22331–22341. [CrossRef] [PubMed]
5. Ieni, A.; Barresi, V.; Caltabiano, R.; Cascone, A.M.; del Sordo, R.; Cabibi, D.; Zeppa, P.; Lanzafame, S.; Sidoni, A.; Franco, V.; et al. Discordance rate of HER2 status in primary breast carcinomas versus synchronous axillary lymph node metastases: A multicenter retrospective investigation. *Onco. Targets Ther.* **2014**, *7*, 1267–1272. [PubMed]
6. Slamon, D.J.; Clark, G.M.; Wong, S.G.; Levin, W.J.; Ullrich, A.; McGuire, W.L. Human breast cancer: Correlation of relapse and survival with amplification of the HER-2/neu oncogene. *Science* **1987**, *235*, 177–182. [CrossRef] [PubMed]
7. Allgayer, H.; Babic, R.; Gruetzner, K.U.; Tarabichi, A.; Schildberg, F.W.; Heiss, M.M. c-erbB-2 is of independent prognostic relevance in gastric cancer and is associated with the expression of tumor-associated protease systems. *J. Clin. Oncol.* **2000**, *18*, 2201–2209. [PubMed]
8. Ieni, A.; Barresi, V.; Rigoli, L.; Caruso, R.A.; Tuccari, G. HER2 Status in Premalignant, Early, and Advanced Neoplastic Lesions of the Stomach. *Dis. Markers* **2015**, *2015*, 234851. [CrossRef] [PubMed]
9. Ieni, A.; Giuffrè, G.; Lanzafame, S.; Nuciforo, G.; Curduman, M.; Villari, L.; Roz, E.; Certo, G.; Cabibi, D.; Salomone, E.; et al. Morphological and biomolecular characteristics of subcentimetric invasive breast carcinomas in Sicily: A multicentre retrospective study in relation to trastuzumab treatment. *Oncol. Lett.* **2012**, *3*, 141–146. [CrossRef] [PubMed]
10. Rakha, E.A.; Starczynski, J.; Lee, A.H.; Ellis, I.O. The updated ASCO/CAP guideline recommendations for HER2 testing in the management of invasive breast cancer: A critical review of their implications for routine practice. *Histopathology* **2014**, *64*, 609–615. [CrossRef] [PubMed]
11. Valtorta, E.; Martino, C.; Sartore-Bianchi, A.; Penaullt-Llorca, F.; Viale, G.; Risio, M.; Rugge, M.; Grigioni, W.; Bencardino, K.; Lonardi, S.; et al. Assessment of a HER2 scoring system for colorectal cancer: Results from a validation study. *Mod. Pathol.* **2015**, *28*, 481–1491. [CrossRef] [PubMed]
12. Baselga, J.; Lewis Phillips, G.D.; Verma, S.; Ro, J.; Huober, J.; Guardino, A.E.; Samant, M.K.; Olsen, S.; de Haas, S.L.; Pegram, M.D. Relationship between Tumor Biomarkers and Efficacy in EMILIA, a Phase III Study of Trastuzumab Emtansine in HER2-Positive Metastatic Breast Cancer. *Clin. Cancer Res.* **2016**, *22*, 3755–3763. [CrossRef] [PubMed]
13. Bang, Y.J.; van Cutsem, E.; Feyereislova, A.; Chung, H.C.; Shen, L.; Sawaki, A.; Lordick, F.; Ohtsu, A.; Omuro, Y.; Satoh, T.; et al. Trastuzumab in combination with chemotherapy versus chemotherapy alone for treatment of HER2-positive advanced gastric or gastro-oesophageal junction cancer (ToGA): A phase 3, open-label, randomised controlled trial. *Lancet* **2010**, *376*, 687–697. [CrossRef]
14. Lemoine, N.K.; Wyllie, F.S.; Lillehaug, J.R.; Staddon, S.L.; Hughes, C.M.; Aasland, R.; Shaw, J.; Varhaug, J.E.; Brown, C.L.; Gullick, W.J.; et al. Absence of abnormalities of the c-erbB-1 and c-erbB-2 proto-oncogenes in human thyroid neoplasia. *Eur. J. Cancer* **1990**, *26*, 777–779. [CrossRef]
15. Haugen, D.R.; Akslen, L.A.; Varhaug, J.E.; Lillehaug, J.R. Expression of c-erbB-2 protein in papillary thyroid carcinomas. *Br. J. Cancer* **1992**, *65*, 832–837. [CrossRef] [PubMed]
16. Sugg, S.L.; Ezzat, S.; Zheng, L.; Rosen, I.B.; Freeman, J.L.; Asa, S.L. Cytoplasmic staining of erbB-2 but not mRNA levels correlates with differentiation in human thyroid neoplasia. *Clin. Endocrinol.* **1998**, *49*, 629–637. [CrossRef]

17. Utrilla, J.C.; Martin-Lacave, I.; San Martin, M.V.; Fernandez-Santos, J.M.; Galera-Davidson, H. Expression of c-erbB-2 oncoprotein in human thyroid tumours. *Histopathology* **1999**, *34*, 60–65. [CrossRef] [PubMed]
18. Kremser, R.; Obrist, P.; Spizzo, G.; Erler, H.; Kendler, D.; Kemmler, G.; Mikuz, G.; Ensinger, C. Her2/neu overexpression in differentiated thyroid carcinomas predicts metastatic disease. *Virchows. Arch.* **2003**, *442*, 322–328. [PubMed]
19. Ensinger, C.; Prommegger, R.; Kendler, D.; Gabriel, M.; Spizzo, G.; Mikuz, G.; Kremser, R. Her2/neu expression in poorly-differentiated and anaplastic thyroid carcinomas. *Anticancer Res.* **2003**, *23*, 2349–2353. [PubMed]
20. Mondi, M.M.; Rich, R.; Ituarte, P.; Wong, M.; Bergman, S.; Clark, O.H.; Perrier, N.D. HER2 expression in thyroid tumors. *Am. Surg.* **2003**, *69*, 1100–1103. [PubMed]
21. Murakawa, T.; Tsuda, H.; Tanimoto, T.; Tanabe, T.; Kitahara, S.; Matsubara, O. Expression of KIT, EGFR, HER2/NEU and tyrosine phosphorylation in undifferentiated thyroid carcinoma: Implication for a new therapeutic approach. *Pathol. Int.* **2005**, *55*, 757–765. [CrossRef] [PubMed]
22. Elliott, D.D.; Sherman, S.I.; Busaidy, N.L.; Williams, M.D.; Santarpia, L.; Clayman, G.L.; El-Naggar, A.K. Growth factor receptors expression in anaplastic thyroid carcinoma: Potential markers for therapeutic stratification. *Hum. Pathol.* **2008**, *39*, 15–20. [CrossRef] [PubMed]
23. Qin, C.; Cau, W.; Zhang, Y.; Mghanga, F.P.; Lan, X.; Gao, Z.; An, R. Correlation of clinic-pathological features and expression of molecular markers with prognosis after 131I treatment of differentiated thyroid carcinoma. *Clin. Nucl. Med.* **2012**, *37*, e40–e46. [CrossRef] [PubMed]
24. Sugishita, Y.; Kammori, M.; Yamada, O.; Poon, S.S.; Kobayashi, M.; Onoda, N.; Yamazaki, K.; Fukumori, T.; Yoshikawa, K.; Onose, H.; et al. Amplification of the human epidermal growth factor receptor 2 gene in differentiated thyroid cancer correlates with telomere shortening. *Int. J. Oncol.* **2013**, *42*, 1589–1596. [PubMed]
25. Mdah, W.; Mzalbat, R.; Gilbey, P.; Stein, M.; Sharabi, A.; Zidan, J. Lack of HER-2 gene amplification and association with pathological and clinical characteristics of differentiated thyroid cancer. *Mol. Clin. Oncol.* **2014**, *2*, 1107–1110. [CrossRef] [PubMed]
26. Wu, G.; Wang, J.; Zhou, Z.; Li, T.; Tang, F. Combined staining for immunohistochemical markers in the diagnosis of papillary thyroid carcinoma: Improvement in the sensitivity or specificity? *J. Int. Med. Res.* **2013**, *41*, 975–983. [CrossRef] [PubMed]
27. Kavanagh, D.O.; McIlroy, M.; Myers, E.; Bane, F.; Crotty, T.B.; McDermott, E.; Hill, A.D.; Young, L.S. The role of oestrogen receptor α in human thyroid cancer: Contributions from coregulatory proteins and the tyrosine kinase receptor HER2. *Endocr. Relat. Cancer* **2010**, *17*, 255–264. [CrossRef] [PubMed]
28. Caria, P.; Cantara, S.; Frau, D.V.; Pacini, F.; Vanni, R.; Dettori, T. Genetic Heterogeneity of HER2 Amplification and Telomere Shortening in Papillary Thyroid Carcinoma. *Int. J. Mol. Sci.* **2016**, *17*, e1759. [CrossRef] [PubMed]
29. The American Joint Committee on Cancer. Thyroid. In *AJCC Cancer Staging Manual*, 7th ed.; Edge, S.B., Byrd, D.R., Compton, C.C., Eds.; Springer: New York, NY, USA, 2010; pp. 87–96.
30. Alonso-Gordoa, T.; Díez, J.J.; Durán, M.; Grande, E. Advances in thyroid cancer treatment: Latest evidence and clinical potential. *Ther. Adv. Med. Oncol.* **2015**, *7*, 22–38. [CrossRef] [PubMed]
31. Bulotta, S.; Celano, M.; Costante, G.; Russo, D. Emerging strategies for managing differentiated thyroid cancers refractory to radioiodine. *Endocrine* **2016**, *52*, 214–221. [CrossRef] [PubMed]
32. Soares, P.; Celestino, R.; Melo, M.; Fonseca, E.; Sobrinho-Simões, M. Prognostic biomarkers in thyroid cancer. *Virchows. Arch.* **2014**, *464*, 333–346. [CrossRef] [PubMed]
33. Ruggeri, R.M.; Campennì, A.; Baldari, S.; Trimarchi, F.; Trovato, M. What is New on Thyroid Cancer Biomarkers. *Biomark. Insights* **2008**, *3*, 237–252. [PubMed]
34. De Lellis, R.A. *WHO Classification of Tumours, Pathology and Genetics of Tumours of Endocrine Organs*, 3rd ed.; Ronald, A., de Lellis, R.A., Riccardo, V.L., Philipp, U.H., Charis, E., Eds.; IARC Press: Lyon, France, 2004; Volume 8, p. 230.
35. La, P.; Salazar, P.A.; Hudis, C.A.; Ladanyi, M.; Chen, B. HER-2 testing in breast cancer using immunohistochemical analysis and fluorescence in situ hybridization: A single-institution experience of 2279 cases and comparison of dual-color and single-color scoring. *Am. J. Clin. Pathol.* **2004**, *121*, 631–636.

International Journal of
Molecular Sciences

MDPI

Article

Genetic Heterogeneity of *HER2* Amplification and Telomere Shortening in Papillary Thyroid Carcinoma

Paola Caria [1], Silvia Cantara [2], Daniela Virginia Frau [1], Furio Pacini [2,†], Roberta Vanni [1,*,†] and Tinuccia Dettori [1]

1   Department of Biomedical Sciences, University of Cagliari, Cittadella Universitaria, Monserrato 09042, Italy; paola.caria@unica.it (P.C.); dvfrau@unica.it (D.V.F.); dettorit@unica.it (T.D.)
2   Department of Medical, Surgical and Neurological Sciences, University of Siena, Siena 53100, Italy; cantara@unisi.it (S.C.); pacini8@unisi.it (F.P.)
*   Correspondence: vanni@unica.it; Tel.: +39-070-675-4123; Fax: +39-070-675-4119
†   These authors contributed equally to this work.

Academic Editor: Daniela Gabriele Grimm
Received: 4 August 2016; Accepted: 12 October 2016; Published: 21 October 2016

Abstract: Extensive research is dedicated to understanding if sporadic and familial papillary thyroid carcinoma are distinct biological entities. We have previously demonstrated that familial papillary thyroid cancer (fPTC) cells exhibit short relative telomere length (RTL) in both blood and tissues and that these features may be associated with chromosome instability. Here, we investigated the frequency of *HER2* (*Human Epidermal Growth Factor Receptor 2*) amplification, and other recently reported genetic alterations in sporadic PTC (sPTC) and fPTC, and assessed correlations with RTL and *BRAF* mutational status. We analyzed *HER2* gene amplification and the integrity of *ALK, ETV6, RET,* and *BRAF* genes by fluorescence in situ hybridization in isolated nuclei and paraffin-embedded formalin-fixed sections of 13 fPTC and 18 sPTC patients. We analyzed $BRAF^{V600E}$ mutation and RTL by qRT-PCR. Significant *HER2* amplification ($p = 0.0076$), which was restricted to scattered groups of cells, was found in fPTC samples. *HER2* amplification in fPTCs was invariably associated with $BRAF^{V600E}$ mutation. RTL was shorter in fPTCs than sPTCs ($p < 0.001$). No rearrangements of other tested genes were observed. These findings suggest that the association of *HER2* amplification with $BRAF^{V600E}$ mutation and telomere shortening may represent a marker of tumor aggressiveness, and, in refractory thyroid cancer, may warrant exploration as a site for targeted therapy.

Keywords: papillary thyroid carcinoma; *HER2* (*Human Epidermal Growth Factor Receptor 2*); Telomere; FISH (fluorescence in situ hybridization)

1. Introduction

The most common histological subtype of non-medullary thyroid carcinoma (NMTC) is papillary thyroid carcinoma (PTC), which represents 75%–85% of all thyroid cancer. Although mostly sporadic (sPTC), there is some evidence for a familial form of PTC (fPTC) not associated with known Mendelian syndromes. Familial PTC is observed in approximately 5%–10% of NMTC cases [1] and, despite the extensive research dedicated to understand if it is a distinct biological entity than sPTC [2], this distinction remains controversial. Indeed, the most common somatic alterations, such as mutations in *RAS* and *BRAF* and rearrangements of *RET/PTC* and *NTRK1*, exhibit similar prevalence and distribution in both sPTC and fPTC [3]. No oncogenic germline mutations of these genes have been detected in fPTC cases [4]. However, there is an ongoing debate on the possible association of *HAPB2* germline mutation to the predisposition to familial forms of NMTC [5–8]. Generally speaking, most PTC can be treated effectively with surgery and radioactive iodine therapy. However, for cases in which these treatments are not effective, targeted drugs might be considered. Kinase inhibitors, such as sorafenib and lenvatinib, are now used as targeted drugs [9,10]. Less frequent genetic alterations, such as rearrangements of *ALK* [11] and *ETV6* [12], identified in sPTC, but not yet investigated in fPTC, also represent new therapeutic targets. In addition, the amplification of the *HER2* gene, reported in highly-malignant PTC nodules [13], might be added to the list of drug-targetable genes. The *HER2* amplification in PTC was observed by fluorescence in situ hybridization (FISH), and average telomere length in *HER2*-positive (*HER2*+) PTC was significantly shorter than *HER2*-negative (*HER2*−) PTC [13]. Of possible significance, shorter average telomere length has also been reported in fPTC compared to sPTC [14]. These observations prompted us to verify *HER2* amplification and telomere length status in 13 fPTC and 18 sPTC. Tumors were also investigated for integrity of the *RET* gene, which is rearranged in 10%–40% of PTC, and *ALK, ETV6*, and *BRAF* genes, which are rearranged in a minority of PTC. $BRAF^{V600E}$ mutation, which is associated with an aggressive biological behavior [15], was also evaluated. Our results indicate an increased prevalence of occasional *HER2* gene intermediate amplification in fPTC compared to sPTC, and shorter telomeres in all fPTC, including those with *HER2* amplification, compared to sPTC. In addition, all *HER2*+ samples invariably possessed the $BRAF^{V600E}$ mutation, but not vice versa. The association might represent a marker of tumor aggressiveness and, in refractory thyroid cancer, may indicate possible exploration for targeted therapy. Additionally, the simultaneous occurrence of these three specific molecular alterations may be suggestive of the existence of a specific fPTC subgroup.

2. Results

2.1. Human Epidermal Growth Factor Receptor 2 (HER2) Amplification in Familial Papillary Thyroid Carcinoma (fPTC) and Sporadic Papillary Thyroid Carcinoma (sPTC)

HER2 amplification was evaluated in 13 fPTC (seven females, mean age at diagnosis of 53.7 ± 14.2; six males, mean age at diagnosis 49.0 ± 21.5) and 18 sPTC (15 females, mean age at diagnosis 46.6 ± 7.8; three males, mean age at diagnosis 43.6 ± 13.6). We found that isolated fPTC and sPTC nuclei were *HER2*− according to the Wolff criteria (originally developed for breast cancer formalin-fixed paraffin-embedded (FFPE) examination) [16], although a number of cases showed scattered *HER2*+ cells, ranging from 1.4% to 9% in fPTC and from 1% to 2.4% in sPTC (Figure 1A,B). Based on these observations, and on the lack of specific criteria for the evaluation of *HER2* amplification in thyroid tumors, we analyzed the distribution of HER2+ cells in these cases to determine if they met the criteria for the presence of genetic heterogeneity (>5% and <50%) according to Vance [17]. We found significant genetic heterogeneity in the distribution of *HER2* amplification in fPTC compared to sPTC (*p* = 0.0076) (Figure 1C–E). We found that 5/13 (38.5%) fPTC cases showed 5.1% to 10% HER2+ cells. FISH on FFPE from sPTC confirmed the findings obtained from isolated nuclei that no sPTC case exceeded the cut-off value. When fPTCs were analyzed by immunohistochemistry using an anti-c-erbB2 antibody [18] to detect HER2 protein expression, inconsistent results compared to the FISH analysis were obtained. However, this result was possibly biased by the age of the available histological sections (7–20 years).

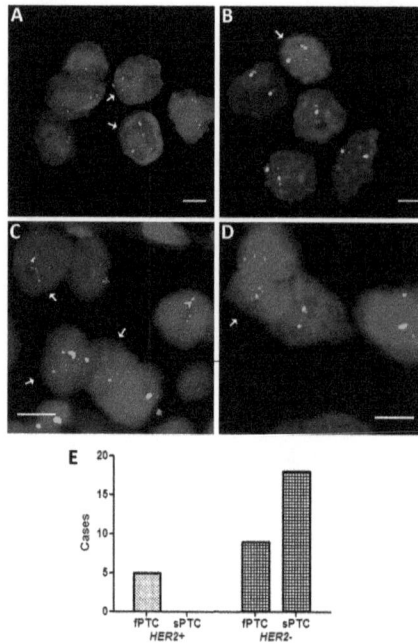

Figure 1. Distribution of *HER2* (*human epidermal growth factor receptor 2*) amplification in familial papillary thyroid carcinoma (fPTC) and sporadic papillary thyroid carcinoma (sPTC) nuclei. Arrows point to isolated nuclei with extra copies of the *HER2* gene (**red spots**) in the presence of disomy 17 (chromosome 17 centromere-specific alphoid repetitive DNA, **green spots**) in fPTC (**A**); and sPTC (**B**); and in nuclei in formalin-fixed paraffin-embedded (FFPE) sections of fPTC (**C**); and sPTC (**D**). The distribution of the amplification in FFPE sections of fPTC versus sPTC was significant ($p = 0.0076$) (Fisher exact test) and indicative of genetic heterogeneity heterogeneity according to Vance criteria [17] (**E**). Scare bar = 10 μm.

2.2. Rearrangements of ALK, BRAF, ETV6, and RET Genes

We found no rearrangements of *ALK*, *BRAF*, and *ETV6* genes by fluorescence in situ hybridization. We found one case of sPTC with a *RET* rearrangement, the remaining cases exhibited neither disruptions nor numerical changes in *RET* gene (see examples in Figure 2A,B).

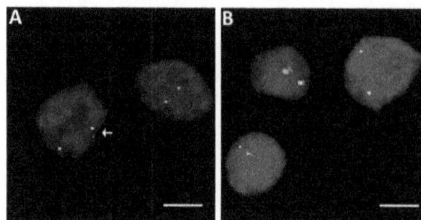

Figure 2. Examples of fluorescence in situ hybridization (FISH) in isolated nuclei for the identification of genes specifically rearranged in papillary thyroid carcinoma (PTC). Arrows point to the split of the red/green signal of a *RET* break-apart [19] probe in the case of sPTC, indicating broken *RET* (**A**); and to un-split red/green signals of an *ALK* break-apart probe in the case of fPTC, indicating unbroken *ALK* (**B**). **Red spot**: 300 kb probe DNA fragment; **Green spot**: 442 kb probe DNA fragment; Scare bar = 10 μm.

2.3. Telomere Length and BRAF^{V600E} Mutation in fPTC and sPTC Patients

Relative telomere length was significantly shorter in fPTC samples than in sPTC samples: median = 0.93 (25th–75th percentile: 0.6–1.2) vs. 1.9 (25th–75th percentile: 1.8–2.3) for fPTC vs. sPTC, respectively ($p < 0.001$) (Figure 3). This result was not due to a difference in the patient's age and sex, as sporadic cases were selected to be age/sex-matched with familial patients (Table 1).

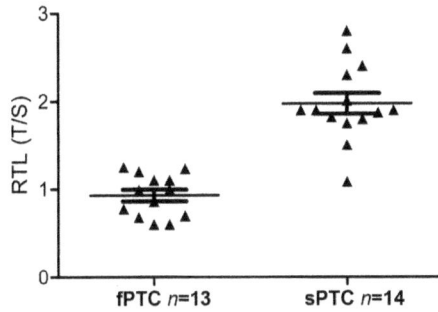

Figure 3. fPTC and sPTC relative telomere length (RTL). RTL was measured by q-PCR, and was expressed as the ratio (T/S) of the telomere (T) repeat copy number to a single-copy gene (S). The difference in RLT between fPTC and sPTC samples was significant ($p < 0.001$) (Mann-Whitney U-test). Triangles represent the RTL of each case; the upper and lower lines represent the interquartile range of the distribution (25th–75th percentile); the middle line represents the median.

BRAF^{V600E} mutation was detected in 9/13 (69.0%) fPTC and 14/18 (78%) sPTC ($p = 0.68$), which was not statistically significant. All *HER2+* fPTC were BRAF^{V600E}-positive (*BRAF+*), although not all *BRAF+* fPTC were *HER2+* (Figure 4).

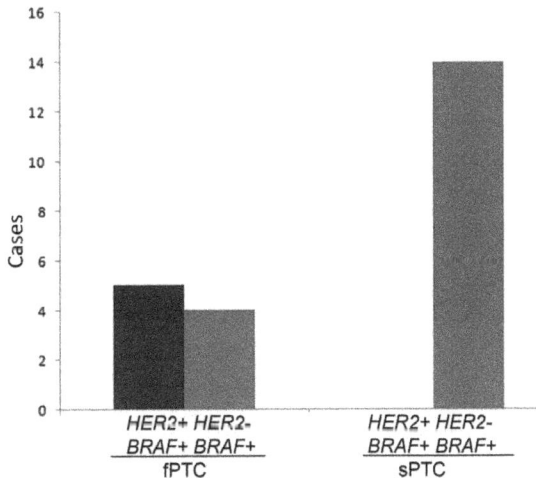

Figure 4. Distribution of *HER2* amplification and BRAF^{V600E} mutation in fPTC and sPTC tumors.

Table 1. Characteristics of patients.

Tumor	Age at Diagnosis (Mean ± SD)	Sex (Males %)	PTC Size (Median/IQR)	TNM	Extrathyroidal Invasion N (%)	Multifocality N (%)	Lymphonode Metastases at Diagnosis N (%)	Final Outcome * N (%)	Follow-up (Mean Years)	Histology
fPTC (*n* = 13)	51.5 ± 17.0	6 (46.1%)	11/11.5	pT1 8 (61.5%) pT2 2 (11.7%) pT3 3 (23.0%)	5 (38.5%)	7 (53.8%)	3 (23%)	*Remission* 9 (69.2%) *Persistent disease* 4 (30.8%)	7.59 ± 3.9	9 CV-PTC 4 FV-PTC
sPTC (*n* = 18)	46.1 ± 8.5	3 (16.6%)	9.5/8.5	pT1 8 (44.4%) pT2 1 (5.5%) pT3 9 (50.%)	9 (50%)	6 (33.3%)	5 (27.7%)	*Remission* 11 (61.1%) *Persistent disease* 2 (11.1%)	5.5 ± 2.8	14 CV-PTC 2 FV-PTC 1 SV-PTC 1 TR-PTC

CV-PTC—classical variant of papillary thyroid carcinoma; fPTC—familial papillary thyroid carcinoma; FV-PTC—follicular variant of papillary thyroid carcinoma; IQR—Inter Quartile Range; N—number of cases; PTC—papillary thyroid carcinoma; sPTC—sporadic papillary thyroid carcinoma; SV-PTC—sclerosing variant of papillary thyroid carcinoma; TNM—(Tumor (limph) Node Metastasis) classification [20]; TR-PTC: trabecular variant of papillary thyroid carcinoma; *—six patients were lost to follow-up.

3. Discussion

Amplification of the *HER2* gene in thyroid cancer was first uncovered by FISH analysis of follicular cells from highly malignant PTC nodules [13]. They observed that *HER2+* PTC exhibited shorter telomeres than *HER2−* PTC. PTC is an entity mostly recognized as sporadic, although the familial form may account for approximately 5% of cases [1,2]. Familial PTC may occur in combination with other Mendelian cancer syndromes (familial adenomatous polyposis, Gardner's syndrome, Peutz-Jeghers syndrome, and Cowden's syndrome) or may be unassociated with other neoplasms in familial aggregates. However, although the risk of developing extra-thyroidal malignancy in non-Cowden's syndrome is documented [21,22], the clinical correlation between sporadic breast cancer (20% of which are *HER2+*) [23] and PTC is still controversial [24,25], and the co-occurrence of both disorders in the same individual is a subject of extensive debate [26,27]. None of our fPTC patients had clinical or pathological evidence of hereditary syndromes associated with NMTC, breast cancer, or other types of sporadic tumors, except for one male patient who had a previous squamous cell carcinoma of the auricle. The thyroid cancer of this patient was not associated with any genetic alterations of the genes examined here. As more than 5%, but fewer than 50%, of nuclei were found to be amplified in our FFPE sections, we use the Vance criteria [17] in the interpretation of our results. Our data indicated a significant difference in *HER2* amplification ($p = 0.019$) in fPTC compared to sPTC. This finding indicates a degree of genetic heterogeneity in the fPTC group and suggests that *HER2+* cells in fPTC possibly undergo apoptosis to a lesser extent than in sPTC. The observation adds thyroid carcinoma to the list of tumors that exhibit *HER2+*. *HER2+* is indeed observed in a growing number of other tumors, including advanced gastric and esophageal cancer [28], ovarian [29], colon [30], bladder, lung, uterine, cervix, head and neck, and endometrial cancer [31]. The extracellular domain of the HER2 receptor has an essential role in cell proliferation and anti-apoptotic processes, making *HER2+* breast and gastric/gastroesophageal cancers more likely to respond to targeted therapies in combination with chemotherapy than *HER2−* tumors. A wide range of solid tumors showing deregulation of *HER2* expression are regarded as biologically aggressive. Familial PTC is often associated with a more severe phenotype than its sporadic counterpart [32], and often harbor *BRAF*[V600E] mutation. *BRAF*[V600E] is considered to have a prognostic value in PTC [33], and usually identifies differentiated thyroid tumors with advanced clinicopathological features. *BRAF*[V600E] is also strongly associated with PTC patient mortality [15]. In contrast to lung adenocarcinoma, in which *HER2* amplification and *BRAF*[V600E] mutation appear to be mutually exclusive events [34], here we found that all *HER2+* fPTC bore the *BRAF*[V600E] mutation, although not all *BRAF*[V600E]-positive nodules had *HER2* amplification. Of significance, none of the *BRAF*[V600E]-positive sPTC were *HER2+*, despite the high frequency of *BRAF*[V600] mutations in this cohort. It is not entirely clear how this discrepancy should be interpreted, considering the limited size of our cohorts. In addition, we do not know whether *HER2* amplification and *BRAF*[V600E] mutation coexist in the same cells within a tumor or if they are segregated in different clones. We do not know, either, if the condition is different in our *HER2+* fPTC versus sPTC with *HER2+* cells <5%.

The other evaluated genes, *ALK*, *BRAF*, *ETV6*, and *RET*, exhibited extensive integrity. The only case bearing a *RET* disruption was *BRAF*[V600E]-negative. Moreover, we found a significant difference ($p < 0.001$) of RTL in fPTC nodules compared to sPTC nodules, in agreement with our previous investigations [14,35]. Telomere length regulation plays a crucial role in genome instability and tumorigenesis [36]. Dysfunctional telomeres can increase chromosome instability by causing either fusion of chromosomes or fusion of sister chromatids, bringing the formation of anaphase bridges and the beginning of the so-called breakage-bridge-fusion cycles [37]. Although biased by the small number of patients investigated in our two cohorts (forced by the low frequency of fPTC), our data stigmatize significantly shorter RTL in fPTC cells versus sPTC cells. This result is in line with the reported predisposition of fPTC patients toward spontaneous chromosome fragility [35]. This observation poses the basis for further investigation exploring the existence of a possible specific three-dimensional (3D) altered telomere organization in fPTC. Telomere remodeling is a feature of cancer cells [38] and may identify tumor subgroups [39,40]. On the other hand, alterations in the telomere 3D profile have been reported in a murine model of thyroid tumors [41]. Recurrent somatic mutations in the promoter of TERT, the catalytic subunit of the enzyme telomerase, has been reported in PTC [42], often in concomitance with mutated *BRAF* [43]. A subclonal distribution in the rare PTC that harbor the alteration, in contrast to a clonal distribution in the poorly-differentiated and anaplastic tumors has been observed [44]. Unfortunately, the scanty material available from our cases prevented the possibility to establish a putative association of TERT promoter mutations with telomere length in our PTCs.

Our finding on RTI, are, substantially, in keeping with the Sugishita data [13]. However, in contrast, 38.5% of our fPTC, but none of our sPTC, showed *HER2* amplification, indicating an apparently preferential association with fPTC. In this regard, the small number of patients investigated in our two cohorts might constitute a bias. Nevertheless, as a whole, the finding of $BRAF^{V600E}$ mutation in association with *HER2*+ genetic heterogeneity, short telomere length, and prevalence of multifocal tumors seems to not be a rare molecular event in fPTC and may characterize a subgroup of fPTC. The response of refractory fPTC patients of this subgroup to target therapy of trastuzumab and lapatinib should be explored.

4. Materials and Methods

4.1. Sample Collection

The PTCs (13 fPTC and 18 sPTC) considered in the present study were selected from the pathological files of the University of Siena. Familial recurrence of the disease was defined as the presence of at least one first-degree relative with differentiated thyroid carcinoma in the absence of any other familial syndrome. None of the fPTC patients presented with any other sporadic tumor, including breast cancer, except for one male patient with a previous squamous cell carcinoma of the auricle. The histology of all tumor samples was classified according to the World Health Organization guidelines [20]. The fPTC cases were selected from 13 families, randomly choosing one affected subject from each family (the oldest affected subject): seven females (mean age at diagnosis of 53.7 ± 14.2, with an age range of 29–69 years) and six males (mean age at diagnosis of 49.0 ± 21.5, range 24–81). Four fPTC tumors were classified as follicular variants and nine were classical PTC. Seven out of the 13 families (53.8%) had three or more affected members, two out of 13 (15.3%) had two members with PTC and at least three members operated for multi-nodular goiter, and four out of 13 had only two members affected by thyroid cancer. In these cases, the phenomenon of genetic anticipation was observed with the second generation acquiring the disease at an earlier age and having more advanced disease at presentation.

The sPTC cases were from 15 females (mean age at diagnosis of 46.6 ± 7.8, range 31–64) and three males (mean age at diagnosis 43.6 ± 13.6, range 31–58). Fourteen were classified as classical PTC, two as follicular variants, one as a diffuse sclerosing variant, and one as a trabecular variant (Table 1). Informed consent was obtained from each patient after a full explanation of the purpose and nature of all of the procedures to be used. All data from the patients were handled in accordance with local ethical committee-approved protocols and in compliance with the Helsinki declaration

4.2. Fluorescence In Situ Hybridization

Thick sections (30 µm) were obtained from formalin-fixed paraffin-embedded (FFPE) tissue blocks of the thyroid nodules or, to assess probe cut-off, from the apparently tumor-free tissue of the contralateral lobe. Nuclei were isolated as reported [45], and were investigated by FISH, using specific probes and a standard protocol [46].

4.3. Detection of HER2 Gene Copy Number Alterations or Amplification

HER2 amplification was determined by counting the total numbers of *HER2* and CEP17 (chromosome 17 centromere-specific alphoid repetitive DNA, Abbott Molecular, Abbott Park, IL, USA) signals per nucleus with a mean of 89.5 (range 34–175) nuclei. The ratio of *HER2* signals to CEP17 (centromeric probe for chromosome 17) signals was calculated according to ASCO/CAP (American Society of Clinical Oncology/College of American Pathologists Guideline) criteria refined for breast cancer [16].

Nodules with, or suspected to have, *HER2* amplification were then re-evaluated by FISH on 4 μm histological sections to assess the distribution of possible clones. The tumor area, marked by the pathologist [47], was entirely scored. As more than 5%, but fewer than 50%, of nuclei were found to be amplified, the Vance criteria were used to define the distribution of abnormal cells [17].

Consecutive histologic sections were used to assess *HER2* expression by immunohistochemistry. The expression of HER2 protein was determined using anti-c-erbB2 antibody (Dako, Glostrup, Denmark) in accordance with the manufacturer's instructions.

Detection of *ALK*, *BRAF*, *ETV6*, and *RET* rearrangements. Commercially available break-apart or single gene probes for *ALK*, *BRAF* (Abbott Molecular, Abbott Park, IL, USA), and *ETV6* (Kreatech Diagnostic, Amsterdam, The Netherlands) were used to verify gene integrity in nuclei isolated from FFPE tissue blocks. An arbitrary cut-off of 3% was employed, as control cells showed no split signal in 200 scored nuclei per sample. For the *RET* gene, a homebrew probe and a previously described cut-off value were used [19]. Microscopic analysis was performed with an Olympus BX41 epifluorescence microscope and a charge-coupled device camera (Cohu, San Diego, CA, USA) interfaced with the CytoVision system (software version 3.9; Applied Imaging, Pittsburg, PA, USA).

4.4. Telomere Length and BRAF^{V600E} Mutation

DNA was extracted from fresh or FFPE tissues, using the QIAamp® DNA Mini Kit (Qiagen, Milano, Italy) following the manufacturer's instructions. RTL of sPTC was determined by quantitative PCR, carried out on 30 ng/μL genomic DNA using an MJ Mini Personal Thermal cycler (Bio-Rad, Milano, Italy) as described [14]. RTL values of the fPTC examined in the present study were reported previously [14]. Relative telomere length was calculated as the ratio of telomere repeats to a single-copy gene in experimental samples using standard curves. This ratio is proportional to the average telomere length. The *36B4* gene, which encodes acidic ribosomal phosphoprotein P0, was used as the single-copy gene [14]. For analysis of the *BRAF*^{V600E} mutation, DNA was amplified in a final volume of 50 μL of 2× PCR Master Mix (AmpliTaq Gold® PCR Master Mix, Applied Biosystems, Milano, Italy) and a final primer concentration of 200 nM. Primer sequences, PCR conditions, and interpretation of results were as previously described [48].

4.5. Statistical Analyses

Mann-Whitney *U*-test (IBM SPSS Statistic version 2.1 software, Armonk, NY, USA) was used for statistical analysis of differences in RTL. Fisher exact test was used to compare *HER2* gene amplification. All *p*-values were two-sided and *p* less than 0.05 was considered significant.

Epidemiological data are presented as the mean ± SD and median when necessary. The *t*-test for independent data was performed for normal variables. To evaluate significant differences in data frequency we analyzed 2 × 2 contingency tables by the Fisher exact test. Tables with sizes larger than 2 × 2 were examined by the Chi-squared test.

Acknowledgments: We thank Sandra Orrù from the Pathology Service Businco Hospital (Cagliari, Italy), Azienda Ospedaliera Brotzu for analysis of specimens by immunohistochemistry. This work was partially supported by the program "Projects of national interest (PRIN)" of the Italian Ministry of Education, University and Research (MIUR) (grant No 20122ZF7HE-002).

Author Contributions: Paola Caria, Daniela Virginia Frau, and Tinuccia Dettori carried out the FISH studies and drafted the manuscript. Silvia Cantara carried out the qPCR studies and the *BRAF^{V600E}* mutational analysis. Silvia Cantara and Furio Pacini participated in the design of the study and took care of the clinical approach to patients. Roberta Vanni and Tinuccia Dettori conceived of the study, participated in its design and coordination, and drafted the manuscript. All authors read and approved the final version.

Conflicts of Interest: The authors declare no conflict of interest.

References

1. Thyroid Disease Manager. Available online: http://www.thyroidmanager.org/ (accessed on 31 July 2016).
2. Bonora, E.; Tallini, G.; Romeo, G. Genetic predisposition to familial nonmedullary thyroid cancer: An update of molecular findings and state-of-the-art studies. *J. Oncol.* **2010**, 385206. [CrossRef] [PubMed]
3. Moses, W.; Weng, J.; Kebebew, E. Prevalence, clinicopathologic features, and somatic genetic mutation profile in familial versus sporadic nonmedullary thyroid cancer. *Thyroid* **2011**, *21*, 367–371. [CrossRef] [PubMed]
4. Hou, P.; Xing, M. Absence of germline mutations in genes within the MAP kinase pathway in familial non medullary thyroid cancer. *Cell Cycle* **2006**, *5*, 2036–2039. [CrossRef] [PubMed]
5. Gara, S.K.; Jia, L.; Merino, M.J.; Agarwa, S.K.; Zhang, L.; Cam, M.; Patel, D.; Kebebew, E. Germline *HABP2* mutation causing familial non medullary thyroid cancer. *N. Engl. J. Med.* **2015**, *373*, 448–455. [CrossRef] [PubMed]
6. Carvajal-Carmona, L.G.; Tomlinson, I.; Sahasrabudhe, R. Re: HABP2 G534E mutation in familial nonmedullary thyroid cancer. *J. Natl. Cancer Inst.* **2016**, *108*, djw108. [CrossRef] [PubMed]
7. Weeks, A.L.; Wilson, S.G.; Ward, L.; Goldblatt, J.; Hui, J.; Walsh, J.P. HABP2 germline variants are uncommon in familial nonmedullary thyroid cancer. *BMC Med. Genet.* **2016**, *17*, 60. [CrossRef] [PubMed]
8. Tomsic, J.; Fultz, R.; Liyanarachchi, S.; He, H.; Senter, L.; de la Chapelle, A. HABP2 G534E variant in papillary thyroid carcinoma. *PLoS ONE* **2016**, *11*, e0146315. [CrossRef] [PubMed]
9. Fallahi, P.; Mazzi, V.; Vita, R.; Ferrari, S.M.; Materazzi, G.; Galleri, D.; Benvenga, S.; Miccoli, P.; Antonelli, A. New therapies for dedifferentiated papillary thyroid cancer. *Int. J. Mol. Sci.* **2015**, *16*, 6153–6182. [CrossRef] [PubMed]
10. Bikas, A.; Vachhani, S.; Jensen, K.; Vasko, V.; Burman, K.D. Targeted therapies in thyroid cancer: An extensive review of the literature. *Expert Rev. Clin. Pharmacol.* **2016**, *15*, 1–15. [CrossRef] [PubMed]
11. Nikiforov, Y.E. Thyroid cancer in 2015: Molecular landscape of thyroid cancer continues to be deciphered. *Nat. Rev. Endocrinol.* **2016**, *12*, 67–68. [CrossRef] [PubMed]
12. Ricarte-Filho, J.C.; Li, S.; Garcia-Rendueles, M.E.; Montero-Conde, C.; Voza, F.; Knauf, J.A.; Heguy, A.; Viale, A.; Bogdanova, T.; Thomas, G.A.; Mason, C.E.; et al. Identification of kinase fusion oncogenes in post-Chernobyl radiation-induced thyroid cancers. *J. Clin. Investig.* **2013**, *123*, 4935–4944. [CrossRef] [PubMed]
13. Sugishita, Y.; Kammori, M.; Yamada, O.; Poon, S.S.; Kobayashi, M.; Onoda, N.; Yamazaki, K.; Fukumori, T.; Yoshikawa, K.; Onose, H.; et al. Amplification of the human epidermal growth factor receptor 2 gene in differentiated thyroid cancer correlates with telomere shortening. *Int. J. Oncol.* **2013**, *42*, 1589–1596. [PubMed]
14. Capezzone, M.; Cantara, S.; Marchisotta, S.; Busonero, G.; Formichi, C.; Benigni, M.; Capuano, S.; Toti, P.; Pazaitou-Panayiotou, K.; Caruso, G.; et al. Telomere length in neoplastic and nonneoplastic tissues of patients with familial and sporadic papillary thyroid cancer. *J. Clin. Endocrinol. Metab.* **2011**, *96*, E1852–E1856. [CrossRef] [PubMed]
15. Xing, M.; Haugen, B.R.; Schlumberger, M. Progress in molecular-based management of differentiated thyroid cancer. *Lancet* **2013**, *381*, 1058–1069. [CrossRef]
16. Wolff, A.C.; Hammond, M.E.H.; Hicks, D.G.; Dowsett, M.; McShane, L.M.; Allison, K.H.; Allred, D.C.; Bartlett, J.M.S.; Bilous, M.; Fitzgibbons, P.; et al. Recommendations for human epidermal growth factor receptor 2 testing in breast cancer: American Society of Clinical Oncology/College of American Pathologists clinical practice guideline update. *J. Clin. Oncol.* **2013**, *31*, 3997–4013. [CrossRef] [PubMed]
17. Vance, G.H.; Barry, T.S.; Bloom, K.J.; Fitzgibbons, P.L.; Hicks, D.G.; Jenkins, R.B.; Persons, D.L.; Tubbs, R.R.; Hammond, M.E.H. Genetic heterogeneity in HER2 testing in breast cancer panel summary and guidelines. *Arch. Pathol. Lab. Med.* **2009**, *133*, 611–612. [PubMed]

18. Wright, C.; Angus, B.; Nicholson, S.; Sainsbury, J.R.; Cairns, J.; Gullick, W.J.; Kelly, P.; Harris, A.L.; Horne, C.H. Expression of c-erbB-2 oncoprotein: A prognostic indicator in human breast cancer. *Cancer Res.* **1989**, *49*, 2087–2090. [PubMed]

19. Caria, P.; Dettori, T.; Frau, D.V.; Borghero, A.; Cappai, A.; Riola, A.; Lai, M.L.; Boi, F.; Calò, P.; Nicolosi, A.; et al. Assessing RET/PTC in thyroid nodule fine-needle aspirates: The FISH point of view. *Endocr. Relat. Cancer* **2013**, *20*, 527–536. [CrossRef] [PubMed]

20. De Lellis, R.A. *WHO Classification of Tumours, Pathology and Genetics of Tumours of Endocrine Organs*, 3rd ed.; Ronald, A., de Lellis, R.A., Riccardo, V.L., Philipp, U.H., Charis, E., Eds.; IARC Press: Lyon, France, 2004; Volume 8, p. 230.

21. Ronckers, C.M.; McCarron, P.; Ron, E. Thyroid cancer and multiple primary tumors in the SEER cancer registries. *Int. J. Cancer* **2005**, *117*, 281–288. [CrossRef] [PubMed]

22. Omür, O.; Ozcan, Z.; Yazici, B.; Akgün, A.; Oral, A.; Ozkiliç, H. Multiple primary tumors in differentiated thyroid carcinoma and relationship to thyroid cancer outcome. *Endocr. J.* **2008**, *55*, 365–372. [CrossRef] [PubMed]

23. Burstein, H.J. The distinctive nature of HER2-positive breast cancers. *N. Engl. J. Med.* **2005**, *353*, 1652–1654. [CrossRef] [PubMed]

24. Joseph, K.R.; Edirimanne, S.; Eslick, G.D. The association between breast cancer and thyroid cancer: A meta-analysis. *Breast Cancer Res. Treat.* **2015**, *152*, 173–181. [CrossRef] [PubMed]

25. Sogaard, M.; Farkas, D.K.; Ehrenstein, V.; Jørgensen, J.O.; Dekkers, O.M.; Sorensen, H.T. Hypothyroidism and hyperthyroidism and breast cancer risk: A nationwide cohort study. *Eur. J. Endocrinol.* **2016**, *174*, 409–414. [CrossRef] [PubMed]

26. Brown, A.P.; Chen, J.; Hitchcock, Y.J.; Szabo, A.; Shrieve, D.C.; Tward, J.D. The risk of second primary malignancies up to three decades after the treatment of differentiated thyroid cancer. *J. Clin. Endocrinol. Metab.* **2008**, *93*, 504–515. [CrossRef] [PubMed]

27. Verkooijen, R.B.; Smit, J.W.; Romijn, J.A.; Stokkel, M.P. The incidence of second primary tumors in thyroid cancer patients is increased, but not related to treatment of thyroid cancer. *Eur. J. Endocrinol.* **2006**, *155*, 801–806. [CrossRef] [PubMed]

28. Nagaraja, V.; Eslick, G.D. HER2 expression in gastric and oesophageal cancer: A metaanalytic review. *J. Gastrointest. Cancer* **2015**, *6*, 143–154.

29. Verri, E.; Guglielmini, P.; Puntoni, M.; Perdelli, L.; Papadia, A.; Lorenzi, P.; Rubagotti, A.; Ragni, N.; Boccardo, F. HER2/neuoncoprotein overexpression in epithelial ovarian cancer: Evaluation of its prevalence and prognostic significance. *Oncology* **2005**, *68*, 154–161. [CrossRef] [PubMed]

30. Seo, A.N.; Kwak, Y.; Kim, D.W.; Kang, S.B.; Choe, G.; Kim, W.H.; Lee, H.S. HER2 status in colorectal cancer: Its clinical significance and the relationship between *HER2* gene amplification and expression. *PLoS ONE* **2014**, *9*, e98528. [CrossRef] [PubMed]

31. Iqbal, N.; Iqbal, N. Human epidermal growth factor receptor 2 (HER2) in cancers: Overexpression and therapeutic implications. *Mol. Biol. Int.* **2014**, *2014*, 1–9. [CrossRef] [PubMed]

32. Capezzone, M.; Marchisotta, S.; Cantara, S.; Busonero, G.; Brilli, L.; Pazaitou-Panayiotou, K.; Carli, A.F.; Caruso, G.; Toti, P.; Capitani, S.; et al. Familial non-medullary thyroid carcinoma displays the features of clinical anticipation suggestive of a distinct biological entity. *Endocr. Relat. Cancer* **2008**, *15*, 1075–1081. [CrossRef] [PubMed]

33. Xing, M.; Alzahrani, A.S.; Carson, K.A.; Shong, Y.K.; Kim, T.Y.; Viola, D.; Elisei, R.; Bendlová, B.; Yip, L.; Mian, C.; et al. Association between BRAF V600E mutation and recurrence of papillary thyroid cancer. *J. Clin. Oncol.* **2015**, *33*, 42–50. [CrossRef] [PubMed]

34. Shan, L.; Qiu, T.; Ling, Y.; Guo, L.; Zheng, B.; Wang, B.; Li, W.; Li, L.; Ying, J. Prevalence and clinicopathological characteristics of HER2 and BRAF mutation in Chinese patients with lung adenocarcinoma. *PLoS ONE* **2015**, *10*, e0130447. [CrossRef] [PubMed]

35. Cantara, S.; Pisu, M.; Frau, D.V.; Caria, P.; Dettori, T.; Capezzone, M.; Capuano, S.; Vanni, R.; Pacini, F. Telomere abnormalities and chromosome fragility in patients affected by familial papillary thyroid cancer. *J. Clin. Endocrinol. Metab.* **2012**, *97*, E1327–E1331. [CrossRef] [PubMed]

36. Meeker, A.K.; Hicks, J.L.; Iacobuzio-Donahue, C.A.; Montgomery, E.A.; Westra, W.H.; Chan, T.Y.; Ronnett, B.M.; De Marzo, A.M. Telomere length abnormalities occur early in the initiation of epithelial carcinogenesis. *Clin. Cancer Res.* **2004**, *10*, 3317–3326. [CrossRef] [PubMed]

37. Gisselsson, D.; Jonson, T.; Petersen, A.; Strombeck, B.; Dal Cin, P.; Hoglund, M.; Mitelman, F.; Mertens, F.; Mandahl, N. Telomere dysfunction triggers extensive DNA fragmentation and evolution of complex chromosome abnormalities in human malignant tumors. *Proc. Natl. Acad. Sci. USA* **2001**, *98*, 12683–12688. [CrossRef] [PubMed]

38. Gadji, M.; Vallente, R.; Klewes, L.; Righolt, C.; Wark, L.; Kongruttanachok, N.; Knecht, H.; Mai, S. *Nuclear Remodeling as a Mechanism for Genomic Instability in Cancer*; Gisselsson, D., Ed.; Academic Press: New York, NY, USA, 2011; Volume 112, pp. 77–126.

39. Gadji, M.; Adebayo Awe, J.; Rodrigues, P.; Kumar, R.; Houston, D.S.; Klewes, L.; Dièye, T.N.; Rego, E.M.; Passetto, R.F.; de Oliveira, F.M.; et al. Profiling three-dimensional nuclear telomeric architecture of myelodysplastic syndromes and acute myeloid leukemia defines patient subgroups. *Clin. Cancer Res.* **2012**, *18*, 3293–3304. [CrossRef] [PubMed]

40. Kuzyk, A.; Gartner, J.; Mai, S. Identification of neuroblastoma subgroups based on three-dimensional telomere organization. *Transl. Oncol.* **2016**, *9*, 348–356. [CrossRef] [PubMed]

41. Wark, L.; Danescu, A.; Natarajan, S.; Zhu, X.; Cheng, S.Y.; Hombach-Klonisch, S.; Mai, S.; Klonisch, T. Three-dimensional telomere dynamics in follicular thyroid cancer. *Thyroid* **2014**, *24*, 296–304. [CrossRef] [PubMed]

42. Liu, R.; Xing, M. TERT promoter mutations in thyroid cancer. *Endocr. Relat. Cancer* **2016**, *23*, R143–R155. [PubMed]

43. Liu, R.; Bishop, J.; Zhu, G.; Zhang, T.; Ladenson, P.W.; Xing, M. Mortality risk stratification by combining BRAF V600E and TERT promoter mutations in papillary thyroid cancer: Genetic duet of BRAF and TERT promoter mutations in thyroid cancer mortality. *JAMA Oncol.* **2016**. [CrossRef] [PubMed]

44. Landa, I.; Ibrahimpasic, T.; Boucai, L.; Sinha, R.; Knauf, J.A.; Shah, R.H.; Dogan, S.; Ricarte-Filho, J.C.; Krishnamoorthy, G.P.; Xu, B.; et al. Genomic and transcriptomic hallmarks of poorly differentiated and anaplastic thyroid cancers. *J. Clin. Investig.* **2016**, *126*, 1052–1066. [CrossRef] [PubMed]

45. Petersen, B.L.; Sorensen, M.C.; Pedersen, S.; Rasmussen, M. Fluorescence in situ hybridization on formalin-fixed and paraffin-embedded tissue: Optimizing the method. *Appl. Immunohistochem. Mol. Morphol.* **2004**, *12*, 259–265. [CrossRef] [PubMed]

46. Caria, P.; Frau, D.V.; Dettori, T.; Boi, F.; Lai, M.L.; Mariotti, S.; Vanni, R. Optimizing detection of RET and PPARg rearrangements in thyroid neoplastic cells using a home-brew tetracolor probe. *Cancer Cytopathol.* **2014**, *122*, 377–385. [CrossRef] [PubMed]

47. Hastings, R.; Bown, N.; Tibiletti, M.G.; Debiec-Rychter, M.; Vanni, R.; Espinet, B.; van Roy, N.; Roberts, P.; van den Berg-de-Ruiter, E.; Bernheim, A.; et al. Guidelines for cytogenetic investigations in tumours. *Eur. J. Hum. Genet.* **2015**, *24*, 1–8. [CrossRef] [PubMed]

48. Cantara, S.; Capezzone, M.; Marchisotta, S.; Capuano, S.; Busonero, G.; Toti, P.; di Santo, A.; Caruso, G.; Carli, A.F.; Brilli, L.; et al. Impact of proto-oncogene mutation detection in cytological specimens from thyroid nodules improves the diagnostic accuracy of cytology. *J. Clin. Endocrinol. Metab.* **2010**, *95*, 1365–1369. [CrossRef] [PubMed]

International Journal of
Molecular Sciences

MDPI

Article

A Comprehensive Characterization of Mitochondrial Genome in Papillary Thyroid Cancer

Xingyun Su [1], Weibin Wang [1], Guodong Ruan [2], Min Liang [3], Jing Zheng [3], Ye Chen [3], Huiling Wu [4], Thomas J. Fahey III [5], Minxin Guan [3,*] and Lisong Teng [1,*]

[1] Department of Surgical Oncology, First Affiliated Hospital, School of Medicine, Zhejiang University, Hangzhou 310003, China; luckymaimai@sina.cn (X.S.); wbwang@zju.edu.cn (W.W.)
[2] Department of Oncology, the Second Hospital of Shaoxing, Shaoxing 312000, China; recardos@163.com
[3] Institute of Genetics, School of Medicine, Zhejiang University, Hangzhou 310058, China; liangmin85685@126.com (M.L.); candy88zj@zju.edu.cn (J.Z.); yechency@zju.edu.cn (Y.C.)
[4] Department of Plastic Surgery, First Affiliated Hospital, School of Medicine, Zhejiang University, Hangzhou 310003, China; whl1616@126.com
[5] Department of Surgery, New York Presbyterian Hospital and Weill Medical College of Cornell University, New York, NY 10021, USA; tjfahey@med.cornell.edu
* Correspondence: gminxin88@zju.edu.cn (M.G.); lsteng@zju.edu.cn (L.T.);
   Tel.: +86-571-8820-6497 (M.G.); +86-571-8706-8873 (L.T.);
   Fax: +86-571-8820-6485 (M.G.); +86-571-8723-6628 (L.T.)

Academic Editor: Daniela Gabriele Grimm
Received: 5 July 2016; Accepted: 8 September 2016; Published: 10 October 2016

Abstract: Nuclear genetic alterations have been widely investigated in papillary thyroid cancer (PTC), however, the characteristics of the mitochondrial genome remain uncertain. We sequenced the entire mitochondrial genome of 66 PTCs, 16 normal thyroid tissues and 376 blood samples of healthy individuals. There were 2508 variations (543 sites) detected in PTCs, among which 33 variations were novel. Nearly half of the PTCs (31/66) had heteroplasmic variations. Among the 31 PTCs, 28 specimens harbored a total of 52 somatic mutations distributed in 44 sites. Thirty-three variations including seven nonsense, 11 frameshift and 15 non-synonymous variations selected by bioinformatic software were regarded as pathogenic. These 33 pathogenic mutations were associated with older age ($p = 0.0176$) and advanced tumor stage ($p = 0.0218$). In addition, they tended to be novel ($p = 0.0003$), heteroplasmic ($p = 0.0343$) and somatic ($p = 0.0018$). The mtDNA copy number increased in more than two-third (46/66) of PTCs, and the average content in tumors was nearly four times higher than that in adjacent normal tissues ($p < 0.0001$). Three sub-haplogroups of N (A4, B4a and B4g) and eight single-nucleotide polymorphisms (mtSNPs) (A16164G, C16266T, G5460A, T6680C, G9123A, A14587G, T16362C, and G709A) were associated with the occurrence of PTC. Here we report a comprehensive characterization of the mitochondrial genome and demonstrate its significance in pathogenesis and progression of PTC. This can help to clarify the molecular mechanisms underlying PTC and offer potential biomarkers or therapeutic targets for future clinical practice.

Keywords: mitochondrial DNA; mitochondrial DNA copy number; haplogroup; papillary thyroid cancer

1. Introduction

Mitochondria are semiautonomous organelles responsible for bioenergetic metabolism, aging and apoptosis [1]. Otto Warburg et al. first proposed that metabolic reprogramming occurred in cancer cells evidenced by highly activated glycolysis even in the presence of oxygen, and this was regarded as a hallmark of cancer [2]. This phenomenon, called the Warburg effect, is probably triggered by insufficient energy supply that is the result of the combination of mitochondrial defects and activated cellular proliferation [3]. Mitochondrial DNA (mtDNA) is a 16,569 bp, double-stranded circular molecule encoding 13 polypeptides, two ribosomal RNAs (rRNAs) and 22 transfer RNAs (tRNAs) for mitochondrial respiration. The replicative origins and transcriptive promoters are located in the non-coding displacement-loop (D-loop) region [4]. Accumulated evidence demonstrates that mtDNA variations and copy number alterations are common in human cancers [5]. Pathogenic mtDNA mutations can severely affect mitochondrial respiration and overproduce endogenous reactive oxygen species (ROS) contributing to anti-apoptosis, proliferation and metastasis of cancer [5,6].

Papillary thyroid cancer (PTC) is the main histological type of thyroid cancer. Most PTC patients have favorable outcome with the 30-year survival rate more than 90% after routine treatment by thyroidectomy with or without radioiodine ablation [7]. However, a small group of PTC patients suffer from tumor persistence, recurrence and even death [8]. Investigating the underlying molecular mechanisms of PTC can provide promising biomarkers and therapeutic targets for early diagnosis and treatment, thus improving prognosis and survival quality of patients, especially those with aggressive tumor behavior and adverse outcomes. The malignant transformation and progression of thyroid cancer is driven by accumulated genetic alterations. Among them, the BRAF[V600E] mutation is the most significant factor for PTC and is associated with high-risk clinicopathological features and unfavorable outcomes [9]. Therefore, many researchers suggest that BRAF[V600E] mutation can be a valuable biomarker and therapeutic target for diagnosis, risk stratification, prognostic prediction and treatment of PTC [9].

In spite of the research achievements in understanding the nuclear genome, the role of mitochondrial genome in pathogenesis and progression of thyroid cancer is still incompletely characterized. Previous researchers have found abnormally excessive mitochondria and prevalent mtDNA alterations in thyroid cancer. However, the majority of these studies are restricted to the oncocytic subtype of thyroid cancer and only focused on mutation hotspots of mtDNA [10–12]. Here we comprehensively characterized the mitochondrial genome in papillary thyroid cancer by sequencing the entire mtDNA of 66 PTCs, 16 normal thyroid tissues and 376 blood samples of healthy individuals. The mtDNA variation distribution, haplogroup and copy number were further analyzed.

2. Results

2.1. Distribution of mtDNA Variations

A total of 2508 variations in 543 sites were identified in 66 PTC cases, and the D-loop region was the hotspot of mtDNA (Figure S1a,b). Single-base substitution was the main component of mtDNA variations, in addition to 76 deletions (13 sites) and 112 insertions (10 sites) (Figure S1c). About 30.9% (101/327) transitions and 60% (12/20) transversions were non-synonymous, suggesting that transversion was more likely to alter the encoded amino-acid and affect the structure or function of protein (Figure S1d). In the protein-coding region, most variations were synonymous (Figure S1e). ATPase6 (14/22, 63.6%), Cytb (20/45, 44.4%), ND4L (3/8, 37.5%) and ND5 (25/71, 35.2%) genes harbored relatively high ratio of nonsynonymous variation (Figure S1e,f). A total of 33 variations—including 11 non-synonymous, seven nonsense and eight frameshift variations—in 25 PTC patients were novel, and all of them were singular (Table 1). Heteroplasmy was one of the most important characteristics of mitochondrial genome, presenting in nearly half of the 66 PTCs (31/66). Among the heteroplasmic variations, 52 somatic mutations (44 sites) in 28 PTC patients and 28 germline variations (20 sites) in 16 patients were detected (Table S1).

Table 1. Novel mtDNA variations in the entire mitochondrial genome.

Position	Gene	Replacement	Amino-Acid Change or Watson-Crick Base-Pairing [a]	Conservation Index (%) [b]	Number of 66 PTC Patients (%)	Number of 376 Healthy Controls (%)	Heter/Homo [c]
			RNA Region				
1629	tRNA^Val	A-T	A-U↓	24.4%	1 (1.52%)	0 (0.00%)	Homo
2274	16S rRNA	A-G		100%	1 (1.52%)	0 (0.00%)	Heter
3275-3276	tRNA^Leu(UUR)	Del CA		-	1 (1.52%)	0 (0.00%)	Heter
4272	tRNA^Ile	T-C	A-U↓	100%	1 (1.52%)	0 (0.00%)	Homo
5835	tRNA^Tyr	Ins T		-	1 (1.52%)	0 (0.00%)	Homo
5881	tRNA^Tyr	G-C	C-G↓	100%	1 (1.52%)	0 (0.00%)	Homo
10040	tRNA^Gly	C-A		43.9%	1 (1.52%)	0 (0.00%)	Homo
			Protein-Coding Region				
4520-4521	ND2	Del AC	-	-	1 (1.52%)	0 (0.00%)	Homo
4875	ND2	C-T	Leu -> Leu	100%	1 (1.52%)	0 (0.00%)	Homo
4969	ND2	G-A	No: Trp -> Ter [d]	100%	1 (1.52%)	0 (0.00%)	Homo
4971	ND2	G-A	No: Gly -> Ser	100%	1 (1.52%)	0 (0.00%)	Homo
5977	COI	G-A	No: Trp -> Ter	100%	1 (1.52%)	0 (0.00%)	Heter
6238	COI	T-C	No: Leu -> Pro	100%	1 (1.52%)	0 (0.00%)	Heter
7104	COI	T-C	No: Ser -> Pro	100%	1 (1.52%)	0 (0.00%)	Heter
7750	COII	C-A	No: Ile -> Met	58.5%	1 (1.52%)	0 (0.00%)	Homo
7928	COII	G-A	No: Gly -> Ter	56.1%	1 (1.52%)	0 (0.00%)	Homo
9253	COIII	G-A	No: Trp -> Ter	100%	1 (1.52%)	0 (0.00%)	Heter
10521	ND4L	G-A	No: Gly -> Ter	100%	1 (1.52%)	0 (0.00%)	Homo
10622	ND4L	C-T	Thr -> Thr	36.6%	1 (1.52%)	0 (0.00%)	Homo
11646	ND4	Ins T	-	-	1 (1.52%)	0 (0.00%)	Homo
11673-11677	ND4	C5-C4		-	1 (1.52%)	0 (0.00%)	Heter
11673-11677	ND4	C5-C6		-	1 (1.52%)	0 (0.00%)	Homo
12794	ND5	T-A	No: Leu -> Ter	100%	1 (1.52%)	0 (0.00%)	Heter
12858	ND5	Ins T		-	1 (1.52%)	0 (0.00%)	Heter
12943	ND5	C-T	No: Leu -> Phe	24.4%	1 (1.52%)	0 (0.00%)	Heter
13128-13132	ND5	Del A		-	1 (1.52%)	0 (0.00%)	Homo
13170	ND5	C-T	No: Leu -> Phe	51.2%	1 (1.52%)	0 (0.00%)	Homo
13621	ND5	G-A	No: Gly -> Ter	100%	1 (1.52%)	0 (0.00%)	Homo
13825	ND5	C-A	No: Gly -> Trp	70.7%	1 (1.52%)	0 (0.00%)	Heter
14310	ND6	(AAAT)2-1	-	-	1 (1.52%)	0 (0.00%)	Homo
14495-14502	ND6	C-A		-	1 (1.52%)	0 (0.00%)	Homo
14774	Cytb	C-A	No: Leu -> Ile	63.4%	1 (1.52%)	0 (0.00%)	Heter
15018	Cytb	T-A	No: Phe -> Tyr	100%	1 (1.52%)	0 (0.00%)	Heter

[a] Watson–Crick base-pairing; abolished (↓); [b] Conservation index denotes the conservative properties of amino-acid or nucleotides in 41 primate species; [c] Heter: Heteroplasmy; Homo: Homoplasmy; [d] Ter: Terminator.

2.2. The mtDNA Variations in Non-Coding Region

There were 103 substitutions and 10 frameshift alterations in D-loop region. Nearly all the insertions and deletions were located in mitochondrial microsatellite instability (mtMSI) regions, such as poly-C in np 303–315 or np 16184–16193 and poly-CA stretch in np 514–523. In the RNA region, 20, 21 and 29 variations were, respectively, identified in 12S rRNA, 16S rRNA and tRNAs. The published secondary structures of RNAs were used to localize the alterations in the stem and loop structure [13]. A total of seven variations in 12S rRNA, one alteration in 16S rRNA and 13 alterations in tRNAs changed the Waston–Crick base-pairing. According to their frequencies in control groups and conservation of the altered nucleotides, 13 variations were identified as potentially deleterious and five of them had been reported in diseases (Table S2, Figure 1).

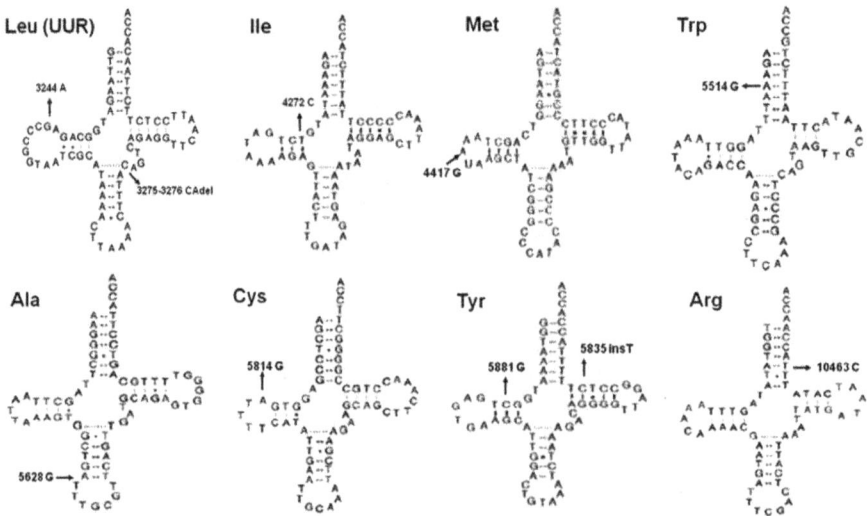

Figure 1. Potential pathogenic tRNA variations in PTC Schematic structures of eight mitochondrial tRNAs are shown. Arrows point out the position of tRNA variation.

2.3. The mtDNA Variations in Protein-Coding Region

A total of 234 synonymous, 113 non-synonymous, seven nonsense and 11 frameshift variations were detected in protein-coding region. All the nonsense and frameshift variations brought in advanced stop-codon (UAG, UGA) and leaded to premature termination of protein synthesis (Table 2, Figure 2). Among the 113 non-synonymous alterations, 26 variations were selected as potentially pathogenic based on their frequencies in control groups and conservation of the altered amino-acid (Table 3). These 26 selected variations were further evaluated by seven bioinformatic programs, and 15 of them were predicted as deleterious by more than half of the programs (Table 3). Therefore, these 33 mutations in 32 patients, including 15 nonsynonymous, seven nonsense and 11 frameshift mutations, were classified as pathogenic mutations. These pathogenic mtDNA mutations were associated with patients' older age ($p = 0.018$) and advanced tumor stage ($p = 0.022$), and tended to be novel ($p < 0.001$), heteroplasmic ($p = 0.034$) and somatic ($p = 0.002$) (Table S3).

Table 2. Nonsense and frameshift mutations identified in protein-coding region.

Position	Gene	Change	Reported [a]	Number of 66 PTC Patients (%)	Number of 16 Normal Thyroid Tissues (%)	Number of 376 Healthy Controls (%)	Heter/Homo [b]
Nonsense Mutation							
4969	ND2	G–A	N	1 (1.52%)	0 (0.00%)	0 (0.00%)	Homo
5977	COI	G–A	N	1 (1.52%)	0 (0.00%)	0 (0.00%)	Heter
7928	COII	G–A	N	1 (1.52%)	0 (0.00%)	0 (0.00%)	Homo
9253	COIII	G–A	N	1 (1.52%)	0 (0.00%)	0 (0.00%)	Heter
10521	ND4L	G–A	N	1 (1.52%)	0 (0.00%)	0 (0.00%)	Homo
12794	ND5	T–A	N	1 (1.52%)	0 (0.00%)	0 (0.00%)	Heter
13825	ND5	G–A	N	1 (1.52%)	0 (0.00%)	0 (0.00%)	Homo
Frameshift Mutation							
4520–4521	ND2	Del AC	N	1 (1.52%)	0 (0.00%)	0 (0.00%)	Homo
10952	ND4	Ins C	Y	1 (1.52%)	0 (0.00%)	0 (0.00%)	Homo
11032–11038	ND4	A7–6	Y	4 (6.06%)	0 (0.00%)	0 (0.00%)	Homo + Heter
11646	ND4	Ins T	N	1 (1.52%)	0 (0.00%)	0 (0.00%)	Homo
11673–11677	ND4	C5–C4	N	1 (1.52%)	0 (0.00%)	0 (0.00%)	Heter
11673–11677	ND4	C5–C6	Y	1 (1.52%)	0 (0.00%)	0 (0.00%)	Homo
12418–12425	ND5	Del A	N	1 (1.52%)	0 (0.00%)	0 (0.00%)	Heter
12858	ND5	Ins T	N	1 (1.52%)	0 (0.00%)	0 (0.00%)	Heter
13128–13132	ND5	C5–4	N	1 (1.52%)	0 (0.00%)	0 (0.00%)	Homo
13170	ND5	Del A	N	1 (1.52%)	0 (0.00%)	0 (0.00%)	Homo
14495–14502	ND6	(AAAT)2–1	N	1 (1.52%)	0 (0.00%)	0 (0.00%)	Homo

[a] According to Mitomap (http://www.mitomap.org); [b] Heter: Heteroplasmy; Homo: Homoplasmy.

Table 3. Potential pathogenic mtDNA variations identified in protein-coding region.

Position	Gene	Change	Amino-Acid Change	Conservation Index (%) [a]	Reported [b]	Number of 66 PTC Patients (%)	Number of 16 Normal Thyroid Tissues (%)	Number of 376 Healthy Controls (%)	Polyphen-2 [c]	SIFT	Mutation Assesor	Provean	SNP&GO	Align GVGD [d]	PANTHER (Pdeleterious) [e]
3392 [f]	ND1	G-A	No: Gly -> Asp	100.00%	Y	1 (1.52%)	0 (0.00%)	0 (0.00%)	Probably	Not Tolerated	High	Deleterious	Disease	C65	NA [g]
3644	ND1	T-C	No: Val -> Ala	97.60%	Y	1 (1.52%)	0 (0.00%)	2 (0.53%)	Benign	Not Tolerated	Medium	Deleterious	Neutral	C65	0.29125
3679	ND1	T-C	No: Ser -> Pro	100.00%	Y	1 (1.52%)	0 (0.00%)	0 (0.00%)	Probably	Not Tolerated	High	Deleterious	Disease	C65	0.74261
3745	ND1	C-A	No: Ala -> Thr	92.70%	Y	1 (1.52%)	0 (0.00%)	0 (0.00%)	Benign	Not Tolerated	Low	Neutral	Neutral	C55	0.21113
4971	ND2	G-A	No: Gly -> Ser	100.00%	N	1 (1.52%)	0 (0.00%)	0 (0.00%)	Probably	Not Tolerated	Medium	Deleterious	Neutral	C55	0.36251
6238	COI	T-C	No: Leu -> Pro	100.00%	N	1 (1.52%)	0 (0.00%)	0 (0.00%)	Probably	Not Tolerated	High	Deleterious	Disease	C65	0.87509
6340	COI	C-T	No: Thr -> Ile	82.90%	Y	1 (1.52%)	0 (0.00%)	0 (0.00%)	Benign	Tolerated	Medium	Neutral	Neutral	C65	0.21096
6681	COI	T-C	No: Tyr -> His	85.40%	N	1 (1.52%)	0 (0.00%)	0 (0.00%)	Benign	Not Tolerated	Neutral	Neutral	Neutral	C65	0.32881
7104	COI	T-C	No: Ser -> Pro	100.00%	N	1 (1.52%)	0 (0.00%)	0 (0.00%)	Possibly	Not Tolerated	Neutral	Neutral	Disease	C65	0.5134
7329	COI	T-C	No: Phe -> Leu	100.00%	N	1 (1.52%)	0 (0.00%)	0 (0.00%)	Benign	Tolerated	Low	Neutral	Neutral	C15	0.16379
8156	COII	C-A	No: Val -> Met	75.61%	Y	1 (1.52%)	0 (0.00%)	0 (0.00%)	Probably	Not Tolerated	Medium	Neutral	Neutral	C15	0.53442
8989	ATP6	G-A	No: Ala -> Thr	100.00%	Y	1 (1.52%)	0 (0.00%)	0 (0.00%)	Probably	Not Tolerated	Low	Deleterious	Neutral	C55	0.47286
9187	ATP6	T-C	No: Tyr -> His	100.00%	Y	1 (1.52%)	0 (0.00%)	0 (0.00%)	Probably	Not Tolerated	High	Deleterious	Disease	C65	NA
9355	COIII	A-G	No: Asn -> Ser	82.90%	Y	1 (1.52%)	0 (0.00%)	0 (0.00%)	Benign	Tolerated	Neutral	Neutral	Neutral	C45	0.14014
10573	ND4L	G-A	No: Gly -> Glu	97.60%	Y	1 (1.52%)	0 (0.00%)	0 (0.00%)	Probably	Not Tolerated	High	Deleterious	Neutral	C65	0.40946
12850	ND5	A-G	No: Ile -> Val	90.20%	Y	1 (1.52%)	0 (0.00%)	0 (0.00%)	Possibly	Tolerated	Neutral	Neutral	Neutral	C25	0.50297
13535	ND5	A-G	No: Asn -> Ser	87.80%	Y	1 (1.52%)	0 (0.00%)	0 (0.00%)	Benign	Not Tolerated	Low	Deleterious	Neutral	C45	NA
13748	ND5	A-G	No: Asn -> Ser	85.40%	Y	1 (1.52%)	0 (0.00%)	0 (0.00%)	Benign	Tolerated	Neutral	Neutral	Neutral	C45	0.5082
14310	ND6	C-A	No: Gly -> Trp	78.05%	N	1 (1.52%)	0 (0.00%)	0 (0.00%)	Probably	Not Tolerated	Medium	Deleterious	Disease	C65	0.71527
14463	ND6	T-C	No: Thr -> Ala	90.20%	Y	1 (1.52%)	0 (0.00%)	0 (0.00%)	Benign	Tolerated	Neutral	Deleterious	Neutral	C55	0.15283
15018	Cytb	T-A	No: Phe -> Tyr	100.00%	N	1 (1.52%)	0 (0.00%)	0 (0.00%)	Possibly	Not Tolerated	High	Deleterious	Disease	C15	0.68543
15045	Cytb	G-A	No: Arg -> Gln	100.00%	Y	1 (1.52%)	0 (0.00%)	0 (0.00%)	Probably	Not Tolerated	High	Deleterious	Disease	C35	0.59378
15090	Cytb	T-C	No: Ile -> Thr	85.40%	Y	1 (1.52%)	0 (0.00%)	1 (0.27%)	Possibly	Tolerated	Low	Deleterious	Neutral	C65	0.42865
15479	Cytb	T-C	No: Phe -> Leu	80.50%	Y	1 (1.52%)	0 (0.00%)	0 (0.00%)	Benign	Tolerated	Low	Deleterious	Neutral	C15	0.39962
15483	Cytb	C-T	No: Ser -> Leu	80.50%	Y	1 (1.52%)	0 (0.00%)	0 (0.00%)	Possibly	Tolerated	Low	Deleterious	Neutral	C65	0.45816

[a] Conservation index denotes the conservative properties of amino-acid or nucleotides in 41 primate species; [b] According to Mitomap (http://www.mitomap.org); [c] Polyphen-2 classified the variations as probably damaging, possibly damaging and benign according to their pathogenic potential; [d] Align GVGD classified the variations as C65, C55, C45, C35, C25, C15 and C0 according to the risk estimates, and here we regarded the C65 as pathogenic; [e] PANTHER predicted the pathogenicity of variations by values of Pdeleterious, and we regarded Pdeleterious >0.5 as deleterious; [f] The variants predicted as by more than half of the bioinformatic software packagess were classified as PTC-associated mutations which were highlighted by bold and italic; [g] NA, not available.

a

ND2 G4969A

```
     I   M   A   G   S   W   G
5' ATC ATA GCA GGC AGT TGA GGT 3'
   4955        4964        4973
   ATC ATA GCA GGC AGT TAA
     I   M   A   G   S   Stop
```

COI G5977A

```
     L   F   G   A   W   A   G
5' TTA TTC GGC GCA TGA GCT GGA 3'
   5966        5975        5984
   TTA TTC GGC GCA TAA
     L   F   G   A   Stop
```

COII G7928A

```
     T   D   Y   G   G   L   I
5' ACC GAC TAC GGC GGA CTA ATC 3'
   7918        7927        7936
   ACC GAC TAC GGC AGA
     T   D   Y   G   Stop
```

COIII G9253A

```
     K   P   S   P   W   P   L
5' AAA CCC AGC CCA TGA CCC CTA 3'
   9242        9251
   AAA CCC AGC CCA TAA
     K   P   S   P   Stop
```

ND4L G10521A

```
     S   L   L   G   M   L   V
5' TCA CTT CTA GGA ATA CTA GTA 3'
   10514       10523       10532
   TCA CTT CTA AGA
     S   L   L   Stop
```

ND5 T12794A

```
     I   M   S   F   L   L   I
5' ATT ATA TCC TTC TTG CTC ATC 3'
   12786       12795
   ATT ATA TCC TTC TAG
     I   M   S   F   Stop
```

ND5 G13825A

```
     V   T   F   L   G   L   L
5' GTC ACT TTC CTA GGA CTT CTA 3'
   13821       13830
   GTC ACT TTC CTA AGA
     V   T   F   L   Stop
```

b

ND4 11032-11038 A7-A6

```
     R   K   K   L   Y   L   S   M   L   I
5' CGA AAA AAA CTC TAC CTC TCT ATA CTA ATC 3'
       11038       11047       11056
   CGA AAA AAC TCT ACC TCT CTA TA C TAA
     R   K   K   L   Y   L   S   M   Stop
```

ND4 11673-11677 C5-C4

```
      T   P   W   S   F   T   G   A   V   I   L   M
5' ACC CCC TGA AGC TTC ACC GGC GCA GTC ATT CTC ATA 3'
        11677       11686       11695       11704
   ACC CCT GAA GCT TCA CCG GCG CAG TCA TTC TCA TAA
      T   P   W   S   F   T   G   A   V   I   L   Stop
```

ND5 12425delA

```
     K   N   S   Y   P   H   Y   V
5' AAA AAC TCA TAC CCC CAT TAT GTA 3'
       12426       12435       12444
   AAA AAC CAT ACC CCC ATT ATG TAA
     K   T   H   T   P   I   M   Stop
```

ND5 12858 Ins T

```
     A   I   Q   A   I   L   Y   N   R   I
5' GCC ATT CAA GCA ATC CTA TAC AAC CGT ATC 3'
   12840       12849       12858       12867
   GCC ATT CAA GCA ATC CTA TAC TAA
     A   I   Q   A   I   L   Y   Stop
```

ND5 13128-13132 C5-C4

```
     H   P   L   A   E   N   S   P   L   I
5' CAC CCC CTA GCA GAA AAT AGC CCA CTA ATC 3'
   13128       13137       13146       13155
   CAC CCC TA G CAG AA A ATA GCC CAC TAA
     H   P   L   A   E   N   S   P   Stop
```

ND5 13170 Del A

```
     L   T   L   C   L   G
5' CTA ACA CTA TGC TTA GGC 3'
   13164       13173
   CTA ACA CTT GCT TAG
     L   T   L   C   Stop
```

ND6 14495-14502 (AAAT)2-1

```
       G   L   Y   I   L   F   V   M   L
3' GGG ATT TAT TTA ATT TTT TTG ATA ATT 5'
       14499       14508       14517
5' CCC TAA ATA AAT AAA AAC TAT TAA 3'

     Stop   I   L   F   V   M   L
3' GAT TTA ATT TTT TTG ATA ATT 5'
   14499       14508       14517
5' CTA AAT TAA AAA AAC TAT TAA 3'
```

Figure 2. The nonsense and framshift mutations: (**a**) Seven nonsense mutations directly introduce stop-codon and thus create premature termination of protein synthesis immediately; and (**b**) Seven frameshift alterations bring stop-codon in the following transcription and induce truncated polypeptide

2.4. The Alteration of mtDNA Copy Number

In comparison with corresponding normal tissues, more than two-thirds (46/66) of the PTCs had increased mtDNA copy number. The average mtDNA content in tumors was nearly four times higher than that in adjacent normal tissues ($p < 0.0001$) (Figure 3). Interestingly, mtDNA content in the tumor of patient No. 48 was more than 38 times higher than the corresponding normal tissue. However, our analysis showed that increased mtDNA content had no significant association with clinicopathological features. No obvious association was observed between mtDNA content with novel or heteroplasmic mtDNA variations, PTC-associated mutations or mtMSIs (309insC/CC and 523del/insCA).

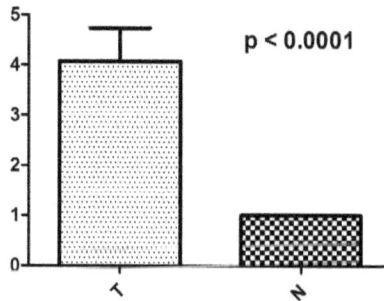

Figure 3. Copy number analysis of mtDNA in thyroid cancer: comparison of the average mtDNA copy number between PTC cases (T) and their corresponding normal tissues (N). Two-sided Mann–Whitney U test was used to analysis the difference, and $p < 0.05$ was considered as significant.

2.5. Analysis of Haplogroup and mtSNP

The entire mtDNA sequences of 66 PTCs were assigned to Asian mtDNA lineage and classified into 11 haplogroups distributed between macro-haplogroups M ($n = 30$) and N ($n = 36$). Sub-haplogroups were descended from macro-haplogroups M (C, D, G and Z) and N (A, B, F, N, R and Y) (Figure 4). Although no statistical significance was found in haplogroup M or N, the sub-haplogroups A4 (OR 3.903, 95% CI 1.070–14.23, $p = 0.027$), B4a (OR 3.903, 95% CI 1.070–14.23, $p = 0.027$) and B4g (OR 11.5, 95% CI 1.027–128.8, $p = 0.013$) descending from haplogroup N tended to be associated with the occurrence of PTC (Table S4). Frequencies of 15 mtSNPs were statistically different between PTC and healthy groups, and eight of them (A16164G, T16362C, C16266T, G5460A, T6680C, G9123A, A14587G, and G709A) may be associated with a predisposition to developing PTC according to their frequencies between PTC and normal thyroid groups (Table S5).

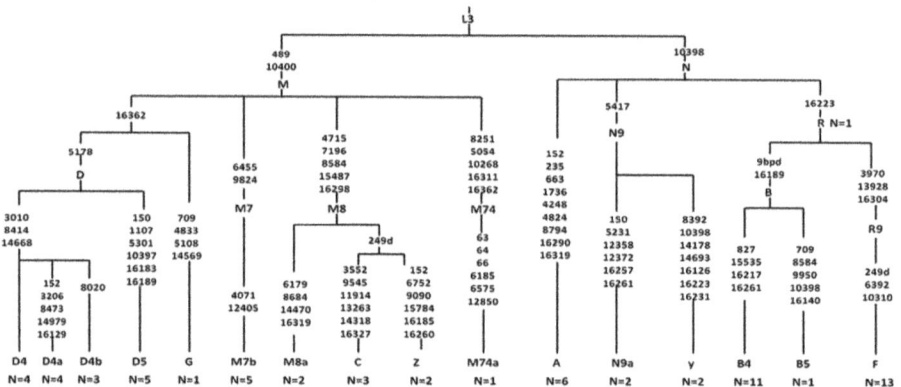

Figure 4. Phylogenetic tree was constructed to reveal the underlying lineages of 16 mtDNA haplogroups in 66 PTC cases.

3. Discussion

In spite of generally indolent behavior and favorable prognosis associated with papillary thyroid cancer, tumor recurrence and distant metastasis are intractable issues in the clinical treatment of a subset of PTC patients [8]. Identifying high-risk patients and offering appropriate, more aggressive therapy in the early stages has been an important goal for clinical researchers. Considering the crucial role of mitochondria in carcinogenesis, investigation of mitochondrial genome may provide potential biomarkers and therapeutic targets for clinical practice. Here we identified 33 pathogenic mtDNA mutations in the protein-coding region, and found three sub-haplogroups and eight mtSNPs that were associated with PTC predisposition. In addition, the average mtDNA copy number in PTCs was significantly higher than that in corresponding normal tissues.

The mutation load of mtDNA is 10–20 times higher than nuclear DNA, probably because the protect and repair system in mitochondria is insufficient and mtDNA is more vulnerable to oxidative stress generated by oxidative phosphorylation [14]. The D-loop region is a mutation hotspot of mtDNA due to the unique triple-stranded DNA structure [15]. The mtMSIs in the D-loop region can modify the binding affinity of transacting elements and direct the formation of persistent RNA-DNA hybrids regulating the efficiency of replication and transcription, which are probably produced by direct oxidative attack, slippage or mis-incorporation during replication and inefficient repair of polymerase. The mtDNA copy number varies in different cell types and microenvironments, and is precisely modulated by alterations in the D-loop region. The content of mtDNA is important for functional maintenance of mitochondria, but alterations in mtDNA and their significance in different types of cancer are still discrepant [16]. Our analysis found excessive replication of mtDNA in PTCs. However, no significant association was presented between mtDNA content and clinicopathological features, and no obvious association was observed between mtDNA content with novel or heteroplasmic mtDNA variations, PTC-associated mutations or mtMSIs. Probably, other factors also take part in the increased copy number of mtDNA. Corver et al. demonstrated that the presence of near-homozygous genome (NHG), rather than damaging or disruptive mtDNA mutations, was correlated with oncocytic phenotype which showed a strikingly mitochondrial proliferation [10]. Interestingly, mtDNA content in tumor of No. 48, a conventional variant PTC, was more than 38 times higher than corresponding normal tissue. In this specimen, we identified a novel frameshift alteration 14495–14502 del (AAAT) in the ND6 gene, which directly resulted in a premature stop-codon (UAG) being introduced and truncated the polypeptide from 175 amino-acid to 58 amino-acid. Thus we speculate that the highly increased mtDNA copy number may have been triggered by defective mitochondrial function caused by this novel deletion [10,16].

Heteroplasmy is a unique characteristic of mitochondrial genome, and also a typical feature of pathogenicity [17]. Once the pathogenic threshold is surpassed, the heteroplasmic level can affect the biochemical and clinical phenotype from mild functional deficiency to complete disassembly of the mitochondrial complex [18]. In our study, nearly half of the PTC cases harbored heteroplasmic variations. Among these heteroplasmic variations, 52 variations were somatic and the majority of them were novel—which dramatically increases their likelihood of being cancer-specific [19]. These somatic variations may confer a neoplastic advantage for tumor cells, and their successive introduction within a developing tumor may provide necessary genetic diversity to satisfy the adaptive evolution and drive tumor progression [20].

Cybrid models have demonstrated that the mitochondrion, but not nuclei, is the master contributor to mitochondrial dysfunction [21]. Pathogenic mtDNA mutations can hamper the electron transport chain (ETC) and generate excessive electrons, which triggers cancer-associated pathways and in turn produces more mutations aggravating the respiratory deficiency. It is reported that more than half of the pathogenic mutations are located in tRNAs which comprise only 10% coding capacity of mitochondrial genome, while the protein-coding region occupying about 70% mtDNA accounts for 40% disease-related mutations. The two rRNAs harbor only about 2% of the pathogenic mutations [20]. In the RNA genes, we identified 13 possibly detrimental variations and five of them had been previously reported in diseases according to the Mitomap database. For example, G3244A in tRNA$^{Leu(UUR)}$, next to the famous pathogenic mutation A3243G, was first detected in mitochondrial myopathy, encephalopathy, lactic acidosis, and stroke-like episodes (MELAS) and later found in several cancers including oncocytic thyroid tumors [22,23]. The A5514G in tRNA^{Trp} damaging an A–U base-pair in ACC-stem was identified in neonatal onset mito-disease and analyzed to be damaging by clinicopathology and biochemistry [24]. The T5628C in tRNA^{Ala} disrupted an extremely conserved

A–U base-pair in the anti-codon stem and resulted in nine unmatched nucleotides (rather than the seven in normal cells) which decreased the energetic stability of tRNA^{Ala} [25].

In the coding region, seven nonsense and 11 frameshift mutations introduced premature stop-codons (UAG, UGA) in protein synthesis and resulted in loss-of-function or even disassembly of the complex. Among them, both 10952insC and 11032–11038delA have been detected in renal oncocytoma [26], and 11032-11038delA was also found in prostate cancer [27]. The 12425delA has been previously identified in a girl having chronic renal failure, persistent lactic acidosis and myopathy [28]. A similar variation 12425insA has been reported in several cancers in a heteroplasmic status [29,30]. These nonsense and frameshift mutations, together with 15 non-synonymous mutations selected by bioinformatics programs, were regarded as pathogenic, and may interfere the OXPHOS system of mitochondrial respiration and contribute to the molecular pathogenesis of thyroid cancer. The association between these pathogenic mtDNA mutations and advanced tumor stage suggests the possible involvement of mtDNA mutations in malignant transformation and progression. Apart from pathogenic mutations, "non-pathologic" mtSNPs can also affect carcinogenesis and progression of cancer in multifactorial manners. Haplotypes, classified by specific combinations of tightly linked mtSNPs, are also correlated with the predisposition to specific cancers [31]. For example, haplogroup U increased the risk of prostate cancer and renal cancer in white North American individuals [32], but decreased the risk of breast cancer in European-American women [33].

The application value of mtDNA variations in early diagnosis, risk stratification, prognostic prediction and disease monitoring of cancer have been widely investigated and discussed. Since mtDNA is a small size and close-circular molecular entity and does not undergo recombination, mtDNA variations are more fixed and persistent than nuclear alterations. Due to the high copy number of mtDNA, detecting mtDNA biomarkers can be more sensitive and powerful than nuclear ones. Therefore, mtDNA biomarkers may have special advantage in samples of limited cellularity including fine-needle aspiration or core-needle aspiration of thyroid nodule. Furthermore, mitochondria are potential therapeutic targets for cancer treatment and can be specifically targeted by antioxidant compounds, selective gene-therapy or approaches changing the mtDNA variation load. Dai et al. demonstrated that mitophagy induced by rapamycin can eliminate pathogenic mtDNA mutations and increase ATP restoration [34]. Recently, several researchers reported that resistance to BRAF inhibition was partly caused by increased mitochondrial biogenesis and oxidative respiration, and therapies inhibiting this metabolic reprogramming restored the function of the BRAF inhibitor and improved treatment efficiency [35,36].

The major limitation of our study is that we do not analyze the mitochondrial genome in anaplastic thyroid cancer (ATC), which has more aggressive behaviors and worse prognosis than PTC. We plan to analyze the mitochondrial characteristics in ATC and evaluate their clinical and prognostic significance. Furthermore, we can compare the role of mitochondrial genome in different histological types of thyroid cancer.

4. Materials and Methods

4.1. Sample Collection

A total of 66 PTC patients underwent primary surgery in the First Affiliated Hospital, Zhejiang University School of Medicine (Hangzhou, China) were enrolled. None of them had a history of cancer or radiotherapy before surgery. Histopathology of tumor specimens was independently evaluated by two experienced pathologists according to the World Health Organization (WHO) classification [37]. Among the 66 PTCs, two were follicular variant and the others were classical variant. Tumors and adjacent normal tissues were immediately frozen in −80 °C after resection. The 16 normal thyroid tissues were used to distinguish tissue-specific variations, and 376 blood samples of healthy individuals from the same geographic region were collected to identify polymorphisms in this population. All the samples were obtained with informed consent. The study was conducted in accordance with the Declaration of Helsinki, and the protocol was approved by the Ethics Committee of the First Affiliated Hospital, College of Medicine, Zhejiang University (2015-443, 30 December 2015).

4.2. Sequencing of the Mitochondrial Genome

Genomic DNA was isolated from frozen tissues and blood samples using a commercial kit (QIAamp DNA Mini Kit from QIAGEN, Hilden, Germany). Concentration and purity of DNA were analyzed by spectrometry. The entire mitochondrial genome was PCR-amplified by 24 pairs of overlapping primer as described previously [38]. The PCR products were detected by electrophoresis in 1% agarose gel and then sequencing by the ABI 3700 automated DNA sequencer (Applied BioSystems, Foster City, CA, USA) using BigDye Terminator v3.1 Cycle Sequencing Kit (Applied BioSystems).

4.3. Sequence Analysis and Haplogroup Classification

The sequences of mtDNA were aligned to the revised Cambridge Reference Sequence (rCRS) (GeneBank accession number: NC_012920) to identify mtDNA variations [39]. The variation load referred to the percentage of variations per gene or complex, which was calculated as follows: total number of altered nucleotides per gene or complex/total number of nucleotides per gene or complex ×100. Variations not recorded in the Mitomap database (http://www.mitomap.org) were regarded as novel. All the heteroplasmic variations were confirmed by repeat analysis of the other strand and compared with the corresponding positions in adjacent normal tissues. The mtDNA haplogroups were classified according to the updated phylogenetic tree of mtDNA (mtDNA tree Built 16) provided by PhyloTree (http://www.phylotree.org) [31].

4.4. Phylogenetic Conservation Analysis and Pathogenic Prediction

Inter-species conservation of the altered amino acids or nucleotides was evaluated by mitochondrial sequences of 41 primates (Table S6). The conservation index (CI) was defined as the percentage of species having wild-type amino-acid or nucleotide by comparing the amino-acid or nucleotide of human with the other 40 species. The higher the conservation of the altered amino-acid or nucleotide was, the greater the pathogenic possibility will be. The variations with potential pathogenicity were selected based on the following criteria: (1) presented in less than 1% of 376 healthy individuals—those variations existed in more than 1% healthy controls were regarded as polymorphisms; (2) were absent in normal thyroid samples, and those variations also identified in normal thyroid samples were regarded as tissue-specific variations; and (3) the altered amino-acids or nucleotides had high conservation (CI > 75%), which indicated the high possibility of functional consequence. Furthermore, the potentially pathogenic variations in protein-coding region were evaluated by 7 bioinformatic programs including PolyPhen-2 (http://genetics.bwh. harvard.edu/pph2/), SIFT (http://sift.jcvi.org/), MutationAssessor (http://mutationassessor.org/), Provean (http://provean.jcvi.org/index.php), SNP & GO (http://snps-and-go.biocomp.unibo.it/), Align GVGD (https://www.biostars.org/) and PANTHER (http://fathmm.biocompute.org.uk/). The variations that were predicted as deleterious by more than half of these 7 programs had high possibility to be "pathogenic" for mitochondrial function and associated with PTC.

4.5. Determination of mtDNA Copy Number

The mtDNA content relative to nuclear encoded 18s RNA was determined by quantitative real-time PCR in ABI Prim 7900HT system using FastStart Universal SYBR Green Master Mix (Roche Diagnostics GmbH, Mannheim, Germany). The primers used for amplification of mtDNA copy number were: the forward primer 5' CACCCAAGAACAGGGTTTGT 3' and the reverse primer 5' TGGCCATGGGTATGTTGTTAA 3'. Another pair of primers was designed to amplify 18s RNA: the forward primer 5' TAGAGGGACAAGTGGCGTTC 3' and the reverse primer 5' CGCTGAGCCAGTCAGTGT 3'. The total volume of PCR mixture was 10 μL including 2 μL DNA (2 ng/μL), 3 μL primers (10 μM) and 5 μL SYBR Green Master Mix. The action was conducted as follows: 50 °C for 2 min, 95 °C for 10 min and followed by 45 cycles of 95 °C for 5 s, 58 °C for 30 s and 72 °C for 1 min. All the reactions were repeated 3 times. Non-template control and a serial dilution of reference DNA were used in each reaction.

4.6. Statistical Analysis

All the statistical analyses were conducted by SPSS software (version 21.0) (SPSS Inc., Chicago, IL, USA). The Pearson chi-square test was performed to analyze the clinicopathological significance of mitochondrial characteristics. Two-sided Mann–Whitney U test was used to analyze the difference of the average mtDNA copy number between PTC cases and their corresponding normal tissues. The odds ratios (ORs) with 95% confidence intervals (CIs) were calculated to clarify the association of haplogroups and single-nucleotide polymorphisms (mtSNPs) with PTC occurrence. For all analyses, $p < 0.05$ was regarded as statistically significant.

5. Conclusions

Here, we have reported a comprehensive characterization of the mitochondrial genome in PTC, and demonstrated that pathogenic mtDNA mutations, as well as some specific mtSNPs and haplogroups, may be involved in the pathogenesis and progression of PTC. These results provide an alternative dimension to clarify the molecular mechanisms underlying PTC carcinogenesis, and present possible novel biomarkers and therapeutic targets for the diagnosis, risk stratification, prognostic prediction and treatment of papillary thyroid cancer.

Supplementary Materials: Supplementary Materials can be found at www.mdpi.com/1422-0067/17/10/1594/s1.

Acknowledgments: This study is supported by Grants from National Natural Science Foundation of China (No. 81202141, and 81272676), the Key Project of Scientific and Technological Innovation of Zhejiang Province (No. 2015C03G2010206), National Science and Technology Major Project of the Ministry of Science and Technology of China (No. 2013ZX09506015), Medical Science and Technology Project of Zhejiang Province (No. 2011ZDA009), and Natural Science Foundation of Zhejiang Province (No. Y2110414).

Author Contributions: Lisong Teng and Minxin Guan conceived and designed the study; Xingyun Su performed the experiment; Weibin Wang and Guodong Ruan contributed the specimens; Min Liang, Jing Zheng, Ye Chen and Huiling Wu contributed reagents/materials/analysis tools; Xingyun Su analyzed the data and wrote the paper; and Thomas J. Fahey III helped to modify the manuscript.

Conflicts of Interest: The authors declare no conflict of interest.

References

1. Chan, D.C. Mitochondria: Dynamic organelles in disease, aging, and development. *Cell* **2006**, *125*, 1241–1252. [CrossRef] [PubMed]
2. Warburg, O. On the origin of cancer cells. *Science* **1956**, *123*, 309–314. [CrossRef] [PubMed]
3. Kroemer, G. Mitochondria in cancer. *Oncogene* **2006**, *25*, 4630–4632. [CrossRef] [PubMed]
4. Wallace, D.C.; Fan, W. Energetics, epigenetics, mitochondrial genetics. *Mitochondrion* **2010**, *10*, 12–31. [CrossRef] [PubMed]
5. Larman, T.C.; DePalma, S.R.; Hadjipanayis, A.G.; Protopopov, A.; Zhang, J.; Gabriel, S.B.; Chin, L.; Seidman, C.E.; Kucherlapati, R.; Seidman, J.G. Spectrum of somatic mitochondrial mutations in five cancers. *Proc. Natl. Acad. Sci. USA* **2012**, *109*, 14087–14091. [CrossRef] [PubMed]
6. Ishikawa, K.; Imanishi, H.; Takenaga, K.; Hayashi, J. Regulation of metastasis; mitochondrial DNA mutations have appeared on stage. *J. Bioenerg. Biomembr.* **2012**, *44*, 639–644. [CrossRef] [PubMed]
7. Markovina, S.; Grigsby, P.W.; Schwarz, J.K.; DeWees, T.; Moley, J.F.; Siegel, B.A.; Perkins, S.M. Treatment approach, surveillance, and outcome of well-differentiated thyroid cancer in childhood and adolescence. *Thyroid* **2014**, *24*, 1121–1126. [CrossRef] [PubMed]
8. Ito, Y.; Miyauchi, A.; Ito, M.; Yabuta, T.; Masuoka, H.; Higashiyama, T.; Fukushima, M.; Kobayashi, K.; Kihara, M.; Miya, A. Prognosis and prognostic factors of differentiated thyroid carcinoma after the appearance of metastasis refractory to radioactive iodine therapy. *Endocr. J.* **2014**, *61*, 821–824. [CrossRef] [PubMed]

Int. J. Mol. Sci. **2016**, *17*, 1594

9.   Xing, M. Molecular pathogenesis and mechanisms of thyroid cancer. *Nat. Rev. Cancer* **2013**, *13*, 184–199. [CrossRef] [PubMed]

10. Corver, W.E.; van Wezel, T.; Molenaar, K.; Schrumpf, M.; van den Akker, B.; van Eijk, R.; Ruano Neto, D.; Oosting, J.; Morreau, H. Near-haploidization significantly associates with oncocytic adrenocortical, thyroid, and parathyroid tumors but not with mitochondrial DNA mutations. *Genes Chromosomes Cancer* **2014**, *53*, 833–844. [CrossRef] [PubMed]

11. Ding, Z.; Ji, J.; Chen, G.; Fang, H.; Yan, S.; Shen, L.; Wei, J.; Yang, K.; Lu, J.; Bai, Y. Analysis of mitochondrial DNA mutations in D-loop region in thyroid lesions. *Biochim. Biophys. Acta* **2010**, *1800*, 271–274. [CrossRef] [PubMed]

12. Gasparre, G.; Porcelli, A.M.; Bonora, E.; Pennisi, L.F.; Toller, M.; Iommarini, L.; Ghelli, A.; Moretti, M.; Betts, C.M.; Martinelli, G.N.; et al. Disruptive mitochondrial DNA mutations in complex I subunits are markers of oncocytic phenotype in thyroid tumors. *Proc. Natl. Acad. Sci. USA* **2007**, *104*, 9001–9006. [CrossRef] [PubMed]

13. Ruiz-Pesini, E.; Wallace, D.C. Evidence for adaptive selection acting on the tRNA and rRNA genes of human mitochondrial DNA. *Hum. Mutat.* **2006**, *27*, 1072–1081. [CrossRef] [PubMed]

14. Cui, H.; Kong, Y.; Zhang, H. Oxidative stress, mitochondrial dysfunction, and aging. *J. Signal Transduct.* **2012**, *2011*. [CrossRef] [PubMed]

15. Kwok, C.S.N.; Quah, T.C.; Ariffin, H.; Tay, S.K.H.; Yeoh, A.E.J. Mitochondrial D-loop polymorphisms and mitochondrial DNA content in childhood acute lymphoblastic leukemia. *J. Pediatr. Hematol. Oncol.* **2011**, *33*, e239–e244. [CrossRef] [PubMed]

16. Yu, M. Generation, function and diagnostic value of mitochondrial DNA copy number alterations in human cancers. *Life Sci.* **2011**, *89*, 65–71. [CrossRef] [PubMed]

17. Chinnery, P.F.; Hudson, G. Mitochondrial genetics. *Br. Med. Bull.* **2013**, *106*, 135–159. [CrossRef] [PubMed]

18. Picard, M.; Zhang, J.; Hancock, S.; Derbeneva, O.; Golhar, R.; Golik, P.; O'Hearn, S.; Levy, S.; Potluri, P.; Lvova, M.; et al. Progressive increase in mtDNA 3243A>G heteroplasmy causes abrupt transcriptional reprogramming. *Proc. Natl. Acad. Sci. USA* **2014**, *111*, E4033–E4042. [CrossRef] [PubMed]

19. He, Y.; Wu, J.; Dressman, D.C.; Iacobuzio-Donahue, C.; Markowitz, S.D.; Velculescu, V.E.; Diaz, L.A., Jr.; Kinzler, K.W.; Vogelstein, B.; Papadopoulos, N. Heteroplasmic mitochondrial DNA mutations in normal and tumour cells. *Nature* **2010**, *464*, 610–614. [CrossRef] [PubMed]

20. Schon, E.A.; DiMauro, S.; Hirano, M. Human mitochondrial DNA: Roles of inherited and somatic mutations. *Nat. Rev. Genet.* **2012**, *13*, 878–890. [CrossRef] [PubMed]

21. Bonora, E.; Porcelli, A.M.; Gasparre, G.; Biondi, A.; Ghelli, A.; Carelli, V.; Baracca, A.; Tallini, G.; Martinuzzi, A.; Lenaz, G.; et al. Defective oxidative phosphorylation in thyroid oncocytic carcinoma is associated with pathogenic mitochondrial DNA mutations affecting complexes I and III. *Cancer Res.* **2006**, *66*, 6087–6096. [CrossRef] [PubMed]

22. Lorenc, A.; Bryk, J.; Golik, P.; Kupryjanczyk, J.; Ostrowski, J.; Pronicki, M.; Semczuk, A.; Szolkowska, M.; Bartnik, E. Homoplasmic melas A3243G mtDNA mutation in a colon cancer sample. *Mitochondrion* **2003**, *3*, 119 124. [CrossRef]

23. Mimaki, M.; Hatakeyama, H.; Ichiyama, T.; Isumi, H.; Furukawa, S.; Akasaka, M.; Kamei, A.; Komaki, H.; Nishino, I.; Nonaka, I.; et al. Different effects of novel mtDNA G3242A and G3244A base changes adjacent to a common A3243G mutation in patients with mitochondrial disorders. *Mitochondrion* **2009**, *9*, 115–122. [CrossRef] [PubMed]

24. Del Mar O'Callaghan, M.; Emperador, S.; López-Gallardo, E.; Jou, C.; Buján, N.; Montero, R.; Garcia-Cazorla, A.; Gonzaga, D.; Ferrer, I.; Briones, P.; et al. New mitochondrial DNA mutations in tRNA associated with three severe encephalopamyopathic phenotypes: Neonatal, infantile, and childhood onset. *Neurogenetics* **2012**, *13*, 245–250. [CrossRef] [PubMed]

25. Spagnolo, M.; Tomelleri, G.; Vattemi, G.; Filosto, M.; Rizzuto, N.; Tonin, P. A new mutation in the mitochondrial tRNAAla gene in a patient with ophthalmoplegia and dysphagia. *Neuromuscul. Disord.* **2001**, *11*, 481–484. [CrossRef]

26. Mayr, J.A.; Meierhofer, D.; Zimmermann, F.; Feichtinger, R.; Kogler, C.; Ratschek, M.; Schmeller, N.; Sperl, W.; Kofler, B. Loss of complex I due to mitochondrial DNA mutations in renal oncocytoma. *Clin. Cancer Res.* **2008**, *14*, 2270–2275. [CrossRef] [PubMed]

27. Jeronimo, C.; Nomoto, S.; Caballero, O.L.; Usadel, H.; Henrique, R.; Varzim, G.; Oliveira, J.; Lopes, C.; Fliss, M.S.; Sidransky, D. Mitochondrial mutations in early stage prostate cancer and bodily fluids. *Oncogene* **2001**, *20*, 5195–5198. [CrossRef] [PubMed]
28. Alston, C.L.; Morak, M.; Reid, C.; Hargreaves, I.P.; Pope, S.A.; Land, J.M.; Heales, S.J.; Horvath, R.; Mundy, H.; Taylor, R.W. A novel mitochondrial MTND5 frameshift mutation causing isolated complex I deficiency, renal failure and myopathy. *Neuromuscul. Disord.* **2010**, *20*, 131–135. [CrossRef] [PubMed]
29. Tseng, L.M.; Yin, P.H.; Yang, C.W.; Tsai, Y.F.; Hsu, C.Y.; Chi, C.W.; Lee, H.C. Somatic mutations of the mitochondrial genome in human breast cancers. *Genes Chromosomes Cancer* **2011**, *50*, 800–811. [CrossRef] [PubMed]
30. Yin, P.H.; Wu, C.C.; Lin, J.C.; Chi, C.W.; Wei, Y.H.; Lee, H.C. Somatic mutations of mitochondrial genome in hepatocellular carcinoma. *Mitochondrion* **2010**, *10*, 174–182. [CrossRef] [PubMed]
31. Van Oven, M.; Kayser, M. Updated comprehensive phylogenetic tree of global human mitochondrial DNA variation. *Hum. Mutat.* **2009**, *30*, E386–E394. [CrossRef] [PubMed]
32. Booker, L.M.; Habermacher, G.M.; Jessie, B.C.; Sun, Q.C.; Baumann, A.K.; Amin, M.; Lim, S.D.; Fernandez-Golarz, C.; Lyles, R.H.; Brown, M.D.; et al. North american white mitochondrial haplogroups in prostate and renal cancer. *J. Urol.* **2006**, *175*, 468–472. [CrossRef]
33. Bai, R.K.; Leal, S.M.; Covarrubias, D.; Liu, A.; Wong, L.J. Mitochondrial genetic background modifies breast cancer risk. *Cancer Res.* **2007**, *67*, 4687–4694. [CrossRef] [PubMed]
34. Dai, Y.; Zheng, K.; Clark, J.; Swerdlow, R.H.; Pulst, S.M.; Sutton, J.P.; Shinobu, L.A.; Simon, D.K. Rapamycin drives selection against a pathogenic heteroplasmic mitochondrial DNA mutation. *Hum. Mol. Genet.* **2014**, *23*, 637–647. [CrossRef] [PubMed]
35. Livingstone, E.; Swann, S.; Lilla, C.; Schadendorf, D.; Roesch, A. Combining BRAF V 600E inhibition with modulators of the mitochondrial bioenergy metabolism to overcome drug resistance in metastatic melanoma. *Exp. Dermatol.* **2015**, *24*, 709–710. [CrossRef] [PubMed]
36. Spagnolo, F.; Ghiorzo, P.; Queirolo, P. Overcoming resistance to BRAF inhibition in BRAF-mutated metastatic melanoma. *Oncotarget* **2014**, *5*, 10206–10221. [CrossRef] [PubMed]
37. Hedinger, C.; Williams, E.D.; Sobin, L.H. The who histological classification of thyroid tumors: A commentary on the second edition. *Cancer* **1989**, *63*, 908–911. [CrossRef]
38. Rieder, M.J.; Taylor, S.L.; Tobe, V.O.; Nickerson, D.A. Automating the identification of DNA variations using quality-based fluorescence re-sequencing: Analysis of the human mitochondrial genome. *Nucleic Acids Res.* **1998**, *26*, 967–973. [CrossRef] [PubMed]
39. Andrews, R.M.; Kubacka, I.; Chinnery, P.F.; Lightowlers, R.N.; Turnbull, D.M.; Howell, N. Reanalysis and revision of the Cambridge Reference Sequence for human mitochondrial DNA. *Nat. Genet.* **1999**, *23*. [CrossRef]

International Journal of
Molecular Sciences

MDPI

Article

Expression of Tenascin C, EGFR, E-Cadherin, and TTF-1 in Medullary Thyroid Carcinoma and the Correlation with RET Mutation Status

Florian Steiner [1], Cornelia Hauser-Kronberger [1], Gundula Rendl [2], Margarida Rodrigues [2] and Christian Pirich [2,*]

[1] Department of Pathology, Paracelsus Medical University Salzburg, Müllner Hauptstrasse 48, A-5020 Salzburg, Austria; f.steiner@salk.at (F.S.); c.hauser-kronberger@salk.at (C.H.-K.)

[2] Department of Nuclear Medicine and Endocrinology, Paracelsus Medical University Salzburg, Müllner Hauptstrasse 48, A-5020 Salzburg, Austria; g.rendl@salk.at (G.R.); rodriguesradischat@hotmail.com (M.R.)

* Correspondence: c.pirich@salk.at; Tel.: +43-57255-26601

Academic Editor: Daniela Gabriele Grimm
Received: 15 May 2016; Accepted: 24 June 2016; Published: 9 July 2016

Abstract: Tenascin C expression correlates with tumor grade and indicates worse prognosis in several tumors. Epidermal growth factor receptor (EGFR) plays an important role in driving proliferation in many tumors. Loss of E-cadherin function is associated with tumor invasion and metastasis. Thyroid transcription factor-1 (TTF-1) is involved in rearranged during transfection (RET) transcription in Hirschsprung's disease. Tenascin C, EGFR, E-cadherin, TTF-1-expression, and their correlations with RET mutation status were investigated in 30 patients with medullary thyroid carcinoma (MTC) ($n = 26$) or C-cell hyperplasia ($n = 4$). Tenascin C was found in all, EGFR in 4/26, E-cadherin in 23/26, and TTF-1 in 25/26 MTC. Tenascin C correlated significantly with tumor proliferation (overall, $r = 0.61$, $p < 0.005$; RET-mutated, $r = 0.81$, $p < 0.01$). E-cadherin showed weak correlation, whereas EGFR and TTF-1 showed no significant correlation with tumor proliferation. EGFR, E-cadherin, and TTF-1 showed weak correlation with proliferation of RET-mutated tumors. Correlation between TTF-1 and tenascin C, E-cadherin, and EGFR was $r = -0.10, 0.37$, and 0.21, respectively. In conclusion, MTC express tenascin C, E-cadherin, and TTF-1. Tenascin C correlates significantly with tumor proliferation, especially in RET-mutated tumors. EGFR is low, and tumors expressing EGFR do not exhibit higher proliferation. TTF-1 does not correlate with RET mutation status and has a weak correlation with tenascin C, E-cadherin, and EGFR expression.

Keywords: tenascin C; epidermal growth factor receptor (EGFR); E-cadherin; thyroid transcription factor-1 (TTF-1); medullary thyroid carcinoma

1. Introduction

Medullary thyroid carcinoma (MTC) may arise sporadically in about 75% of cases or as part of multiple endocrine neoplasia type 2 (MEN2) syndrome in 20%–25% of cases [1]. MEN 2 syndromes are caused by activating mutations of the proto-oncogene rearranged during transfection (RET) [2]. On the other hand, a loss of function mutation of RET leads to Hirschsprung's disease [3].

Tenascin C is an extracellular glycoprotein complex expressed by a variety of cells including epithelial, stromal, and tumor cells [4]. It is overexpressed in a wide variety of tumors including gliomas, where it was originally discovered [5]. In most cases, the expression of tenascin C correlates with the tumor grade and is indicative of a worse prognosis [6]. Koperek et al. [7] found tenascin C expression in medullary microcarcinoma and C-cell hyperplasia and suggested that stromal tenascin C expression seems to be an indicator of a further step in carcinogenesis of MTC, irrespective of a RET germ-line mutation.

Mutations of epidermal growth factor receptor (EGFR) have been found in several tumor entities including gliomas, breast cancer, and non-small lung cancer [8]. In the case of MTC, mutations are rarely found, and their significance is unknown [9]. Rodríguez-Antona et al. [9] showed that EGFR overexpression in MTC is seen in as many as 13% of tumors and that metastases show stronger positivity than primary tumors. Furthermore, EGFR overexpression is linked to RET activation. However, in the presence of RET, EGFR does not appear to play an important role in signaling [10].

Loss of function of the molecule E cadherin in tumors is associated with invasion and metastasis [11,12]. Naito et al. [13] found that expression of E-cadherin was reduced or absent in 50% or more of thyroid cancer cases, and concluded that this loss of E-cadherin expression may be involved in regional lymph node metastasis and in malignant potential of thyroid neoplasms.

Thyroid transcription factor-1 (TTF-1) is involved in gene expression of thyroperoxidase [14] and thyreoglobulin [15]. TTF-1 expression is seen in follicular cell neoplasms [16] as well as in MTC [17]. In the parafollicular cells of MTC, TTF-1 modulates the activity of genes involved in calcium homeostasis [18]. It was recently shown that TTF-1 is also involved in the transcription of human RET in Hirschsprung's disease [19].

In MTC, the Ki-67 index correlates with the stage of the disease [20]. Primary tumors that had metastasized were found to have higher Ki-67 indices than primary tumors that had not metastasized. Recurrent lymph node metastases were shown to have higher Ki-67 indices than the primary tumors. The Ki-67 index can therefore be used as a prognostic marker in MTC.

In this study, we investigated the expression of tenascin C, EGFR, E-cadherin, and TTF-1 in MTC, and their correlation with RET mutation status. Furthermore, EGFR mutation status in MTC was evaluated.

2. Results

Tenascin C showed positive staining results in all the 26 tumors (Figure 1). In contrast, all four cases of C-cell hyperplasia stained negative for tenascin C. The tumor-staining pattern was homogeneously located in all areas of the tumor. However, 14 out of 26 tumors showed expression in the whole tumor field, with the remaining 12 tumors showing partial expression. Except for one case that showed much stronger staining in the periphery, no predominance for tumor center or invasion front could be detected.

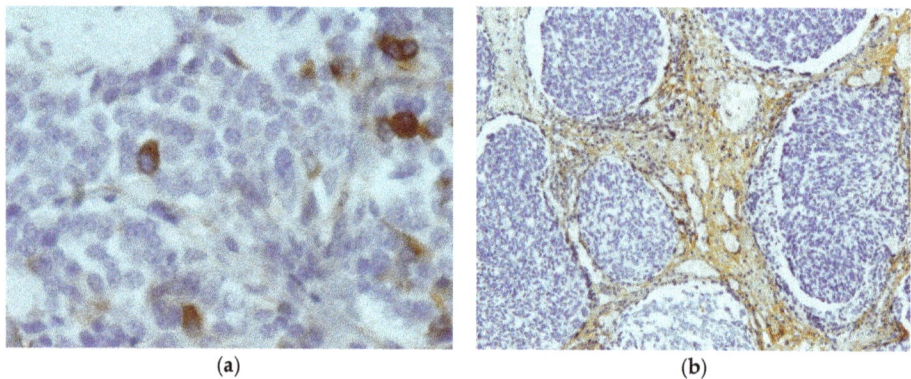

(a)   (b)

Figure 1. Staining results of tenascin C: (**a**) 40× magnification, showing cytoplasmic staining; (**b**) 10× magnification, depicting staining of the extracellular matrix and the lymph follicle-like accumulation of tumor cells.

Expression of tenascin C was primarily located in the extracellular matrix but also in the plasma membrane and the cytoplasm of parafollicular cells. In areas of lymphocyte infiltration expression levels of tenascin C were particularly high. In non-pathological areas, staining was observed in endothelial cells of blood vessels. The average immunoreactivity score for tenascin C staining was 4.69 ± 2.18. The score ranged from 0 (all negative stained samples) to a maximum score of 7.5.

Staining with E-cadherin showed positive expression in 27 out of 30 cases. However, highest expression was observed in three cases of C-cell hyperplasia, while one case of C-cell hyperplasia showed no staining. Expression of E-cadherin was particularly high in the thyroid follicles. Altogether staining in all areas of the tumor was observed in 15 cases of MTC, with 8 MTC cases showing partial expression and 3 MTC cases showing no expression at all. There were no significant differences in E-cadherin expression between MTC and C-cell hyperplasia. As expected form a membrane bound protein, E-cadherin expression was primarily observed in the plasma membrane of cells (Figure 2). Immunoreactivity scores in E-cadherin samples ranged from 1 to 9 (full range) with a mean score of 4.69 ± 2.4.

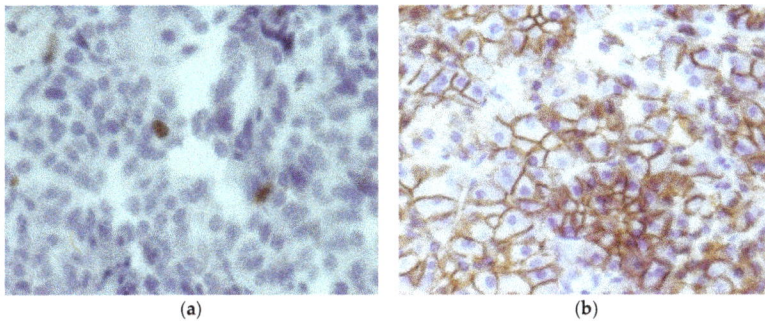

Figure 2. (a) $40\times$ magnification, depicting the staining results of Ki-67 (note the proliferating cell in the center); (b) $40\times$ magnification, showing strong plasma-membrane expression of E-cadherin.

EGFR expression was very weak, with six positively stained cases consisting of four cases of MTC and two cases of C-cell hyperplasia. Staining was primarily found in the cytoplasm of cells and endothelial cells. The highest staining intensity was found in non-neoplastic follicular cells scattered amid the tumor mass (Figure 3). Inside the tumor area, staining was relatively weak. The mean immunoreactivity score was 1.58 ± 1.20.

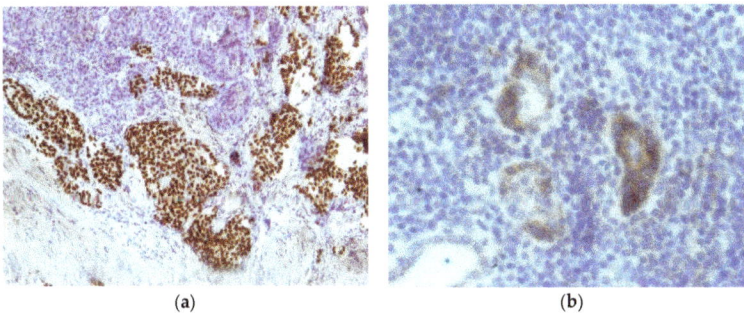

Figure 3. (a) $10\times$ magnification, showing thyroid transcription factor-1 (TTF-1) expression in a metastasis in a Meckel's diverticulum; (b) $40\times$ magnification, illustrating epidermal growth factor receptor (EGFR) expression in follicular cells scattered between medullary thyroid carcinoma (MTC) cells.

With exception of the metastasis to the adrenal gland all tissue samples, including the metastasis in a Meckel's diverticulum showed TTF-1 expression (Figure 3). The entire tumor area and all of the follicles showed strong staining with the TTF-1 antibody. The samples only differed in the staining intensity, which was moderate to strong. The mean immunoreactivity score was 7.77 ± 1.89.

The staining results for the proliferation marker Ki-67 showed positive staining results in all 30 samples. As expected, protein expression was only seen in the nucleus (Figure 2). Immunoreactivity scores ranged from 1.5 to 6.75 after correction with the correlation coefficient. The mean score was 3.35 ± 1.57. The four cases of C-cell hyperplasia had the lowest Ki-67 expression ($p < 0.001$) with only 1–2 cells per high power field. Tenascin C expression correlated moderately to strongly with the level of the proliferation marker Ki-67 in the tumor tissue. A weak correlation could be observed with E-cadherin, whereas EGFR and TTF-1 showed no significant correlation (Table 1).

Table 1. Correlation of tenascin C, EGFR, E-cadherin, and TTF-1 expression with the proliferation marker Ki-67.

MTC	Tenascin C	EGFR	E-Cadherin	TTF-1
Overall MTC				
r-value	0.61	−0.04	−0.19	0.13
p-value	<0.005	ns	<0.05	ns
RET-mutated MTC				
r-value	0.81	0.14	−0.11	−0.12
p-value	<0.01	ns	ns	ns
Wild-type MTC				
r-value	0.08	–	−0.40	0.72
p-value	ns	–	ns	<0.001

EGFR: epidermal growth factor receptor; TTF-1: thyroid transcription factor-1; MTC: medullary thyroid carcinoma; RET: rearranged during transfection; *r*: Pearson correlation coefficient; *p*: probability of obtaining a positive test result; ns: not significant.

All 15 tumors that showed RET mutation were analyzed regarding their expression of tenascin C, EGFR, E-cadherin, and TTF-1. They were then correlated with the proliferation marker Ki-67. Tenascin C expression showed a very strong correlation with the proliferation of RET-mutated tumors, while EGFR, E-cadherin, and TTF-1 showed a very weak correlation (Table 1).

The group of RET-mutated tumors was then split in germ-line-mutated ($n = 7$) and somatic-mutated ($n = 8$) tumors. Expression profiles of both groups were then correlated with proliferation in those tumors. In the case of the germ-line-mutated tumors, tenascin C expression correlated highly ($r = 0.86$) with proliferation. A weak correlation could be observed with E-cadherin and TTF-1 ($r = -0.26$ and -0.33, respectively), whereas EGFR only showed a very weak correlation ($r = -0.11$). In the case of MTC with somatic RET mutation, tenascin C still showed a moderate-to-strong correlation with proliferation ($r = 0.67$). EGFR correlation with proliferation was moderate ($r = 0.51$), while E-cadherin and TTF-1 showed low ($r = 0.39$) and very low correlations ($r = 0.02$), respectively.

MTC with RET wild-type were also investigated. Tenascin C showed a very weak correlation, E-cadherin a weak to moderate correlation, and TTF-1 a strong correlation with tumor proliferation (Table 1). EGFR analysis was not performed in this group because none of the specimens showed positivity for EGFR. Only EGFR expression differed significantly between RET-mutated and RET wild-type tumors ($r = 0.51$, $p = 0.001$). Tenascin C, E-cadherin, and TTF-1 did not differ in their respective expression levels.

RET-mutated and wild-type tumors were compared to evaluate whether the mutation status of RET affects TTF-1 expression. No significant difference in TTF-1 expression was found between both groups. RET-mutated MTC showed no correlation with TTF-1 expression for germ-line- and somatic-mutated tumors ($r = -0.33$ and 0.02, respectively, *p*-value is not significant (p ns)). TTF-1 expression correlated with tenascin C, EGFR, and E-cadherin expression. Tenascin C correlation was very weak ($r = -0.10$, p ns), while the EGFR and E-cadherin correlation was weak ($r = 0.37$ and 0.21, respectively, p ns).

EGFR positively stained tumors (all RET wild-type) did not show a significantly higher Ki-67 index, as compared with EGFR negatively stained tumors.

A weak to moderate correlation ($r = 0.08$–0.40, p ns) between calcitonin levels and Ki-67 was found. Preoperative calcitonin levels only showed a weak correlation with tenascin C expression ($r = 0.18$, $p < 0.05$) and Ki-67 ($r = 0.10$, p ns). Post-operative calcitonin levels correlated moderately with tenascin C expression ($r = 0.53$, $p < 0.005$) and Ki-67 ($r = 0.40$, p ns). Except for the inverse correlation for EGFR ($r = -0.38$, $p < 0.05$), post-operative calcitonin levels showed weak to no correlation with E-cadherin, TTF-1, and EGFR expression ($r = 0.22$, 0.08 and 0.07, respectively, p ns).

Both pre-operative and post-operative calcitonin levels were not significantly different between RET-mutated and wild-type tumors.

3. Discussion

In this study, we found tenascin C expression in all MTC, but in none of the C-cell hyperplasia cases. Tenascin C was primarily located in the stromal areas of tumors, but could also be detected in the cytoplasm and plasma membrane. Our results are in agreement with the findings by Koperek et al. [7] of tenascin C expression in all cases of MTC and in only 52% of C-cell hyperplasia cases. The difference in tenascin C expression in C-cell hyperplasia is most likely due to the larger study group used by Koperek et al. Our study cohort included only 30 patients because of the rarity of the disease. Furthermore, we investigated only four C-cell hyperplasia and two MTC metastases. The relationship between tenascin C expression and tumor proliferation needs to be further investigated. It seems that RET mutation is associated with a higher level of tenascin C expression, even though we found no significant difference between RET-mutated and wild-type MTC. It might be that, with a larger study cohort, a significant difference between RET-mutated and wild-type MTC could be established. Furthermore, it seems that the bc-24 clone used for tenascin C staining does not uniquely bind to tenascin C, but to other tenascin subtypes over the EGF-like repeats. Therefore, the results might not solely represent tenascin C, but also the expression of other tenascin isoforms.

EGFR plays an important role in driving proliferation in a variety of tumors [8]. In the study of Rodriguez-Antona et al. [9], EGFR expression was shown in a subset of 18 tumors, and it was thus concluded that EGFR might be a target for drug therapy. We therefore evaluated if MTC expresses EGFR and, if so, to what degree. In our study, EGFR expression could be detected in six cases (15%), with few staining cells and scattered expression. Our results are consistent with the reported EGFR expression of 9% and 35% in primary MTC and metastasis, respectively. Additionally, it seems that EGFR expression is significantly higher in MTC carrying a RET mutation [9]. Due to these reports, we performed EGFR mutation analysis on three cases with the highest EGFR expression. However, no mutations could be detected using the Cobas® EGFR mutation analysis kit. Our results are consistent with the finding by Rodriguez-Antona et al. [9] of nucleotide changes of unknown significance in only one sample. It thus seems that, although some EGFR expression can be detected, the role of EGFR in MTC is of a minor nature. Therefore, the absence of activating mutations questions the use of EGFR inhibitor drugs. This suggestion is further backed by Vitagliano et al. [21], who found that EGFR downstream signaling is of minor significance in the presence of active RET.

We also looked at the expression of E-cadherin, a plasma membrane protein important for cell–cell adhesion [11,12]. In our findings, E-cadherin showed staining in 26 cases of MTC (87%). The remaining samples showed no staining including one case of C-cell hyperplasia. Naito et al. [13] reported that low E-cadherin expression was associated with a higher malignant potential as well as regional lymph node metastasis. We therefore compared the expression levels of E-cadherin in C-cell hyperplasia and MTC, which showed no statistical significance. This is probably due to the small number of cases with C-cell hyperplasia in our study cohort.

In our study, with the exception of a metastasis in the adrenal gland, TTF-1 staining was moderate to strong in all tissue samples. We also found TTF-1 expression in a metastasis in a Meckel's diverticulum. These data seem to indicate that TTF-1 can be used as a useful marker for detecting primary MTC or metastasis, as previously suggested by Katho et al. [17].

The expression of the proliferation marker Ki-67 was generally low in our study cohort. As expected, C-cell hyperplasia showed the lowest Ki-67 indices, which were significantly lower than those found in MTC. Ishihara et al. [22] reported that breast cancers staining positive for tenascin carried a less favorable prognosis. We therefore evaluated if the expression of tenascin C in MTC correlates with tumor proliferation. We found that the Ki-67 index correlated moderately to strong with tenascin C expression. It might therefore be that tenascin C expression can be used as a marker for the malignant potential of a MTC. On the other hand, we observed that E-cadherin shows weak inverse correlation to tumor proliferation. As previously found by Naito et al. [13], low E-cadherin expression correlates with higher malignant potential of the tumor. This might also be true for our study group, but the size of our cohort may be a limiting factor.

The RET proto-oncogene is an important molecule in the development of MTC. We investigated if RET mutation correlates with a higher expression of tenascin C, EGFR, E-cadherin, or TTF-1. Furthermore, we evaluated whether proliferation is higher in RET-mutated MTC. We found that tenascin C expression in RET-mutated tumors showed a high correlation to proliferation. However, except for a significantly higher degree of EGFR expression in RET wild-type tumors, no significant difference in the expression of E-cadherin or TTF-1 could be detected between RET-mutated and wild-type MTC. Rodriguez-Antona et al. [9] also found that EGFR expression was higher in RET-mutated tumors, depending on the localization of the mutation.

We thereafter investigated RET-mutated tumors where the mutation was germ-line-derived or a somatic mutation. The expression profiles of the tumors in each group were then correlated with the proliferation marker Ki-67. Tenascin C correlated highly to proliferation in the germ-line-mutated group, whereas EGFR, E-cadherin, and TTF-1 showed a weak correlation. In the somatic-mutated tumors, tenascin C correlation was lower but showed a higher correlation to EGFR.

Calcitonin has proven to be a useful marker in the diagnosis and prognosis of MTC [23]. We found that both basal and pentagastrin stimulated calcitonin levels did not differ significantly between RET-mutated and wild-type MTC. Furthermore, no correlation between basal calcitonin levels and the Ki-67 index, tenascin C, EGFR, E-cadherin, or TTF-1 was observed. A moderate correlation was found between post-operative calcitonin levels and both Ki-67 index and tenascin C expression. However, due to the low level of correlation, it is possible that these results are stochastic.

The role of TTF-1 in the development of Hirschsprung's disease by RET interaction has been recently outlined [19]. Furthermore, not only papillary thyroid carcinoma but also MTC show expression of TTF-1 [17]. Garcia-Barceló et al. [24] found that mutations in single nucleotide polymorphisms (SNPs) of NKX2 (codes for TTF-1) and the RET promoter region correlated with the decreased TTF-1 binding and activation of RET, leading to Hirschsprung's disease. It is known that a loss of RET activation leads to Hirschsprung's disease [3], whereas a gain in function leads to MTC [2]. It is possible therefore that TTF-1 expression in RET-mutated MTCs might be higher, leading to consecutive RET activation. However, we found no significant difference in TTF-1 levels between RET-mutated (germ-line- and somatic-mutated) and RET wild-type tumors. Our data seem thus to indicate that TTF-1 does not play a role in the consecutive activation of RET. Moreover, we observed that TTF-1 has a weak correlation with EGFR and E-cadherin, but no correlation with tenascin-C or the Ki-67 index. However, the role of TTF-1 in MTC has yet to be established by a study with a larger cohort.

4. Materials and Methods

In the present study, 30 patients (16 females, 14 males; age: 2–81 years, mean age: 51 ± 18 years) with diagnosed MTC ($n = 26$) or C-cell hyperplasia ($n = 4$) at the Medical University of Salzburg were investigated. Eight patients showed MEN (MEN2A, 7 patients; MEN2B, 1 patient). All subjects gave their informed consent for inclusion before they participated in the study. The study was conducted in accordance with the Declaration of Helsinki, and the protocol (Approval: 14 February 2014) was approved by an institutional review board.

Routinely performed formalin-fixed paraffin embedded (FFPE) tissue was obtained from the primary thyroid site in 22 patients, lymph node metastasis in 6 patients, metastasis in a Meckel's diverticulum in 1 patient, and metastasis in the adrenal gland in 1 patient.

Genetic analysis of RET mutations was carried out in 21 patients, 6 of them with MEN2. RET gene mutations were detected in 15 patients (Table 2), while 6 patients showed RET wild-type.

Table 2. Rearranged during transfection (RET) mutations detected in the study group.

Mutation Detected	Sporadic MTC/MEN2
Codon 769 on Exon 13 ($n = 5$)	Sporadic
Codon 904 on Exon 15 ($n = 3$)	Sporadic
Codon L790F on Exon 13 + Codon 769 on Exon 13 ($n = 3$)	MEN2A (familial)
Codon L790F on Exon 13 + Codon 904 on Exon 15 ($n = 1$)	MEN2A
Codon 790 on Exon 13 ($n = 1$)	MEN2A
Codon 634 on Exon 11 ($n = 1$)	MEN2A
Codon 836 on Exon 14 ($n = 1$)	Sporadic

n, number of patients; MEN, multiple endocrine neoplasia.

Preoperative serum calcitonin levels (2.2–3293.4 ng/L, mean: 596.4 ng/L) were measured in 23 patients and pentagastrin tests (calcitonin: 17.7–2936.7 ng/L; mean: 708.3 ng/L) were performed in 11 patients. At time of the study, serum calcitonin levels (0.7–289,951.0 ng/L, mean: 11,056.6 ng/L) and pentagastrin test results (calcitonin: 2.6–971.7 ng/L; mean: 188.9 ng/L) were available in 29 patients and 16 patients, respectively. The normal calcitonin levels were <15 ng/L for males and <5 ng/L for females.

The expression of tenascin C, EGFR, E-cadherin, TTF-1, and Ki-67 was evaluated by immunohistochemistry. The primary antibodies used, with the working dilutions and pH of antigen retrieval buffers, are listed in Table 3.

Table 3. List of primary antibodies, working dilutions and pH of antigen retrieval buffers used.

Antibody	Source	Clone	Type	Species	pH-Retrieval	Working Dilution
Tenascin C	Sigma Aldrich™	bc-24	mc	Mouse	pH 6	1:4000
	Santa Cruz™	bc-24	mc	Mouse	pH 6	1:4000
EGFR	Dako™	E30	mc	Mouse	pH 6	1:20
E-Cadherin	Thermo Scientific™	SPM471	mc	Mouse	pH 9	1:100
TTF-1	Novocastra™	SPT24	mc	Mouse	pH 9	1:50
Ki-67	Dako™	MIB-1	mc	Mouse	pH 9	1:500

EGFR: epidermal growth factor receptor; TTF-1: thyroid transcription factor-1; mc: monoclonal antibody.

EGFR mutation analysis was performed using the Roche™ Cobas® EGFR mutation kit (Roche Molecular Systems, Inc., Branchburg, NJ, USA) on a Cobas® 4800 platform, v2.0 (Roche Molecular Systems, Inc.).

Statistical Analysis

Excel® software (Microsoft Corporation, Vienna, Austria) was used for the statistical evaluation of results.

Correlation analysis of tenascin C, EGFR, E-cadherin, and TTF-1 with the Ki-67 index was done by using the Pearson correlation coefficient test. For the assessment of statistical significance, the *t*-test for unpaired variance was used. Statistical significance was defined as $p < 0.05$.

5. Conclusions

MTC express tenascin C, E-cadherin, and TTF-1. Tenascin C expression correlates significantly with tumor proliferation, especially in RET-mutated tumors. EGFR expression is low in MTC and tumors showing EGFR expression do not exhibit higher proliferation. However, EGFR expression is significantly higher in MTC with RET mutation. No EGFR mutation was found in MTC. TTF-1 expression does not correlate with RET mutation status. TTF-1 expression has a weak correlation with tenascin C, E-cadherin, and EGFR expression.

Acknowledgments: The authors declare that no funds or grants were received.

Author Contributions: Cornelia Hauser-Kronberger and Christian Pirich conceived and designed the experiments; Florian Steiner and Gundula Rendl performed the experiments; Cornelia Hauser-Kronberger, Margarida Rodrigues, and Christian Pirich analyzed the data; Florian Steiner and Margarida Rodrigues wrote the paper.

Conflicts of Interest: The authors declare no conflict of interest.

References

1. Lairmore, T.C.; Wells, S.A.; Moley, J.F. Molecular biology of endocrine tumors. In *Cancer: Principles and Practice of Oncology*, 6th ed.; DeVita, V.T., Jr., Hellman, S., Rosenberg, S.A., Eds.; Lippincott: Philadelphia, PA, USA, 2001; pp. 1727–1740.
2. Edery, P.; Eng, C.; Munnich, A.; Lyonnet, S. RET in human development and oncogenesis. *Bioessays* **1997**, *19*, 389–395. [CrossRef] [PubMed]
3. Pasini, B.; Borrello, M.G.; Greco, A.; Bongarzone, I.; Luo, Y.; Mondellini, P.; Alberti, L.; Miranda, C.; Arighi, E.; Bocciardi, R.; et al. Loss of function effect of RET mutations causing Hirschsprung disease. *Nat. Genet.* **1995**, *10*, 35–40. [CrossRef] [PubMed]
4. Yoshida, T.; Matsumoto, E.; Hanamura, N.; Kalembeyi, I.; Katsuta, K.; Ishihara, A.; Sakakura, T. Co-expression of tenascin and fibronectin in epithelial and stromal cells of benign lesions and ductal carcinomas in the human breast. *J. Pathol.* **1997**, *182*, 421–428. [CrossRef]
5. Bourdon, M.A.; Wikstrand, C.J.; Furthmayr, H.; Matthews, T.J.; Bigner, D.D. Human glioma-mesenchymal extracellular matrix antigen defined by monoclonal antibody. *Cancer Res.* **1983**, *43*, 2796–2805. [PubMed]
6. Herold-Mende, C.; Mueller, M.M.; Bonsanto, M.M.; Schmitt, H.P.; Kunze, S.; Steiner, H.H. Clinical impact and functional aspects of tenascin-C expression during glioma progression. *Int. J. Cancer* **2002**, *98*, 362–369. [CrossRef] [PubMed]
7. Koperek, O.; Prinz, A.; Scheuba, C.; Niederle, B.; Kaserer, K. Tenascin C in medullary thyroid microcarcinoma and C-cell hyperplasia. *Virchows Arch.* **2009**, *455*, 43–48. [CrossRef] [PubMed]
8. Wikstrand, C.J.; Hale, L.P.; Batra, S.K.; Hill, M.L.; Humphrey, P.A.; Kurpad, S.N.; McLendon, R.E.; Moscatello, D.; Pegram, C.N.; Reist, C.J.; et al. Monoclonal antibodies against EGFRvIII are tumor specific and react with breast and lung carcinomas and malignant gliomas. *Cancer Res.* **1995**, *55*, 3140–3148. [PubMed]
9. Rodriguez-Antona, C.; Pallares, J.; Montero-Conde, C.; Inglada-Pérez, L.; Castelblanco, E.; Landa, I.; Leskelä, S.; Leandro-García, L.J.; López-Jiménez, E.; Letón, R.; et al. Overexpression and activation of EGFR and VEGFR2 in medullary thyroid carcinomas is related to metastasis. *Endocr. Relat. Cancer* **2010**, *17*, 7–16. [CrossRef] [PubMed]
10. Croyle, M.; Akeno, N.; Knauf, J.A.; Fabbro, D.; Chen, X.; Baumgartner, J.E.; Lane, H.A.; Fagin, J.A. RET/PTC-induced cell growth is mediated in part by epidermal growth factor receptor (EGFR) activation: Evidence for molecular and functional interactions between RET and EGFR. *Cancer Res.* **2008**, *68*, 4183–4191. [CrossRef] [PubMed]
11. Shimoyama, Y.; Hirohashi, S. Expression of E- and P-cadherin in gastric carcinomas. *Cancer Res.* **1991**, *51*, 2185–2192. [PubMed]

12. Vleminckx, K.; Vakaet, L.; Mareel, M.; Fiers, W.; van Roy, F. Genetic manipulation of E-cadherin expression by epithelial tumor cells reveals an invasion suppressor role. *Cell* **1991**, *66*, 107–119. [CrossRef]

13. Naito, A.; Iwase, H.; Kuzushima, T.; Nakamura, T.; Kobayashi, S. Clinical significance of E-cadherin expression in thyroid neoplasms. *J. Surg. Oncol.* **2001**, *76*, 176–180. [CrossRef] [PubMed]

14. Francis-Lang, H.; Price, M.; Polycarpou-Schwarz, M.; Di Lauro, R. Cell-type-specific expression of the rat thyroperoxidase promoter indicates common mechanisms for thyroid-specific gene expression. *Mol. Cell. Biol.* **1992**, *12*, 576–588. [CrossRef] [PubMed]

15. Civitareale, D.; Lonigro, R.; Sinclair, A.J.; di Lauro, R. A thyroid-specific nuclear protein essential for tissue-specific expression of the thyroglobulin promoter. *EMBO J.* **1989**, *8*, 2537–2542. [PubMed]

16. Fabbro, D.; di Loreto, C.; Beltrami, C.A.; Belfiore, A.; di Lauro, R.; Damante, G. Expression of thyroid-specific transcription factors TTF-1 and PAX-8 in human thyroid neoplasms. *Cancer Res.* **1994**, *54*, 4744–4749. [PubMed]

17. Katoh, R.; Miyagi, E.; Nakamura, N.; Li, X.; Suzuki, K.; Kakudo, K.; Kobayashi, M.; Kawaoi, A. Expression of thyroid transcription factor-1 (TTF-1) in human C cells and medullary thyroid carcinomas. *Hum. Pathol.* **2000**, *31*, 386–393. [CrossRef]

18. Suzuki, K.; Lavaroni, S.; Mori, A.; Okajima, F.; Kimura, S.; Katoh, R.; Kawaoi, A.; Kohn, L.D. Thyroid transcription factor 1 is calcium modulated and coordinately regulates genes involved in calcium homeostasis in C cells. *Mol. Cell. Biol.* **1998**, *18*, 7410–7422. [CrossRef] [PubMed]

19. Zhu, J.; Garcia-Barcelo, M.M.; Tam, P.K.H.; Lui, V.C.H. HOXB5 cooperates with NKX2–1 in the transcription of human RET. *PLoS ONE* **2011**, *6*, e20815. [CrossRef] [PubMed]

20. Tisell, L.E.; Oden, A.; Muth, A.; Altiparmak, G.; Mölne, J.; Ahlman, H.; Nilsson, O. The Ki67 index a prognostic marker in medullary thyroid carcinoma. *Br. J. Cancer* **2003**, *89*, 2093–2097. [CrossRef] [PubMed]

21. Vitagliano, D.; de Falco, V.; Tamburrino, A.; Coluzzi, S.; Troncone, G.; Chiappetta, G.; Ciardiello, F.; Tortora, G.; Fagin, J.A.; Ryan, A.J.; et al. The tyrosine kinase inhibitor ZD6474 blocks proliferation of RET mutant medullary thyroid carcinoma cells. *Endocr. Relat. Cancer* **2011**, *18*, 1–11. [CrossRef] [PubMed]

22. Ishihara, A.; Yoshida, T.; Tamaki, H.; Sakakura, T. Tenascin expression in cancer cells and stroma of human breast cancer and its prognostic significance. *Clin. Cancer Res.* **1995**, *1*, 1035–1041. [PubMed]

23. Kloos, R.T.; Eng, C.; Evans, D.B.; Francis, G.L.; Gagel, R.F.; Gharib, H.; Moley, J.F.; Pacini, F.; Ringel, M.D.; Schlumberger, M.; et al. Medullary thyroid cancer: Management guidelines of the American Thyroid Association. *Thyroid* **2009**, *19*, 565–612. [CrossRef] [PubMed]

24. Garcia-Barcelo, M.; Ganster, R.W.; Lui, V.C.; Leon, T.Y.; So, M.T.; Lau, A.M.; Fu, M.; Sham, M.H.; Knight, J.; Zannini, M.S.; et al. TTF-1 and RET promoter SNPs: Regulation of RET transcription in Hirschsprung's disease. *Hum. Mol. Genet.* **2005**, *14*, 191–204. [CrossRef] [PubMed]

International Journal of
Molecular Sciences

MDPI

Review

Differentiated Thyroid Cancer—Treatment: State of the Art

Benedikt Schmidbauer [†], Karin Menhart [†], Dirk Hellwig and Jirka Grosse *

Department of Nuclear Medicine, University of Regensburg, 93053 Regensburg, Germany;
benedikt.schmidbauer@ukr.de (B.S.); karin.menhart@ukr.de (K.M.); dirk.hellwig@ukr.de (D.H.)
* Correspondence: jirka.grosse@ukr.de; Tel.: +49-941-944-7510
† These authors contributed equally to this work.

Received: 6 April 2017; Accepted: 5 June 2017; Published: 17 June 2017

Abstract: Differentiated thyroid cancer (DTC) is a rare malignant disease, although its incidence has increased over the last few decades. It derives from follicular thyroid cells. Generally speaking, the prognosis is excellent. If treatment according to the current guidelines is given, cases of recurrence or persistence are rare. DTC requires special expertise by the treating physician. In recent years, new therapeutic options for these patients have become available. For this article we performed a systematic literature review with special focus on the guidelines of the American Thyroid Association, the European Association of Nuclear Medicine, and the German Society of Nuclear Medicine. For DTC, surgery and radioiodine therapy followed by levothyroxine substitution remain the established therapeutic procedures. Even metastasized tumors can be cured this way. However, in rare cases of radioiodine-refractory tumors, additional options are to be discussed. These include strict suppression of thyroid-stimulating hormone (also known as thyrotropin, TSH) and external local radiotherapy. Systemic cytostatic chemotherapy does not play a significant role. Recently, multikinase or tyrosine kinase inhibitors have been approved for the treatment of radioiodine-refractory DTC. Although a benefit for overall survival has not been shown yet, these new drugs can slow down tumor progression. However, they are frequently associated with severe side effects and should be reserved for patients with threatening symptoms only.

Keywords: differentiated thyroid cancer; radioiodine therapy; targeted therapy; tyrosine kinase inhibitors

1. Introduction

Patients with differentiated thyroid carcinoma have an excellent prognosis. The multimodal therapeutic approach is risk-adapted to achieve optimal treatment of differentiated thyroid cancer (DTC) and to minimize treatment-related morbidity. The treatment includes surgery (near-/total thyroidectomy) usually followed by remnant ablation using radioiodine according to the guidelines of the American Thyroid Association (ATA) and European Association of Nuclear Medicine (EANM) as well as a risk-stratified follow-up including hormone substitution.

However, in patients with primary or secondary radioiodine-refractory thyroid carcinoma the prognosis becomes significantly poorer. External beam irradiation may be used for locoregional control. Receptor tyrosine kinase inhibitors (TKIs) have shown clinical effectiveness in iodine-refractory DTC.

In this review, we present the current state of treatment of DTC.

2. Epidemiology and Classification

DTC is a rare disease with mostly excellent prognosis. The appearance of DTC depends on age, sex, family history, radiation exposure and many other factors [1]. DTC occurs in 7–15% of patients with thyroid surgery. In the year 2014, approximately 63,000 new cases of DTC were diagnosed in the

US [2] compared to 2009 with only 31,200 new cases. In Germany there are about 6000 new cases of DTC per year. The growing incidence of thyroid cancer and the tumor shift to diagnosis of smaller tumors is due to the increased usage of diagnostic methods, such as ultrasound of the neck [3].

Differentiated thyroid cancer includes papillary and follicular cancer that derive from thyrocytes and express the sodium iodine symporter. DTC represents the majority (90%) of all types of thyroid cancer [4]. One study predicts that papillary thyroid cancer will become the third most expensive cancer in women, with costs of US$ 19–21 billion in the US in 2019 [5].

Worldwide, there are many clinical practice guidelines for diagnosis, therapy and follow-up of DTC. The European Thyroid Association (ETA) published new guidelines for the management of DTC in 2013 [6]. The Society for Nuclear Medicine and Molecular Imaging and European Association of Nuclear Medicine published their most recent guidelines for radioiodine therapy of differentiated thyroid cancer in 2012 and 2008, respectively [7,8]. The Japanese Association of Endocrine Surgeons and the Japanese Society of Thyroid Surgeons recently reviewed their guidelines in 2014 [9]. The new ATA guidelines for management of differentiated thyroid cancer for adults were published in 2015 [10]. The updated ATA guidelines for management of DTC for children were also published in 2015 [11].

The risk classification of DTC using multiple staging systems is based on a combination of the size of the primary tumor, specific histology, extrathyroidal spread of the tumor and the age at diagnosis. It helps to predict the risk of local recurrence and developing metastases and the mortality in patients with DTC. The TNM classification depends on the size of primary tumor, the number and localization of metastatic lymph nodes and number of distant metastases (Table 1) [12]. The American Joint Committee on Cancer (AJCC) uses the combination of TNM Classification and an age of more than 55 years at diagnosis as risk factor [13]. The differentiation of lymphatic invasion and angioinvasion is of high importance, because angioinvasion is associated with an intermediate risk of recurrence. A common risk-stratification of DTC is based on the TNM classification (see also Section 4.2) [14]:

- high-risk group: pT3, pT4, each N1, all M1;
- low-risk group: pT1b, pT2, cN0/pN0, cM0;
- very low risk-group: pT1a, cN0/pN0, cM0.

Table 1. TNM Classification of thyroid cancer, 8th edition (modified from [12]).

TX	Primary Tumor Cannot be Assessed
T0	No evidence of primary tumor
T1	Tumor size maximum 2 cm, limited to the thyroid
T1a	Tumor size maximum 1 cm, limited to the thyroid
T1b	Tumor size >1 cm up to a maximum of 2 cm, limited to the thyroid
T2	Tumor size >2 cm up to 4 cm, limited to the thyroid
T3	Tumor size >4 cm, limited to the thyroid, or any tumor with macroscopic extrathyroidal extension (*Musculus sternohyoideus, Musculus sternothyreoideus, Musculus omohyoideus*)
T3a	Tumor size >4 cm, limited to the thyroid
T3b	Any tumor with macroscopic extrathyroidal extension (*M. sternohyoideus, M. sternothyreoideus, M. omohyoideus*)
T4a	Any tumor size with extrathyroidal extension beyond the thyroid capsule and invasion of subcutaneous soft tissue, larynx, trachea, esophagus and/or recurrent laryngeal nerve
T4b	Any tumor size with invasion of prevertebral fascia, mediastinal vessels or carotid artery
NX	Regional lymph nodes cannot be assessed
N0	No regional lymph node metastases
N1	Regional lymph node metastases
N1a	Lymph node metastases unilateral in level VI or upper mediastinum
N1b	Metastases in other unilateral, bilateral or contralateral cervical lymph nodes (level I, II, III, IV and V) or retropharyngeal
M0	No distant metastases
M1	Distant metastases

The American Thyroid Association defines in their current guideline a stratification based on the risk of structural disease recurrence [10]:

- high-risk group: gross extrathyroidal extension, incomplete tumor resection, distant metastases, or lymph node >3 cm;
- intermediate-risk: aggressive histology, minor extrathyroidal extension, vascular invasion, or >5 involved lymph nodes (0.2–3 cm);
- low-risk: intrathyreoidal DTC, ≤5 lymph nodes micrometastases (<0.2 cm).

In the last few years new molecular and genetic biomarkers, such as BRAF (V600E), phosphatidylinositol 4,5-bisphosphate 3-kinase catalytic subunit α (PIK3CA), tumor protein p53 (TP53), RAC-α serine/threonine-protein kinase 1 (AKT1) and telomerase reverse transcriptase (TERT) became more important for the management of diagnosis, therapy and observing of DTC. The role of RAS is discussed controversially. Table 2 shows the impact of the two well-evaluated molecular markers BRAF and TERT [15]. Some of these alterations might be interesting molecular targets for new therapies.

Table 2. Mutations of BRAF and TERTp in follicular-derived thyroid carcinoma and clinicopathological impact (modified from [15]).

Mutation	Histology	Clinicopathological Associations
BRAF	papillary thyroid carcinoma (PTC)	recurrence, multifocality, extrathyreoidal extension, lymph nodes metastasis, advanced stage, absence of capsule, vascular invasion, more aggressive histological subtype
BRAF	micro PTC	multifocality, extrathyreoidal extension, advanced stage, lymph node metastasis
BRAF	thyroid carcinoma derived from follicular cells	no association
TERT	papillary thyroid carcinoma	more advanced stage by tall cell variant, higher tumor size, vascular invasion, older age, poor outcome, lymph node and distant metastasis
TERT	thyroid carcinoma derived from follicular cells	more aggressive histologic variants, concomitant presence of mutated RAS/BRAF, age > 45, higher tumor size, vascular invasion, persistent or recurrent disease, lymph node metastasis

2.1. Papillary Thyroid Cancer

Papillary thyroid carcinoma (PTC) is the most common form of DTC. Histologically it is a tumor of follicular cells of the thyroid gland with characteristic nuclear signs. There are more than 10 histological variants of papillary thyroid cancer documented, can be seen in Table 3 [16,17]. Due to this microscopic diversity, different risk stratifications are needed.

The tall cell variant is one of the tumor entities with unfavorable outcome. This type of thyroid cancer is presented in tall columnar cells and occurs in older age showing a higher rate of lymph node metastases. In nearly 80% of these tumors the BRAF (V600E) mutation is found [18]. A new aggressive variant of papillary thyroid carcinoma, which is characterized by cells with hobnail appearance and apically placed nuclei, was described recently. The BRAF (V600E) mutation is found frequently and associated with distant metastases [19]. In children and adults affected by the Chernobyl incident the solid variant of PTC appears predominantly. Mortality within the first 10 years after initial diagnosis and treatment is low (<1%) [20,21]. It is very important to recognize that there are histological differences compared to poorly differentiated carcinomas, because of the very different therapy strategy. In poorly differentiated thyroid cancer the capability to take up (radio) iodine is clearly reduced (e.g., decreased expression of sodium iodine symporter) and therefore not sufficient to achieve a significant therapeutic effect. Another form of PTC is the diffuse sclerosing variant. It is characterized by a higher incidence of lymph node and distant metastases. Nevertheless, overall mortality appears low. The encapsulated follicular variant of papillary carcinoma very rarely shows capsular or vascular invasion. Histologically it is characterized by follicular growth, typical nuclear features of papillary carcinoma and total tumor encapsulation. RAS mutations can be detected frequently. The non-encapsulated follicular variant of papillary cancer shows BRAF (V600E) mutations quite often [22,23]. This tumor is associated with lymph node metastases in about 25–30% and low rates of distant metastases.

Table 3. WHO classification of papillary and follicular carcinoma of the thyroid (modified from [17]).

Histology	Histological Variants
Papillary carcinoma	Classic (usual)
	Clear cell variant
	Columnar cell variant
	Cribriform-morular variant
	Diffuse sclerosing variant
	Follicular variant
	Macrofollicular variant
	Microcarcinoma (occult, latent, small, microtumor)
	Oncocytic or oxyphilic variant (follicular/nonfollicular variant)
	Solid variant
	Tall cell variant
	Warthin-like variant
Follicular carcinoma	Clear cell variant
	Oncocytic (Hürthle cell) variant
	Mucinous variant
	With signet-ring cells

PTC presents distant metastases mainly in bones or lungs.

Papillary microcarcinoma is a PTC < 1 cm corresponding to the classification of the World Health Organization (WHO) which is often found incidentally. In some autopsy studies the papillary microcarcinoma was found in 6–35% of the thyroids by incident [10]. Papillary microcarcinoma may also exhibit RET proto-oncogene (RET)/PTC-rearrangements or BRAF (V600E) mutations.

2.2. Follicular Thyroid Cancer

Follicular thyroid carcinoma (FTC) is a malignant tumor, histologically derived from follicular thyroid cells, showing transcapsular or vascular invasion and missing the typical nuclear signs of papillary carcinoma. In the traditional classification of FTC there are two groups: minimally invasive and widely invasive [24 26]. The widely invasive FTC shows an extensive vascular invasion, often also associated with extrathyroidal growth.

Oncocytic follicular carcinoma is a special form of FTC with some microscopic differences compared to conventional FTC. One of them is the accumulation of innumerable mitochondria. Due to its histological differences, oncocytic carcinoma shows some different biological behavior with a higher ability to metastasize to lymph nodes and a possibly higher rate of recurrence and tumor-related mortality [27–29].

2.3. Familial Tumor Syndromes

Some of the histopathological variants of DTC are associated with familial tumor syndromes. For example, the cribriform-morular form of papillary thyroid cancer is frequently seen in patients with a germline mutation in adenomatous polyposis coli gene [30,31]. About 40% of patients with this special histological form of papillary thyroid carcinoma show simultaneously a familial adenomatous polyposis (FAP) [32]. Due to this high rate of association of cribriform-morular PTC and FAP it is very important to complete the diagnostic work-up with colonoscopy and genetic counseling.

Another type of FTC is associated with the germline mutation of the phosphatase and tensin homolog (PTEN) gene [33–35]. The follicular variant of thyroid carcinoma is in this case very characteristic and should be known by pathologists. The syndrome is associated with a high risk of appearance of other tumors, such as colon hamartomas or breast and endometrium tumors. Genetic counseling is recommended.

3. Diagnostic Approach to Thyroid Nodules

The prevalence of sonographically-detected thyroid nodules in the U.S. is described between 19% and 35% [36]. Toxic adenomas are found in up to 4 percent of the population. In Europe the incidence of thyroid nodules is higher in some areas. In Germany, a country with relative iodine deficiency, nodules are found in 33% of the population.

Risk factors for malignancy are exposure to ionizing radiation through radiotherapy or fallout especially in younger years, familial thyroid carcinoma, or syndromes that are associated with thyroid cancer like PTEN, Cowdens disease or multiple endocrine neoplasia type 2 (MEN2). Warning signs in clinical examination are rapid nodule growth, fixation in the surrounding tissue, vocal cord paralysis, possibly accompanied by hoarseness.

The diagnostic cornerstone of thyroid nodules remains the ultrasound examination. It should be performed in any case of known or suspected thyroid nodules or cervical lymphadenopathy to assess if further diagnostic is needed. Sonographic patterns suspicious of malignancy are microcalcifications, irregular margins, solid consistency, hypoechogenity, extrathyroidal extension (ETE) and a tall shape rather than a wide one. Intranodulary vascularization does not seem to have a clear correlation with malignancy [10].

Roughly one third of thyroid nodules are larger than 1 cm and eligible for scintigraphy [37]. The guidelines of the German Society of Nuclear Medicine recommend a scintigraphic examination of every thyroid nodule >1 cm. By routinely performing a Tc-99m thyroid scan autonomous adenomas that have not yet an impact on thyroid-stimulating hormone (TSH) level can be detected without subjecting the patient to the risks and stress of fine-needle aspiration (FNA). This applies especially to groups at increased risk for complications like patients that are treated with coagulation inhibitors. The diagnostic algorithm for evaluation of thyroid nodules according to the German guidelines that was recently published by Feldkamp et al. is shown in Figure 1 [38].

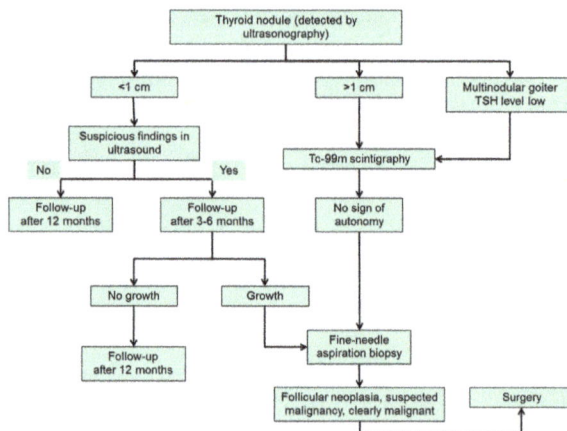

Figure 1. Diagnostic algorithm for the evaluation of thyroid nodules (modified from [38]). TSH: thyroid-stimulating hormone.

The ATA guidelines recommend measurement of the TSH level if a thyroid nodule is found. A radionuclide scan (Tc-99m, preferable I-123) should be performed only if TSH level is subnormal [10]. Nevertheless, the American guidelines cannot be applied to the rest of the world without adjustment for differences in the patient populations. Except for clearly benign cysts, for every lesion of a certain size ultrasound guided fine-needle aspiration biopsy (FNA) is recommended (Table 4) [10]. Furthermore, the measurement of serum calcitonin is recommended when new thyroid nodules are detected time to rule out medullary thyroid cancer that derives from c-cells and is not added to group of DTC.

Table 4. Sonographic patterns and risk of malignancy (modified from [10]).

Ultrasound Features	Estimated Risk of Malignancy	Sonographic Pattern	FNA Size Cutoff
Solid hypoechogenic nodule or solid hypoechogenic component of a partially cystic nodule with one or more of the following features: irregular margins, microcalcification, taller rather than wide shape, rim calcifications with small extrusive soft tissue component, evidence of extrathyroidal extension (ETE)	>70–90%	Highly suspicious	>1 cm
Solid hypoechogenic nodule with smooth margins without microcalcification, taller rather than wide shape or signs of ETE	10–20%	Intermediate suspicion	>1 cm
Isoechogenic solid nodule or partially cystic nodule with eccentric solid areas without microcalcification, taller rather than wide shape or signs of ETE	5–10%	Low suspicion	>1.5 cm
Spongiform or partially cystic nodule without any of the sonographic features described above	<3%	Very low suspicion	>2 cm, alternative: observation without fine needle aspiration (FNA)
Purely cystic nodules without solid components	<1%	Benign	No biopsy

Cytological analysis is performed according to the Bethesda System for Reporting Thyroid Cytopathology. The findings are graded into six categories:

I: nondiagnostic/unsatisfactory;

II: benign;

III: atypia of undetermined significance/follicular lesion of undetermined significance;

IV: follicular neoplasm/suspicious for follicular neoplasm;

V: suspicious for malignancy;

VI: malignant.

If the FNA biopsy is graded non-diagnostic/unsatisfactory, biopsy should be repeated. Numerous molecular tests can be applied to distinguish malignant from benign lesions, such as BRAF (V600E), PIK3CA and TERT promoter, AKT1, and TP53, although there is no explicit recommendation in the current guidelines. Accordingly, adjustments are to be expected in the future [15]. For more non-diagnostic biopsies in a row, the decision for close surveillance without intervention or for surgery should be made in dependence of the sonographic pattern [10].

4. Therapy of Differentiated Thyroid Carcinoma

DTC should be treated interdisciplinary in facilities with an appropriate expertise in order to ensure an optimal long-term treatment quality. Specialists in surgery, endocrinology, pathology and nuclear medicine should be available. The therapeutic approach is individualized and risk-adapted.

4.1. Surgery

For widely invasive follicular thyroid carcinomas and FTC with vascular infiltration, thyreoidectomy is recommended. Lymph node dissection is recommended if lymph node metastases can be detected pre- or intraoperatively by sonographic examination and/or palpation. The solitary minimally invasive FTC without vascular invasion does not require a second surgical intervention as completion, if the tumor has been completely removed (R0). Thyreoidectomy and lymph node dissection of the central compartment are recommended for prognostically unfavorable variants.

For all papillary thyroid carcinomas >1 cm and/or for all metastasized or macroscopically invasive PTC irrespective of size, thyreoidectomy is recommended [10,39]. If lymph node metastases have been detected sonographically or intraoperatively, lymph node dissection in the affected compartment should be done to reduce the risk of (local) recurrence. On the other hand, the diagnostic or therapeutic extirpation of only single lymph nodes as a part of the primary intervention is not recommended. Although at present the importance of central lymph dissection with prophylactic intention is still unclear, the high probability of lymph node metastases is a substantial argument to expand the surgical procedure. Furthermore, it is difficult to exclude lymph node metastases pre- or intraoperatively. After all, the increased risk of a local recurrence associated with an increased morbidity due to the surgical intervention in the situation of relapse should be mentioned. On the other hand, the main arguments against a prophylactic dissection are the lack of evidence regarding a better outcome of the patients and the remarkably higher complication rate due to the more extensive intervention (e.g., vocal cord paralysis, parathyreoprival tetany). Specifically, papillary thyroid microcarcinoma that is found incidentally does not require a further surgical treatment.

After all, accurate histopathological examination of the specimen after (hemi)thyroidectomy and lymphadenectomy (if done) is regarded as the gold standard and is indispensable for the management and further diagnostic and therapeutic approach.

4.2. Adjuvant Radioiodine Therapy

Radioiodine therapy (RIT) has been established for more than 60 years. The benefit was demonstrated in DTC patients with a high risk for recurrence. In patients with very low-risk DTC a positive effect of a RIT on tumor-free and overall survival has not been proven by prospective clinical trials.

RIT is defined as the systemic administration of I-131 (radioiodine as sodium iodide or potassium iodide) to irradiate thyroid remnants as well as non-resectable or incompletely resected DTC.

Adjuvant ablative RIT of thyroid remnants or tumor tissue is the optimal precondition for the follow-up including determination of serum thyroglobulin (Tg) and I-131 whole-body scans. The rationale that underlies this approach is to detect a local recurrence or distant metastases in an early and potentially curable stage to minimize mortality. However, regional or distant metastases frequently are only detectable by rising Tg levels after a successful remnant-ablation. It was shown that an ablative RIT decreases the rate of recurrence and mortality over a follow-up period of more than 10 years [40–43]. RIT is indicated in high-risk DTC (pT3, pT4, each N1, every M1), in low-risk DTC (pT1b, pT2, cN0, pN0, M0) and in small papillary thyroid carcinoma (very low-risk DTC), if there are risk factors (see Section 4.7) [44,45]. Furthermore, RIT can be used for the treatment of radioiodine-positive tumor residues, lymph node and distant metastases with curative or palliative intention. In the case of tumor activity shown by an increasing serum level of thyroglobulin without a macroscopically detectable tumor using morphological and functional imaging RIT can be carried out after carefully weighing risks and benefits [14].

To ensure a high uptake of radioiodine (I-131) in remnant tissue, (suspected) tumor, or metastases, an elevated serum level of TSH is required (>30 mU/L). This level is believed to increase the expression of the sodium iodine symporter (NIS) in benign and malignant follicular cells of the thyroid [46]. According to the guidelines of the ATA and EANM [8,10] this TSH level can be reached by waiting not less than 3 weeks after thyroidectomy or after a withdrawal (4–5 weeks) of levothyroxine (LT4). The subsequent period of hypothyroidism decreases the quality of life significantly in many patients. The physical and psychological symptoms of hypothyroidism include gain of weight, impaired renal function, cardiovascular abnormalities, dyslipidemia (exacerbation), constipation, dry skin, hoarseness, fatigue, sleep disturbance, impaired ability to concentrate and depression [47].

Alternatively, recombinant TSH (rhTSH) can be administered intramuscularly (2 times 0.9 mg rhTSH) to avoid inconvenience and morbidity due to the lack of thyroid hormone. This drug is approved for radioiodine ablation (without known distant metastases) of T1–4 tumors, diagnostic whole-body scan and preparation for testing of serum Tg in adults [48–51].

Absolute contraindications of RIT are pregnancy and breastfeeding. Relative contraindications include depression of the bone marrow (especially if the administration of high activities of I-131 is planned), a restriction of salivary gland function, pulmonary function restriction (if a high accumulation of I-131 in lung metastases is possible) and symptomatic metastases of the central nervous system, because local edema and inflammation caused by RIT and hypothyroidism can lead to severe compression effects [8].

The activity of I-131 for remnant ablation is still discussed controversially. The HiLo study (Great Britain) and the ESTIMABL study (France) both compared ablative RIT with 1.1 GBq I-131 versus administration of 3.7 GBq (100 mCi) I-131 after thyroid hormone withdrawal or stimulation with rhTSH in patients with low-risk carcinoma [52,53]. Both studies showed that RIT with only 1.1 GBq (30 mCi) I-131 is not inferior compared to the higher activity in regard to the success of ablation. However, the definition of "success of ablation" used in both studies is not accepted by all departments and associations. Several authors report that the rate of a second RIT increases when low therapy activities were used initially [54,55].

In the ESTIMABL study the diagnostic I-131-whole-body scan 8 months after I-131 ablation was limited to patients with elevated Tg antibodies and disturbed Tg recovery. Even in this subgroup this concept was not consistently implemented. Based on an observation study, iodine-accumulating metastases are possible with a measurable Tg level of up to 1 ng/mL [56]. Due to the open question of the "optimal" activity for remnant ablation, the German Society of Nuclear Medicine for example recommends a single administration of 1 to 3.7 GBq (about 30–100 mCi) I-131 [14]. Preablation scanning with Tc-99m pertechnetate on the day of ablation (as used in the HiLo trial [52]) can give very useful information in clinical decision making. In low-risk DTC patients with a large remnant (multiple foci or one large focus) ablation with 3.7 GBq (30 mCi) may be prefered.

Although RIT is generally well tolerated, the procedure has some potential short- and long-term side effects [8]. Short-term risks/side effects are: thyroiditis due to irradiation, swelling of the tumor or metastases (including compression symptoms), gastritis and nausea, sialadenitis and abnormalities of taste and smell, bone marrow depression, and hypospermia.

Long-term risks and side effects include permanent bone marrow depression, second primary malignancy after RIT with a high cumulative activity (leukemia and solid tumors) [57], chronic sialadenitis (including abnormalities of taste and smell, xerostomia,) and pulmonary fibrosis (in patients with diffuse iodine-avid pulmonary metastases). Due to the risk of chronic hypospermia or azoospermia, sperm banking should be considered if high cumulative activities are expected [58]. These risks have to be weighed against the expected benefits of the RIT.

4.3. Metastatic Differentiated Thyroid Carcinoma

Distant metastases occur in patients with differentiated thyroid carcinoma with a prevalence of up to 10%. In particular, they affect lung and bone [59]. If a sufficient uptake of I-131 in metastases is measurable, different therapeutic approaches are to be weighed regarding risk and benefit.

If locoregional lymph node metastases are detectable, surgery should be performed. I-131 is used for iodine-avid metastases for treatment control after surgery or as an alternative therapy if no surgery is possible/planned (e.g., additional detection of distant metastases requiring RIT, previously performed radiotherapy, previous lymph node dissection).

In the case of micronodular metastases of the lung RIT is carried out as a treatment with curative intent. Macronodular pulmonary metastases should also be treated with I-131 in a curative intention but a complete remission is unlikely. Alternatively, (or in combination) the resectability can be evaluated.

The complete surgical resection of isolated bone metastases leads to an improved outcome. A combination of different therapeutic approaches like percutaneous radiotherapy, RIT and local interventional therapy could be helpful if symptomatic metastases of the bone cannot be (completely) resected. The same strategy is applied to brain metastases [14].

For the treatment of metastases, standard activities of 4–11 GBq (about 100–300 mCi) I-131 are given depending on individual patient characteristics like age, renal function, bone marrow depression and tumor load.

4.4. Thyroid Hormone Treatment

After thyroidectomy, life-long thyroid hormone therapy is required, usually as monotherapy with levothyroxine (LT4). Since TSH is able to promote the growth of remaining DTC cells, the dosage of LT4 should initially be high enough to achieve a suppression of thyrotropin. The thyroid function should be checked after 6 to 8 weeks. Depending on the result the dosage should be adjusted. An elevated level of triiodothyronine has to be avoided.

A long-term suppression of TSH to values <0.1 mU/L is currently only recommended for high-risk patients and patients with persistent disease indefinitely in the absence of specific contraindications [10]. In these cases, a better prognosis was demonstrated for the suppression of thyrotropin. No evidence-based data are available for optimal duration of TSH suppression.

According to the guidelines of the ATA, serum TSH should be maintained between 0.1 and 0.5 mU/L in patients with high-risk disease but excellent or intermediate response to therapy for up to 5 years and also in patients with a biochemical incomplete response taking into account the initial ATA risk classification published by Haugen et al. [10]. This recommendation is rated as weak with low-quality evidence. If the response to therapy is excellent biochemically and clinically in patients with a low risk for recurrence and there is no evidence of disease in the course of time, the serum level of TSH may be kept in a range of 0.5–2.0 mU/L, because there is no data showing a benefit of TSH suppression for low-risk patients.

Individual patient-related factors such as osteoporosis or osteopenia and cardiac co-morbidities like atrial fibrillation should always be taken into account during thyroid hormone therapy and weighed against the risk of recurrence. Especially in elderly patients >60 years, the use of TSH suppressive therapy should be carefully considered since the risk of such complications is significantly increased [60].

4.5. Follow-Up

Although the cumulative relapse rate is up to 30%, the life expectancy of DTC patients (pT1-3, pN0-1, M0) is not significantly different from the general population after therapy according to the current guidelines. Lifelong follow-up examinations should be carried out because relapses can occur even after decades and may be cured again. Initial checks should be carried out every six months (e.g., for the first 5 years after diagnosis). If there are no pathological findings later on, annual examinations are adequate [61]. The follow-up examination is based on the medical interview, clinical examination, cervical sonography, determination of TSH, triiodothyronine, levothyroxine, and thyroglobulin including Tg antibodies. In the case of postoperative hypoparathyroidism, the substitution therapy (cholecalciferol, calcium) should be checked and adapted (if necessary) to minimize the risk of osteoporosis.

A diagnostic whole-body scan is obligatory 6–12 months after initial RIT, a second scan is only needed in the case of relapse [10,62].

The criteria for a disease-free stage 6–12 months after primary therapy of DTC with total thyroidectomy ± radioiodine therapy are no clinical signs of DTC, no pathological uptake in the I-131 whole-body scan (only after remnant ablation) and a serum Tg below the detection limit (under suppression and after TSH stimulation, with absence of Tg antibodies) [10,62,63]. Under these conditions patients have a very low probability of relapse. If there are signs of relapse (e.g., elevated/rising serum levels of Tg) and no radioiodine-accumulating tumor tissue is detectable, clinical diagnostics should include the search for non-radioiodine-avid tumor tissue using F-18-fluorodeoxy-glucose positron emission tomography (FDG-PET) combined with computed tomography ideally under TSH-stimulation.

4.6. Tyrosine Kinase Inhibitors

For poorly differentiated thyroid carcinoma without relevant iodine metabolism and therefore very low radioiodine uptake, RIT is not a therapeutic option. Radioiodine resistance is currently defined as lesions without iodine uptake under TSH stimulation, progression in size in the year following RIT or persistent metastases after a cumulative dose of 22 GBq (600 mCi) I-131 of radioiodine. In these cases, complimentary diagnostic using FDG-PET/computed tomography (CT) is essential. FDG uptake is typically increased in poorly differentiated lesions that can be overlooked on radioiodine scans. Prognosis for radioiodine resistant thyroid cancer with distant metastases is very poor, with an estimated median survival time of about 2.5 to 3.5 years [64,65].

Chemotherapy comes at high toxicity with disappointing response rates [66]. For these patients a strict LT4 regime with TSH suppression is the best way to go. On showing rapid progression under such a regime, therapy options were few until recently.

Tyrosine kinase inhibitors like vandetanib, sorafenib and lenvatinib are a relatively new approach to systemic therapy in these cases. Tyrosine kinase receptors, the target structure of TKI, are trans-membrane proteins that mediate cell survival and proliferation [67]. If mutated, they can cause uncontrolled cell proliferation, dedifferentiation and apoptosis reduction. A large part of DTC show at least one mutation of RAF, RET or paired box gene 8 (PAX8)/peroxisome proliferator-activated receptor gamma (PPARγ) which makes them targets for TKI therapy. Furthermore, TKIs block receptors of the vascular endothelial growth factor (VEGF), fibroblast growth factor receptors and platelet-derived growth factor and thus inhibit tumor angiogenesis and lymphangiogenesis and cause hypoxia in malignant tissue [68]. TKIs, already approved for the treatment of irresectable liver cancer and renal carcinoma, promise to be an effective new tool for the treatment of poorly differentiated thyroid carcinoma (PDTC) [66].

A recent review on the use of sorafenib, sunitinib and lenvatinib showed a benefit for progression-free survival of up to five months [69]. While initially showing partial response or at least disease stabilization after sorafenib, the first TKI approved for thyroid cancer, patients almost always develop resistance over the course of the following one to two years. Switching to another TKI is possible at this point [70]. A study using lenvatinib was able to indicate prolonged progression-free survival regardless of BRAF or RAS mutation status, suggesting a diminished role of these pathways [71]. However, a benefit in overall survival could not be found.

Therapy with kinase inhibitors may be accompanied by severe side effects. Induced hypertension is one of the most common; the underlying pathophysiology is yet unclear. Vasoconstriction following reduced nitric oxide production via inhibition of the VEGF-PI3K pathway is discussed. A reduction of peripheral arterioles due to antiangiogenic effects resulting in increased peripheral resistance and an activation of the endothelin-1-system causing vasoconstriction have also been suggested [72]. Other side effects may include diarrhea, fatigue, hepatotoxicity, skin changes, nausea, increased LT4 dosage requirement, changes in taste and weight loss and associated with a severe decrease of quality of life [69].

Keeping this in mind kinase inhibitors can certainly not be considered as a standard regime or an alternative to TSH-suppression. For patients with radioiodine-refractory DTC they can be a useful complementation to standard therapy.

While undergoing TKI therapy patients can display somewhat inconclusive lab results. Normally a reliable parameter, thyroglobulin levels can fluctuate under TKI treatment. These changes do not necessarily represent the actual course of the disease as it is monitored in anatomical imaging. Sufficient therapeutic monitoring not only by relying on lab tests but also on CT or PET/CT diagnosis to determine a morphologic or metabolic response is essential [73].

Considering the extensive side effects, this therapy should be reserved for patients with rapid tumor progression and severe to life threatening symptoms. In these cases, the decision for TKI therapy should be made in a interdisciplinary manner, carefully weighed against local strategies like radiotherapy and local surgery [74]. TKI treatment should only be performed by a team of physicians experienced with side effects management.

The European guidelines for treating differentiated thyroid carcinoma are from 2008. Kinase inhibitors are therefore not considered there [8] and a unanimous European recommendation is still awaited.

4.7. Papillary Microcarcinoma

Because of the excellent prognosis of papillary microcarcinoma (PTMC), a hemithyroidectomy without RIT is regarded as sufficient therapy, if there is no sign of local invasion, lymph node and/or distant metastases. The substitution of LT4 should keep the serum level of TSH in a euthyroid metabolic state.

In a meta-analysis, PTMC showed a prevalence of distant metastases of 0.4%, a probability of locoregional relapse of 2.5% but also a prevalence of micrometastases in locoregional lymph nodes of 12–50% [45]. The risk of lymphogenic micrometastases increases with increasing tumor diameter [75]. Using single photon emission computed tomography (SPECT) combined with CT, other studies showed a prevalence of lymph node metastases up to 57% [76,77]. The relapse-free survival in patients with PTMC after 5 years was 78.6% without RIT compared to 95.0% in patients that have had a remnant ablation with RIT [78]. The recommendation for a RIT in PTMC is based on the extent of resection and the individual risk profile. Risk factors are multifocality, infiltration of the thyroid gland, histological variants of papillary thyroid carcinoma, low degree of differentiation, tumor diameter 6–10 mm, molecular markers like BRAF-V600E mutation, infiltrative tumor growth, surrounding desmoplastic fibrosis and previous percutaneous irradiation of the neck [14]. In patients with a residual thyroid gland (e.g., after lobectomy) an ablative RIT is not indicated.

5. Summary and Conclusion

Differentiated thyroid cancer is a rare tumor entity but shows a strongly increasing incidence over the last decades. It derives from the follicular epithelium of the thyroid and shows basic biological characteristics of healthy thyroid tissue. The expression of the sodium iodide symporter is the key feature for specific iodine uptake. Patients with DTC have an excellent prognosis.

The therapeutic approach including surgery and remnant ablation with radioiodine should be risk-adapted to achieve an optimal treatment and to minimize treatment-related morbidity. Overtreatment should be avoided.

With regard to so-called low-risk carcinoma defined by the ATA there are controversial therapeutic approaches. The guidelines of the ATA recommend a lobectomy under certain conditions. Following the guidelines of the EANM a thyreodectomy with RIT should be performed (except PTC pT1a). However, long-term studies are currently not available. These studies are certainly necessary (against the background of the slow growth of the well-differentiated thyroid carcinoma) to decide which approach is appropriate. A risk-stratified follow-up is required since recurrences can occur over years. Furthermore, thyroid hormone substitution must be controlled.

The life span of most DTC patients does not differ from general population when appropriate treatment is given. The prognosis becomes poorer in patients with radioiodine refractory thyroid carcinoma. TKI have shown clinical effectiveness in iodine-refractory DTC with regard to progression free survival. A positive effect on overall survival could not be shown yet and has to be evaluated in further studies. However, therapy should be carried out in centers with special expertise.

In the current guidelines of the ATA and EANM there is no evidence-based treatment concept (or strong recommendation) for every situation. There are still open questions:

- The value of RIT under the condition of increasing serum level of Tg without a detectable correlatation in the morphological or functional imaging (i.e. iodine-negative whole-body scan);
- The benefit of a remnant ablation in patients with papillary microcarcinoma (very low risk of relapse, lymph node metastasis possible);
- Optimal activities of I-131 for safe and effective radioiodine ablation;
- The role of rhTSH as preparation for RIT to treat incomplete or non-resectable local recurrence or metastases;
- The role of a short LT4 withdrawal to reduce blood levels of iodine before RIT or diagnostic whole-body scan.

An analysis by the Cancer Genome Atlas Research Network identifies previously unknown genetic alterations and molecular subtypes of PTC. These alterations may lead to a more accurate diagnosis of tumors and potentially more targeted treatment [79]. Although in the current guidelines no explicit recommendation concerning the determination of molecular markers in the cyto-/histopathological specimen is made, further adjustments are to be expected in the future.

Author Contributions: The manuscript was written by Benedikt Schmidbauer, Karin Menhart and Jirka Grosse, the systematic literature research and corrections were done by Dirk Hellwig and Jirka Grosse, Jirka Grosse concieved the manuscript.

Conflicts of Interest: The authors declare no conflict of interest.

Abbreviations

AKT1	RAC-α serine/threonine-protein kinase 1
ATA	American Thyroid Association
CT	computed tomography
DTC	differentiated thyroid carcinoma
EANM	European Association of Nuclear Medicine
ETE	extrathyroidal extension
ETA	European Thyroid Association
FAP	familial adenomatous polyposis
FNA	fine-needle aspiration biopsy
FTC	follicular thyroid carcinoma
FDG-PET	F-18-fluorodeoxy-glucose positron emission tomography
LT4	levothyroxine
MEN2	multiple endocrine neoplasia type 2
PAX8	paired box gene 8
PDTC	poorly differentiated thyroid carcinoma
PIK3CA	phosphatidylinositol 4,5-bisphosphate 3-kinase catalytic subunit α
PPARγ	peroxisome proliferator-activated receptor gamma
PTC	papillary thyroid carcinoma
PTEN	phosphatase and tensin homolog
PTMC	papillary microcarcinoma
RET	RET proto-oncogene
rhTSH	recombinant thyrotropin
RIT	radioiodine therapy
SPECT	single photon emission computed tomography
TERT	telomerase reverse transcriptase
Tg	serum thyroglobulin
TKI	receptor tyrosine kinase inhibitors
TP53	tumor protein p53
TSH	thyroid-stimulating hormone (also known as thyrotropin)
VEGF	vascular endothelial growth factor
WHO	World Health Organization

References

1. Hegedüs, L. Clinical practice. The thyroid nodule. *N. Engl. J. Med.* **2004**, *351*, 1764–1771. [CrossRef] [PubMed]
2. Siegel, R.; Ma, J.; Zou, Z.; Jemal, A. Cancer statistics, 2014. *CA Cancer J. Clin.* **2014**, *64*, 9–29. [CrossRef] [PubMed]
3. Leenhardt, L.; Bernier, M.O.; Boin-Pineau, M.H.; Conte Devolx, B.; Maréchaud, R.; Niccoli-Sire, P.; Nocaudie, M.; Orgiazzi, J.; Schlumberger, M.; Wémeau, J.L.; et al. Advances in diagnostic practices affect thyroid cancer incidence in France. *Eur. J. Endocrinol.* **2004**, *150*, 133–139. [CrossRef] [PubMed]
4. Sherman, S.I. Thyroid carcinoma. *Lancet* **2003**, *361*, 501–511. [CrossRef]
5. Aschebrook-Kilfoy, B.; Schechter, R.B.; Shih, Y.C.; Kaplan, E.L.; Chiu, B.C.; Angelos, P.; Grogan, R.H. The clinical and economic burden of a sustained increase in thyroid cancer incidence. *Cancer Epidemiol. Biomark. Prev.* **2013**, *22*, 1252–1259. [CrossRef] [PubMed]
6. Leenhardt, L.; Erdogan, M.F.; Hegedus, L.; Mandel, S.J.; Paschke, R.; Rago, T.; Russ, G. 2013 European thyroid association guidelines for cervical ultrasound scan and ultrasound-guided techniques in the postoperative management of patients with thyroid cancer. *Eur. Thyroid J.* **2013**, *2*, 147–159. [CrossRef] [PubMed]
7. Silberstein, E.B.; Alavi, A.; Balon, H.R.; Clarke, S.E.; Divgi, C.; Gelfand, M.J.; Goldsmith, S.J.; Jadvar, H.; Marcus, C.S.; Martin, W.H.; et al. The SNMMI practice guideline for therapy of thyroid disease with 131I 3.0. *J. Nucl. Med.* **2012**, *53*, 1633–1651. [CrossRef] [PubMed]
8. Luster, M.; Clarke, S.E.; Dietlein, M.; Lassmann, M.; Lind, P.; Oyen, W.J.; Tennvall, J.; Bombardieri, E.; European Association of Nuclear Medicine (EANM). Guidelines for radioiodine therapy of differentiated thyroid cancer. *Eur. J. Nucl. Med. Mol. Imaging* **2008**, *35*, 1941–1959. [CrossRef] [PubMed]
9. Takami, H.; Ito, Y.; Okamoto, T.; Onoda, N.; Noguchi, H.; Yoshida, A. Revisiting the guidelines issued by the Japanese Society of Thyroid Surgeons and Japan Association of Endocrine Surgeons: A gradual move towards consensus between Japanese and western practice in the management of thyroid carcinoma. *World J. Surg.* **2014**, *38*, 2002–2010. [CrossRef] [PubMed]
10. Haugen, B.R.; Alexander, E.K.; Bible, K.C.; Doherty, G.M.; Mandel, S.J.; Nikiforov, Y.E.; Pacini, F.; Randolph, G.W.; Sawka, A.M.; Schlumberger, M.; et al. 2015 American Thyroid Association Management Guidelines for Adult Patients with Thyroid Nodules and Differentiated Thyroid Cancer: The American Thyroid Association Guidelines Task Force on Thyroid Nodules and Differentiated Thyroid Cancer. *Thyroid* **2016**, *26*, 1–133. [CrossRef] [PubMed]
11. Francis, G.L.; Waguespack, S.G.; Bauer, A.J.; Angelos, P.; Benvenga, S.; Cerutti, J.M.; Dinauer, C.A.; Hamilton, J.; Hay, I.D.; Luster, M.; et al. Management guidelines for children with thyroid nodules and differentiated thyroid cancer. *Thyroid* **2015**, *25*, 716–759. [CrossRef] [PubMed]
12. Brierley, J.D.; Gospodarowicz, M.K.; Wittekind, C. *TNM Classification of Malignant Tumours*, 8th ed.; John Wiley & Sons: Weinheim, Germany, 2017; pp. 69–71.
13. Armin, M.B.; Edge, S.; Greene, F.; Byrd, D.R.; Brookland, R.K.; Washington, M.K.; Gershenwald, J.E.; Compton, C.C.; Hess, K.R.; Sullivan, D.C.; et al. *AJCC Cancer Staging Manual*, 8th ed.; Springer: New York, NY, USA, 2017; pp. 1–19.
14. Dietlein, M.; Eschner, W.; Grünwald, F.; Lassmann, M.; Verburg, F.A.; Luster, M. Procedure guidelines for radioiodine therapy of differentiated thyroid cancer. Version 4. *Nuklearmedizin* **2016**, *55*, 77–89. [CrossRef] [PubMed]
15. Penna, G.C.; Vaisman, F.; Vaisman, M.; Sobrinho-Simões, M.; Soares, P. Molecular Markers Involved in Tumorigenesis of Thyroid Carcinoma: Focus on Aggressive Histotypes. *Cytogenet. Genome Res.* **2016**, *150*, 194–207. [CrossRef] [PubMed]
16. Nikiforov, Y.E.; Ohori, N.P. Papillary Carcinoma. In *Diagnostic Pathology and Molecular Genetics of the Thyroid*, 1st ed.; Nikiforov, Y.E., Biddinger, P.W., Thompson, L.D.R., Eds.; Lippincott: Philadelphia, PA, USA, 2012; pp. 183–262.
17. Seethala, R.R.; Asa, S.L.; Carty, S.E.; Hodak, S.P.; McHugh, J.B.; Richardson, M.S.; Shah, J.; Thompson, L.D.R.; Nikiforov, Y.E. For the Members of the Cancer Committee, College of American Pathologists. Protocol for the Examination of Specimens From Patients With Carcinomas of the Thyroid Gland, Based on AJCC/UICC TNM, 7th edition. Version: Thyroid 3.2.0.0. Available online: http://www.cap.org/ShowProperty?nodePath=/UCMCon/Contribution%20Folders/WebContent/pdf/cp-thyroid-16protocol-3200.pdf (accessed on 2 June 2017).

18. Nikiforova, M.N.; Kimura, E.T.; Gandhi, M.; Biddinger, P.W.; Knauf, J.A.; Basolo, F.; Zhu, Z.; Giannini, R.; Salvatore, G.; Fusco, A.; et al. BRAF mutations in thyroid tumors are restricted to papillary carcinomas and anaplastic or poorly differentiated carcinomas arising from papillary carcinomas. *J. Clin. Endocrinol. Metab.* **2003**, *88*, 5399–5404. [CrossRef] [PubMed]

19. Asioli, S.; Erickson, L.A.; Sebo, T.J.; Zhang, J.; Jin, L.; Thompson, G.B.; Lloyd, R.V. Papillary thyroid carcinoma with prominent hobnail features: A new aggressive variant of moderately differentiated papillary carcinoma. A clinicopathologic, immunohistochemical, and molecular study of eight cases. *Am. J. Surg. Pathol.* **2010**, *34*, 44–52. [CrossRef] [PubMed]

20. Cardis, E.; Howe, G.; Ron, E.; Bebeshko, V.; Bogdanova, T.; Bouville, A.; Carr, Z.; Chumak, V.; Davis, S.; Demidchik, Y.; et al. Cancer consequences of the Chernobyl accident: 20 years on. *J. Radiol. Prot.* **2006**, *26*, 127–140. [CrossRef] [PubMed]

21. Nikiforov, Y.E. Radiation-induced thyroid cancer: What we have learned from Chernobyl. *Endocr. Pathol.* **2006**, *17*, 307–317. [CrossRef] [PubMed]

22. Howitt, B.E.; Jia, Y.; Sholl, L.M.; Barletta, J.A. Molecular alterations in partially-encapsulated or well-circumscribed follicular variant of papillary thyroid carcinoma. *Thyroid* **2013**, *23*, 1256–1262. [CrossRef] [PubMed]

23. Liu, J.; Singh, B.; Tallini, G.; Carlson, D.L.; Katabi, N.; Shaha, A.; Tuttle, R.M.; Ghossein, R.A. Follicular variant of papillary thyroid carcinoma: A clinicopathologic study of a problematic entity. *Cancer* **2006**, *107*, 1255–1264. [CrossRef] [PubMed]

24. Brennan, M.D.; Bergstralh, E.J.; van Heerden, J.A.; McConahey, W.M. Follicular thyroid cancer treated at the Mayo Clinic, 1946 through 1970: Initial manifestations, pathologic findings, therapy, and outcome. *Mayo Clin. Proc.* **1991**, *66*, 11–22. [CrossRef]

25. Collini, P.; Sampietro, G.; Pilotti, S. Extensive vascular invasion is a marker of risk of relapse in encapsulated non-Hürthle cell follicular carcinoma of the thyroid gland: A clinicopathological study of 18 consecutive cases from a single institution with a 11-year median follow-up. *Histopathology* **2004**, *44*, 35–39. [CrossRef] [PubMed]

26. Lang, W.; Choritz, H.; Hundeshagen, H. Risk factors in follicular thyroid carcinomas. A retrospective follow-up study covering a 14-year period with emphasis on morphological findings. *Am. J. Surg. Pathol.* **1986**, *10*, 246–255. [CrossRef] [PubMed]

27. Hundahl, S.A.; Fleming, I.D.; Fremgen, A.M.; Menck, H.R. A National Cancer Data Base report on 53,856 cases of thyroid carcinoma treated in the U.S., 1985–1995. *Cancer* **1998**, *83*, 2638–2648. [CrossRef]

28. Haigh, P.I.; Urbach, D.R. The treatment and prognosis of Hürthle cell follicular thyroid carcinoma compared with its non-Hürthle cell counterpart. *Surgery* **2005**, *138*, 1152–1157. [CrossRef] [PubMed]

29. Shaha, A.R.; Loree, T.R.; Shah, J.P. Prognostic factors and risk group analysis in follicular carcinoma of the thyroid. *Surgery* **1995**, *118*, 1131–1136. [CrossRef]

30. Cetta, F.; Montalto, G.; Gori, M.; Curia, M.C.; Cama, A.; Olschwang, S. Germline mutations of the APC gene in patients with familial adenomatous polyposis-associated thyroid carcinoma: Results from a European cooperative study. *J. Clin. Endocrinol. Metab.* **2000**, *85*, 286–292. [PubMed]

31. Harach, H.R.; Williams, G.T.; Williams, E.D. Familial adenomatous polyposis associated thyroid carcinoma: A distinct type of follicular cell neoplasm. *Histopathology* **1994**, *25*, 549–561. [CrossRef] [PubMed]

32. Ito, Y.; Miyauchi, A.; Ishikawa, H.; Hirokawa, M.; Kudo, T.; Tomoda, C.; Miya, A. Our experience of treatment of cribriform morular variant of papillary thyroid carcinoma; difference in clinicopathological features of FAP-associated and sporadic patients. *Endocr. J.* **2011**, *58*, 685–689. [CrossRef] [PubMed]

33. Hollander, M.C.; Blumenthal, G.M.; Dennis, P.A. PTEN loss in the continuum of common cancers, rare syndromes and mouse models. *Nat. Rev. Cancer* **2011**, *11*, 289–301. [CrossRef] [PubMed]

34. Laury, A.R.; Bongiovanni, M.; Tille, J.C.; Kozakewich, H.; Nosé, V. Thyroid pathology in PTEN-hamartoma tumor syndrome: Characteristic findings of a distinct entity. *Thyroid* **2011**, *21*, 135–144. [CrossRef] [PubMed]

35. Nosé, V. Familial thyroid cancer: A review. *Mod. Pathol.* **2011**, *24*, S19–S33. [CrossRef] [PubMed]

36. Dean, D.S.; Gharib, H. Epidemiology of thyroid nodules. *Best Pract. Res. Clin. Endocrinol. Metab.* **2008**, *22*, 901–911. [CrossRef] [PubMed]

37. Vanderpump, M.P. The epidemiology of thyroid disease. *Br. Med. Bull.* **2011**, *99*, 39–51. [CrossRef] [PubMed]

38. Feldkamp, J.; Führer, D.; Luster, M.; Musholt, T.J.; Spitzweg, C.; Schott, M. Fine Needle Aspiration in the Investigation of Thyroid Nodules. *Dtsch. Arztebl. Int.* **2016**, *113*, 353–359. [PubMed]

39. Dralle, H.; Musholt, T.J.; Schabram, J.; Steinmüller, T.; Frilling, A.; Simon, D.; Goretzki, P.E.; Niederle, B.; Scheuba, C.; Clerici, T.; et al. German Association of Endocrine Surgeons practice guideline for the surgical management of malignant thyroid tumors. *Langenbecks Arch. Surg.* **2013**, *398*, 347–375. [CrossRef] [PubMed]

40. Mazzaferri, E.L.; Jhiang, S.M. Long-term impact of initial surgical and medical therapy on papillary and follicular thyroid cancer. *Am. J. Med.* **1994**, *97*, 418–428. [CrossRef]

41. Samaan, N.A.; Schultz, P.N.; Hickey, R.C.; Goepfert, H.; Haynie, T.P.; Johnston, D.A.; Ordonez, N.G. The results of various modalities of treatment of well differentiated thyroid carcinomas: A retrospective review of 1599 patients. *J. Clin. Endocrinol. Metab.* **1992**, *75*, 714–720. [PubMed]

42. Sawka, A.M.; Thephamongkhol, K.; Brouwers, M.; Thabane, L.; Browman, G.; Gerstein, H.C. Clinical review 170: A systematic review and metaanalysis of the effectiveness of radioactive iodine remnant ablation for well-differentiated thyroid cancer. *J. Clin. Endocrinol. Metab.* **2004**, *89*, 3668–3676. [CrossRef] [PubMed]

43. Sawka, A.M.; Brierley, J.D.; Tsang, R.W.; Thabane, L.; Rotstein, L.; Gafni, A.; Straus, S.; Goldstein, D.P. An updated systematic review and commentary examining the effectiveness of radioactive iodine remnant ablation in well-differentiated thyroid cancer. *Endocrinol. Metab. Clin. N. Am.* **2008**, *37*, 457–480. [CrossRef] [PubMed]

44. Mehanna, H.; Al-Maqbili, T.; Carter, B.; Martin, E.; Campain, N.; Watkinson, J.; McCabe, C.; Boelaert, K.; Franklyn, J.A. Differences in the recurrence and mortality outcomes rates of incidental and nonincidental papillary thyroid microcarcinoma: A systematic review and meta-analysis of 21,329 person-years of follow-up. *J. Clin. Endocrinol. Metab.* **2014**, *99*, 2834–2843. [CrossRef] [PubMed]

45. Perros, P.; Boelaert, K.; Colley, S.; Evans, C.; Evans, R.M.; Gerrard, B.G.; Gilbert, J.; Harrison, B.; Johnson, S.J.; Giles, T.E.; et al. British Thyroid Association. Guidelines for the management of thyroid cancer. *Clin. Endocrinol.* **2014**, *81*, 1–122. [CrossRef] [PubMed]

46. Cooper, D.S.; Doherty, G.M.; Haugen, B.R.; Kloos, R.T.; Lee, S.L.; Mandel, S.J.; Mazzaferri, E.L.; McIver, B.; Sherman, S.I.; Tuttle, R.M.; et al. Management guidelines for patients with thyroid nodules and differentiated thyroid cancer. *Thyroid* **2006**, *16*, 109–142. [CrossRef] [PubMed]

47. Luster, M.; Felbinger, R.; Dietlein, M.; Reiners, C. Thyroid hormone withdrawal in patients with differentiated thyroid carcinoma: A one hundred thirty-patient pilot survey on consequences of hypothyroidism and a pharmacoeconomic comparison to recombinant thyrotropin administration. *Thyroid* **2005**, *15*, 1147–1155. [CrossRef] [PubMed]

48. Pacini, F.; Ladenson, P.W.; Schlumberger, M.; Driedger, A.; Luster, M.; Kloos, R.T.; Sherman, S.; Haugen, B.; Corone, C.; Molinaro, E.; et al. Radioiodine ablation of thyroid remnants after preparation with recombinant human thyrotropin in differentiated thyroid carcinoma: Results of an international, randomized, controlled study. *J. Clin. Endocrinol. Metab.* **2006**, *91*, 926–932. [CrossRef] [PubMed]

49. Ladenson, P.W.; Braverman, L.E.; Mazzaferri, E.L.; Brucker-Davis, F.; Cooper, D.S.; Garber, J.R.; Wondisford, F.E.; Davies, T.F.; DeGroot, L.J.; Daniels, G.H.; et al. Comparison of administration of recombinant human thyrotropin with withdrawal of thyroid hormone for radioactive iodine scanning in patients with thyroid carcinoma. *N. Engl. J. Med.* **1997**, *337*, 888–896. [CrossRef] [PubMed]

50. Luster, M. Acta Oncologica Lecture. Present status of the use of recombinant human TSH in thyroid cancer management. *Acta Oncol.* **2006**, *45*, 1018–1030. [CrossRef] [PubMed]

51. Schlumberger, M.; Ricard, M.; De Pouvourville, G.; Pacini, F. How the availability of recombinant human TSH has changed the management of patients who have thyroid cancer. *Nat. Clin. Pract. Endocrinol. Metab.* **2007**, *3*, 641–650. [CrossRef] [PubMed]

52. Mallick, U.; Harmer, C.; Yap, B.; Wadsley, J.; Clarke, S.; Moss, L.; Nicol, A.; Clark, P.M.; Farnell, K.; McCready, R.; et al. Ablation with low-dose radioiodine and thyrotropin alfa in thyroid cancer. *N. Engl. J. Med.* **2012**, *366*, 1674–1685. [CrossRef] [PubMed]

53. Schlumberger, M.; Catargi, B.; Borget, I.; Deandreis, D.; Zerdoud, S.; Bridji, B.; Bardet, S.; Leenhardt, L.; Bastie, D.; Schvartz, C.; et al. Strategies of radioiodine ablation in patients with low-risk thyroid cancer. *N. Engl. J. Med.* **2012**, *366*, 1663–1673. [CrossRef] [PubMed]

54. Kukulska, A.; Krajewska, J.; Gawkowska-Suwińska, M.; Puch, Z.; Paliczka-Cieslik, E.; Roskosz, J.; Handkiewicz-Junak, D.; Jarzab, M.; Gubała, E.; Jarzab, B. Radioiodine thyroid remnant ablation in patients with differentiated thyroid carcinoma (DTC): Prospective comparison of long-term outcomes of treatment with 30, 60 and 100 mCi. *Thyroid Res.* **2010**, *3*, 9. [CrossRef] [PubMed]

55. Fallahi, B.; Beiki, D.; Takavar, A.; Fard-Esfahani, A.; Gilani, K.A.; Saghari, M.; Eftekhari, M. Low versus high radioiodine dose in postoperative ablation of residual thyroid tissue in patients with differentiated thyroid carcinoma: A large randomized clinical trial. *Nucl. Med. Commun.* **2012**, *33*, 275–282. [CrossRef] [PubMed]

56. Robbins, R.J.; Chon, J.T.; Fleisher, M.; Larson, S.M.; Tuttle, R.M. Is the serum thyroglobulin response to recombinant human thyrotropin sufficient, by itself, to monitor for residual thyroid carcinoma? *J. Clin. Endocrinol. Metab.* **2002**, *87*, 3242–3247. [CrossRef] [PubMed]

57. Rubino, C.; de Vathaire, F.; Dottorini, M.E.; Hall, P.; Schvartz, C.; Couette, J.E.; Dondon, M.G.; Abbas, M.T.; Langlois, C.; Schlumberger, M. Second primary malignancies in thyroid cancer patients. *Br. J. Cancer* **2003**, *89*, 1638–1644. [CrossRef] [PubMed]

58. Wichers, M.; Benz, E.; Palmedo, H.; Biersack, H.J.; Grünwald, F.; Klingmüller, D. Testicular function after radioiodine therapy for thyroid carcinoma. *Eur. J. Nucl. Med.* **2000**, *27*, 503–507. [CrossRef] [PubMed]

59. Benbassat, C.A.; Mechlis-Frish, S.; Hirsch, D. Clinicopathological characteristics and long-term outcome in patients with distant metastases from differentiated thyroid cancer. *World J. Surg.* **2006**, *30*, 1088–1095. [CrossRef] [PubMed]

60. Abonowara, A.; Quraishi, A.; Sapp, J.L.; Alqambar, M.H.; Saric, A.; O'Connell, C.M.; Rajaraman, M.M.; Hart, R.D.; Imran, S.A. Prevalence of atrial fibrillation in patients taking TSH suppression therapy for management of thyroid cancer. *Clin. Investig. Med.* **2012**, *35*, E152–E156. [CrossRef] [PubMed]

61. Tiedje, V.; Schmid, K.W.; Weber, F.; Bockisch, A.; Führer, D. Differentiated thyroid cancer. *Internist* **2015**, *56*, 153–166. [CrossRef] [PubMed]

62. Pacini, F.; Schlumberger, M.; Dralle, H.; Elisei, R.; Smit, J.W.; Wiersinga, W.; European Thyroid Cancer Taskforce. European consensus for the management of patients with differentiated thyroid carcinoma of the follicular epithelium. *Eur. J. Endocrinol.* **2006**, *154*, 787–803. [CrossRef] [PubMed]

63. Tuttle, R.M.; Tala, H.; Shah, J.; Leboeuf, R.; Ghossein, R.; Gonen, M.; Brokhin, M.; Omry, G.; Fagin, J.A.; Shaha, A. Estimating risk of recurrence in differentiated thyroid cancer after total thyroidectomy and radioactive iodine remnant ablation: Using response to therapy variables to modify the initial risk estimates predicted by the new American Thyroid Association staging system. *Thyroid* **2010**, *20*, 1341–1349. [PubMed]

64. Durante, C.; Haddy, N.; Baudin, E.; Leboulleux, S.; Hartl, D.; Travagli, J.P.; Caillou, B.; Ricard, M.; Lumbroso, J.D.; De Vathaire, F.; et al. Long-term outcome of 444 patients with distant metastases from papillary and follicular thyroid carcinoma: Benefits and limits of radioiodine therapy. *J. Clin. Endocrinol. Metab.* **2006**, *91*, 2892–2899. [CrossRef] [PubMed]

65. Robbins, R.J.; Wan, Q.; Grewal, R.K.; Reibke, R.; Gonen, M.; Strauss, H.W.; Tuttle, R.M.; Drucker, W.; Larson, S.M. Real-time prognosis for metastatic thyroid carcinoma based on 2-[18F]fluoro-2-deoxy-D-glucose-positron emission tomography scanning. *J. Clin. Endocrinol. Metab.* **2006**, *91*, 498–505. [CrossRef] [PubMed]

66. Sherman, S.I. Early clinical studies of novel therapies for thyroid cancers. *Endocrinol. Metab. Clin. N. Am.* **2008**, *37*, 511–524. [CrossRef] [PubMed]

67. Fassnacht, M.; Kreissl, M.C.; Weismann, D.; Allolio, B. New targets and therapeutic approaches for endocrine malignancies. *Pharmacol. Ther.* **2009**, *123*, 117–141. [CrossRef] [PubMed]

68. Lorusso, L.; Pieruzzi, L.; Biagini, A.; Sabini, E.; Valerio, L.; Giani, C.; Passannanti, P.; Pontillo-Contillo, B.; Battaglia, V.; Mazzeo, S.; et al. Lenvatinib and other tyrosine kinase inhibitors for the treatment of radioiodine refractory, advanced, and progressive thyroid cancer. *Onco Targets Ther.* **2016**, *9*, 6467–6477. [CrossRef] [PubMed]

69. Laursen, R.; Wehland, M.; Kopp, S.; Pietsch, J.; Infanger, M.; Grosse, J.; Grimm, D. Effects and Role of Multikinase Inhibitors in Thyroid Cancer. *Curr. Pharm. Des.* **2016**, *22*, 5915–5926. [CrossRef] [PubMed]

70. Pitoia, F.; Jerkovich, F. Selective use of sorafenib in the treatment of thyroid cancer. *Drug Des. Dev. Ther.* **2016**, *10*, 1119–1131. [CrossRef] [PubMed]

71. Schlumberger, M.; Tahara, M.; Wirth, L.J.; Robinson, B.; Brose, M.S.; Elisei, R.; Habra, M.A.; Newbold, K.; Shah, M.H.; Hoff, A.O.; et al. Lenvatinib versus placebo in radioiodine-refractory thyroid cancer. *N. Engl. J. Med.* **2015**, *372*, 621–630. [CrossRef] [PubMed]

72. Ancker, O.V.; Wehland, M.; Bauer, J.; Infanger, M.; Grimm, D. The Adverse Effect of Hypertension in the Treatment of Thyroid Cancer with Multi-Kinase Inhibitors. *Int. J. Mol. Sci.* **2017**, *18*, 625. [CrossRef] [PubMed]

73. Werner, R.A.; Lückerath, K.; Schmid, J.S.; Higuchi, T.; Kreissl, M.C.; Grelle, I.; Reiners, C.; Buck, A.K.; Lapa, C. Thyroglobulin fluctuations in patients with iodine-refractory differentiated thyroid carcinoma on lenvatinib treatment—initial experience. *Sci. Rep.* **2016**, *6*, 28081. [CrossRef] [PubMed]

74. Kreissl, M.C.; Fassnacht, M.; Mueller, S.P. Systemic treatment of advanced differentiated and medullary thyroid cancer. Overview and practical aspects. *Nuklearmedizin* **2015**, *54*, 88–93. [PubMed]

75. Machens, A.; Holzhausen, H.J.; Dralle, H. The prognostic value of primary tumor size in papillary and follicular thyroid carcinoma. *Cancer* **2005**, *103*, 2269–2273. [CrossRef] [PubMed]

76. Gallicchio, R.; Giacomobono, S.; Capacchione, D.; Nardelli, A.; Barbato, F.; Nappi, A.; Pellegrino, T.; Storto, G. Should patients with remnants from thyroid microcarcinoma really not be treated with iodine-131 ablation? *Endocrine* **2013**, *44*, 426–433. [CrossRef] [PubMed]

77. Avram, A.M.; Fig, L.M.; Frey, K.A.; Gross, M.D.; Wong, K.K. Preablation 131-I scans with SPECT/CT in postoperative thyroid cancer patients: What is the impact on staging? *J. Clin. Endocrinol. Metab.* **2013**, *98*, 1163–1171. [CrossRef] [PubMed]

78. Creach, K.M.; Siegel, B.A.; Nussenbaum, B.; Grigsby, P.W. Radioactive iodine therapy decreases recurrence in thyroid papillary microcarcinoma. *ISRN Endocrinol.* **2012**, *2012*, 816386. [CrossRef] [PubMed]

79. Cancer Genome Atlas Research Network. Integrated genomic characterization of papillary thyroid carcinoma. *Cell* **2014**, *159*, 676–690.

International Journal of
Molecular Sciences

MDPI

Review

The Adverse Effect of Hypertension in the Treatment of Thyroid Cancer with Multi-Kinase Inhibitors

Ole Vincent Ancker [1], Markus Wehland [2], Johann Bauer [3], Manfred Infanger [2] and Daniela Grimm [1,2,*

[1]   Department of Biomedicine, Aarhus University, Wilhelm Meyers Allé 4, 8000 Aarhus C, Denmark;
      ole.vincent.ancker@post.au.dk
[2]   Clinic and Policlinic for Plastic, Aesthetic and Hand Surgery, Otto von Guericke University, Leipziger Str. 44,
      39120 Magdeburg, Germany; markus.wehland@med.ovgu.de (M.W.); manfred.infanger@med.ovgu.de (M.I.)
[3]   Max-Planck-Institute for Biochemistry, Am Klopferspitz 18, 82152 Martinsried, Germany;
      jbauer@biochem.mpg.de
*   Correspondence: dgg@biomed.au.dk; Tel.: +45-8716-7693; Fax: +45-8612-8804

Academic Editor: Harry A. J. Struijker-Boudier
Received: 16 January 2017; Accepted: 9 March 2017; Published: 14 March 2017

Abstract: The treatment of thyroid cancer has promising prospects, mostly through the use of surgical or radioactive iodine therapy. However, some thyroid cancers, such as progressive radioactive iodine-refractory differentiated thyroid carcinoma, are not remediable with conventional types of treatment. In these cases, a treatment regimen with multi-kinase inhibitors is advisable. Unfortunately, clinical trials have shown a large number of patients, treated with multi-kinase inhibitors, being adversely affected by hypertension. This means that treatment of thyroid cancer with multi-kinase inhibitors prolongs progression-free and overall survival of patients, but a large number of patients experience hypertension as an adverse effect of the treatment. Whether the prolonged lifetime is sufficient to develop sequelae from hypertension is unclear, but late-stage cancer patients often have additional diseases, which can be complicated by the presence of hypertension. Since the exact mechanisms of the rise of hypertension in these patients are still unknown, the only available strategy is treating the symptoms. More studies determining the pathogenesis of hypertension as a side effect to cancer treatment as well as outcomes of dose management of cancer drugs are necessary to improve future therapy options for hypertension as an adverse effect to cancer therapy with multi-kinase inhibitors.

Keywords: thyroid cancer; hypertension; vascular endothelial growth factor; multi-kinase inhibitors; lenvatinib; sorafenib; sunitinib

1. Introduction

The most common and effective strategies to treat thyroid cancer are surgery, radioactive iodine (RAI) therapy and thyroid-stimulating hormone (TSH) suppression treatment. This therapy regimen shows good results in patients affected by differentiated thyroid carcinoma (DTC) as well as a long-term survival rate of up to 90% [1]. The therapy options for de-differentiated thyroid cancers or for recurrent thyroid cancer are extremely limited. Poorly differentiated thyroid cancer types (PDTC) do not respond to RAI treatment and have a remarkably reduced survival rate. Under these circumstances, multi-kinase inhibitors, such as lenvatinib, sorafenib and sunitinib, may be useful. The multi-kinase inhibitors target an important step in the development of tumors. When a tumor reaches a critical level in its development, oxygen must be delivered through blood vessels and not simply by diffusion. At this point, the tumor produces new blood vessels and thereby obtains the required oxygen and nutrition to grow. The multi-kinase inhibitors work anti-angiogenically by preventing the transmission of signals from multiple tyrosine kinases, which are essential for the development of a new vasculature [2].

Along with their effects as cancer drugs, multi-kinase inhibitors have been shown to cause several unwanted side effects; examples are proteinuria, stomatitis, diarrhea and hypertension, the latter of which had been observed in up to half of the treated patients [3].

Hypertension, or elevated blood pressure, is a physical condition in which the pressure in the blood vessels is persistently raised and the heart must labor against higher systolic and/or higher diastolic pressure. Hypertension exists per definition when the systolic blood pressure (SBP) equals or exceeds 140 mmHg and/or the diastolic pressure (DBP) equals or exceeds 90 mmHg, whereas normal blood pressure is defined as 120 mmHg systolic and 80 mmHg diastolic [4].

Hypertension can physically be described by Ohm's law:

$$blood\ pressure = cardiac\ output \times total\ periphery\ resistance$$

Isolated hypertension, when not extremely elevated, is not dangerous and many people live with raised blood pressure without even being aware of it. However, hypertension can have severe impacts on overall health, numerous studies have shown that patients with hypertension have a higher risk of cardiovascular and renal diseases [5].

The aim of this review is to create an overview of hypertension as an adverse effect (AE) of multi-kinase inhibitors when treating metastatic RAI-refractory thyroid cancer. In addition, this review will focus on the function of multi-kinase inhibitors, and on the mechanisms of the development of hypertension. It will reflect the importance of hypertension as an AE.

This review will consider and address the following questions: (1) How do multi-kinase inhibitors cause hypertension? (2) How can we manage hypertension induced by tyrosine kinase inhibitor (TKI)-treatment? (3) Is the relationship between the efficacy of cancer treatment and the AE of hypertension favorable? (4) Is hypertension as a side effect of the multi-kinase inhibitors a severe concern?

2. Background

2.1. Thyroid Cancer

The thyroid gland is located in front of the tracheal tube. The function of the thyroid gland is to produce the thyroid hormones T3 and T4, which stimulate a great number of processes in the human body, such as metabolic rate, protein synthesis, development, and they also influence the cardiovascular system. Furthermore, the thyroid produces calcitonin, which plays a role in calcium homeostasis. The thyroid gland can be enlarged both by benign and malignant causes: it is often enlarged due to a dietary iodine deficiency that is not cancer associated (struma), but other tumors of the thyroid are caused by malignant alterations [6]. Thyroid cancer can be classified into several categories: differentiated (DTC), covering papillary (PTC) and follicular (FTC), medullary (MTC) and anaplastic thyroid cancer (ATC). The cancer cells in DTC appear similar to normal thyroid cells, whereas poorly differentiated thyroid cancer (PDTC) is comprised of cancer cells that do not share the same characteristics or abilities as normal thyroid cells [7].

PDTC is accountable for up to 10% of thyroid cancer forms and is more aggressive than DTC. Unfortunately, the prognosis for PDTC is not encouraging: the 5-, 10- and 15-year survival rates are 50%, 34% and 0%, respectively [8]. MTC has a 10-year survival rate of 75%–80% [9], whereas the 5-year survival rate for DTC is as high as 98% due to successful treatment, such as surgery, RAI ablation and treatment with thyroid stimulating hormone [10]. Regrettably, about 20% of patients experience recurrence of the disease. The recurrent form of thyroid cancer is poorly differentiated and thereby more malignant; this makes it for example resistant to RAI ablation because due to an inability to take up iodine [11].

The incidence of thyroid cancer in Denmark in 2012 was estimated to be 220 new cases, both sexes included, which places it in the intermediary group below the most common cancer types as colorectal, breast and prostate cancer [12]. In 2012, the incidence worldwide was 298,102 new cases for both sexes representing 2.1% of all cancers [13].

2.2. Multi-Kinase Inhibitors

Most thyroid cancer types have promising prospects as a result of surgical treatment and radioactive iodine ablation. PDTC, RAI-refractory carcinomas and tumors showing resistance to various forms of available treatment must be treated with alternatives in the hope of a good result [14].

Angiogenesis is the formation of new blood vessels from pre-existing vasculature and is a normal physiological process that starts during fetal development and persists in adults during inflammation and vascular or wound healing [15]. Angiogenesis is utilized by tumors to create new blood supply, so forming a path for the delivery of oxygen and nutrients, and in turn supporting tumor growth. Several factors play a role in the creation of new blood vessels, both in normal physiology and in pathophysiology. The growth of a tumor is determined by its nutrient supply. By diffusion alone, only very limited amounts can reach the tumor, and especially its core, so that for continued expansion, an increased internal blood supply is necessary. The central hypothesis is that an increase in tumor size must be preceded by expansion of tumor vasculature, which is stimulated by the tumor. Tumor cells take advantage of the normal physiological process involving the secretion of vascular endothelial growth factor (VEGF). VEGF is of high importance in the induction of new vessel formation and in the survival of endothelial cells [16–18]. Therefore, angiogenesis is a critical process for the development and subsequent growth of tumors.

The superfamily of VEGF comprises VEGF-A, VEGF-B, VEGF-C, VEGF-D and placental growth factor (PGF). Their corresponding tyrosine kinase receptors, the vascular endothelial growth factor receptors (VEGFRs) VEGFR-1, -2 and -3, are distinguished according to their affinities to different VEGFs. VEGFR-1 and -2 are found in endothelial cells, while VEGFR-3 is expressed in lymphatic endothelial cells. VEGFR-1 binds VEGF-A, VEGF-B and PGF, whereas VEGFR-2 binds VEGF-A and proteolytically modified VEGF-C and VEGF-D. Finally, VEGFR-3 is activated by VEGF-C and VEGF-D [19–21].

In addition, growth factors platelet derived growth factor (PDGF), epidermal growth factor (EGF), and fibroblast growth factor (FGF) also play a significant role in angiogenesis [19,20]. The ligand FGF and its receptor FGFR play an important role in cell growth, proliferation, differentiation and survival of thyroid cancer cells, where FGFR-1, -3 and -4 are overexpressed and expression of FGFR-2 is reduced. Binding of a growth factor to one of the receptors leads to a tyrosine kinase activation of either the mitogen activated protein kinase (MAPK) or the phosphatidylinositol-3-kinase (PI3K) pathway that eventually affect oncogenic gene expression [22].

There are two main pathways by which VEGF signaling can be interfered pharmacologically. Direct inhibition of VEGF is one possibility: an immunoglobulin designed specifically for VEGF targets and binds before the interaction with the corresponding receptor [23]. Another pathway focuses on inhibition of the phosphorylation cascade triggered after the binding of ligand and receptor by blocking the signal from the tyrosine receptor and thereby preventing the oncogenic features such as angiogenesis, proliferation and growth. Lenvatinib, sorafenib and sunitinib (Table 1) are drugs used in cancer therapy that inhibit multiple tyrosine kinases in thyroid cancer treatment [22].

Table 1. Characteristics of lenvatinib, sorafenib, and sunitinib.

Drug	Targets	Half-Life	Bioavailability	Metabolism
Lenvatinib	VEGF-R1-3, FGFR1-4, PDGF-RA, c-KIT, RET	28 h	85%	Hepatic CYP3A4
Sorafenib	VEGF-R1-3, PDGF-RA-D, C-RAF, B-RAF	25–48 h	38%–49%	Hepatic CYP3A4
Sunitinib	VEGF-R1-3, PDGF-RA-D, c-KIT, RET, CD114, CD135	40–60 h	50%	Hepatic CYP3A4

Lenvatinib is an orally taken TKI that targets VEGF-R1/-3, FGFR1-4, ret proto-oncogene (RET), and platelet derived growth factor receptor (PDGFRβ). By blocking these receptors, lenvatinib disturbs angiogenesis of the tumor, invasion of tissue and metastasis. By inhibiting VEGF-R1/-3, lenvatinib disrupts angiogenic processes in the tumor. Furthermore, lenvatinib inhibits RET, which is important for controlling tumor growth and, by hitting FGFR and PDGFRβ, it also influences the tumor's microenvironment, as presented in Figure 1 [22].

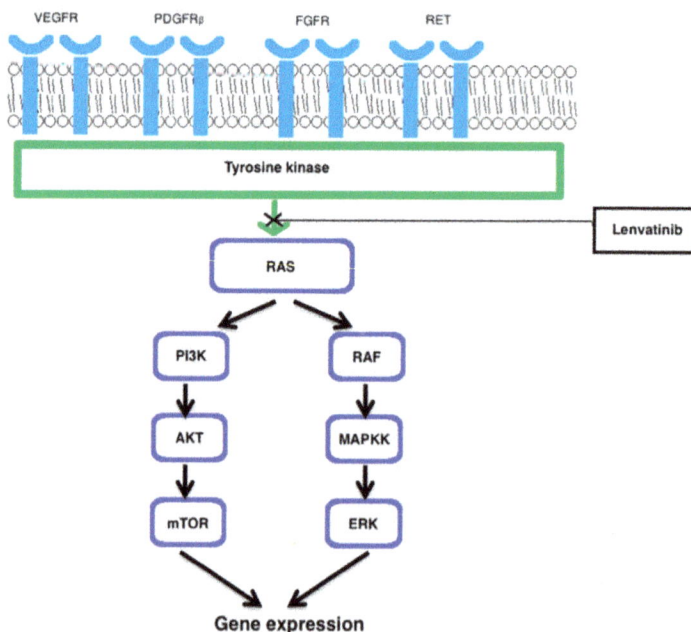

Figure 1. Lenvatinib inhibits signaling from VEGFR, PDGFRβ, FGFR and RET. It decreases angiogenesis and lymphogenesis, stunts tumor growth and damages the tumor's microenvironment [11,22]. VEGFR (vascular endothelial growth factor receptor), PDGFR (platelet derived growth factor receptor), FGFR (fibroblast growth factor receptor), RET (rearranged during transfection), RAS (rat sarcoma), PI3K (phosphatidylinositol-3-kinase), AKT (protein kinase B), mTOR (mammalian target of rapamycin), RAF (rapidly accelerated fibrosarcoma kinase), MAPKK (mitogen activated protein kinase kinase), ERK (extracellular signal regulated kinase).

Sunitinib (Sunitinib malate; Sutent; Pfizer, New York, USA) is a multi-targeted TKI used in the treatment of metastatic renal cell carcinomas (RCC) and gastrointestinal stromal tumors, and is under evaluation for other malignancies [11]. It inhibits VEGFR-1 and -2, platelet-derived growth factor receptors, stem cell factor receptor (c-KIT), FLT3 as well as RET kinases. Sunitinib acts on VEGF receptors and on RET and is therefore a suitable drug to treat RAI-refractory thyroid cancer. Sunitinib is still not approved for thyroid cancer therapy by the FDA, and therefore therapy approaches applying sunitinib are still off-label [11].

Sorafenib (NEXAVAR®) is widely used in cancer therapy and is an oral serine-threonine TKI that targets VEGFR-1/−3, PDGFR, BRAF, RET/PTC, and c-kit. The agent has an anti-proliferative effect and an anti-angiogenic activity by blocking the intracellular signal transduction of VEGFR2 in endothelial cells [11]. Sorafenib is a FDA-approved drug for patients with RAI-refractory metastatic thyroid cancer. The approval of sorafenib is based on the results of the randomized DECISION (stuDy of sorafEnib in loCally advanced or metastatIc patientS with radioactive Iodine refractory thyrOid caNcer) trial.

This was an international, multi-center, placebo-controlled study involving 417 thyroid cancer patients. The patients (400 mg oral sorafenib twice daily) lived on average nearly 11 months longer without disease progression compared to the placebo group [24–26].

2.3. Hypertension

The effects of VEGF are not only present in cancer cells: healthy endothelial cells also express VEGFRs and, because of this property, unwanted consequences may appear [20,27,28]. Hypertension is the most common adverse effect in the treatment of the tyrosine kinase inhibitors. VEGF is known to regulate the vasomotor tonus and maintains blood pressure by dilating small arterioles and venules. In case of an anti-VEGF therapy the result is a reduced density of microvessels (Figure 2). Hypertension has been reported to occur at a higher incidence in patients with DTC and treated with sorafenib [29]. Hypertension usually occurs in the first six weeks of treatment with sorafenib; therefore, blood pressure (BP) should be monitored regularly (at least once a week) at the start of sorafenib therapy [30].

In the SELECT trial (ClinicalTrials.gov number, NCT01321554), lenvatinib, compared to placebo, revealed significant improvements in progression-free survival and the response rate in patients with RAI-refractory thyroid cancer, but it also induced more adverse effects [31]. Treatment-related adverse effects (TEAE) of any grade, occurring in more than 40% of lenvatinib-treated patients were hypertension (in 67.8% of the patients), diarrhea (in 59.4%), fatigue or asthenia (in 59.0%), decreased appetite (in 50.2%), decreased weight (in 46.4%), and nausea (in 41.0%) [31]. Cabanillas et al. [32] investigated 58 patients with advanced, progressive, RAI-refractory DTC, receiving lenvatinib 24 mg once daily in 28-day cycles until disease progression, unmanageable toxicity, withdrawal, or death. TEAE were evaluated: 44 patients had hypertension (all grades 76%) and six patients had grade 3 TEAE (10%). Most patients with hypertension and proteinuria were managed successfully without lenvatinib dose adjustments [32].

There is no unanimous agreement on how these cancer drugs result in hypertension, but some hypotheses have gained a footing in explaining why. One explanation depends on reduced production of the vasodilator, nitric oxide (NO). Blockage of VEGF induces vasoconstriction. VEGFR-2 signaling generates nitric oxide (NO) and prostaglandin I2, which induces endothelial cell-dependent vasodilatation in arterioles and venules. Inhibition of VEGFR-2 signaling reduces NO synthase expression and NO synthesis. Normally, activation of VEGFR-2, by VEGF or shear stress in the vessel walls, induces the PI3K pathway, resulting in an increased production of the vasodilator NO and hence a reduction of peripheral resistance and blood pressure. VEGF inhibition results in an increase in vascular resistance, followed by hypertension. The multi-kinase inhibitors prevent the phosphorylation cascade and thus the formation of NO, leading to a rise in blood pressure [33]. In addition, an increase in blood pressure may also result from the VEGF/VEGFR inhibition in the kidney. VEGF and VEGFR are also expressed in podocytes. Electron microscopy images revealed glomerular lesions associated with VEGF-targeted therapies [34] The reduced VEGF activity influences renal endothelial cells and podocytes and results in a dysregulation of VEGF expression and a downregulation of tight junction proteins with the consequence of proteinuria. As an alternative explanation to the theory of the lack of NO, an increased amount of the vasoconstrictor endothelin-1 (ET-1) has been suggested. ET-1 binds to its receptors in endothelial cells causing the smooth muscle cells to contract and thus increase resistance in the vessels and raise the blood pressure [35,36].

A study has considered a third theory that suggests hypertension is due to a decrease in the number of small arterioles and capillaries, leading to higher peripheral resistance and thereby to increased blood pressure. The authors found that patients treated with anti-angiogenic medicaments showed fewer mucosal capillaries. However, the study was not able to determine whether the observed effects were a consequence of a direct lack of small arterioles or simply a hypo-perfusion of these, since the technique used for recording the numbers of vessels depended on perfusion. Both a decreased number of arterioles or a stopping of perfusion of existing ones could explain a rise in blood pressure, since the blood is distributed in fewer vessels, increasing resistance inside them [37].

Some studies have shown a rapid increase in blood pressure, which challenges the understanding of the mechanisms giving rise to hypertension, and argues against a structural or anatomical explanation of acute induced hypertension. However, rapid rises of hypertension make a theory of active vasomodulators more likely [38]. Risk factors for hypertension occurring upon TKI therapy are of older age, obesity, high sodium intake, alcohol abuse, smoking or reduced physical activity. A pre-existing high blood pressure or certain VEGF polymorphisms might be associated with a lower risk of grade 3 or 4 hypertension.

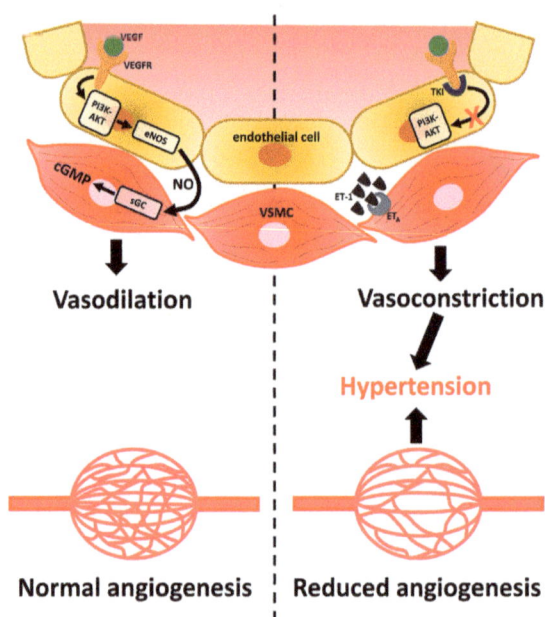

Figure 2. The effects of VEGF on blood pressure and capillary vascularization under: physiological conditions (**left**); and TKI therapy (**right**) (adapted from [39]). VEGF (vascular endothelial growth factor), VEGFR (vascular endothelial growth factor receptor), PI3K (phosphatidylinositol-3-kinase), AKT (protein kinase B), eNOS (endothelial nitric oxide synthase), sGC (soluble guanylyl cyclase), ET-1 (endothelin-1), ET_A (endothelin receptor type A), NO (nitric oxide), cGMP (cyclic guanosine monophosphate).

2.4. Efficacy of Cancer Drug Treatment

A recent study [31] has monitored the efficacy of lenvatinib, compared with placebo, in adults with DTC, RAI-refractory cancer and without prior treatment with a multi-kinase inhibitor. The primary endpoints were either progression of the cancer disease or death. Patients treated with lenvatinib had a median progression-free survival of 18.3 months, whereas placebo-treated patients had a progression-free survival of only 3.6 months. The six month progression-free survival rate for the patients treated with lenvatinib was 77.5%, while the rate for placebo-treated patients was 25.4%. Patients receiving lenvatinib also had more AE. Treatment-related AE of any grade occurred in more than 40% of patients in the lenvatinib group; for example, hypertension (in 67.8% of the patients), diarrhea (in 59.4%), and others [31].

Another study published in December 2015 showed similar results: Japanese patients with DTC, RAI-refractory disease and a progression in disease in the last 13 months were eligible. Patients treated with lenvatinib experienced a median progression-free survival of 16.5 months, while the placebo group had a progression-free survival of 3.7 months. The six-month progression-free survival rate was 70.0% for lenvatinib-treated patients and 31.7% for placebo-treated patients [40]. The most common AE (any grade) in Japanese patients from the SELECT trial was hypertension (86.7%).

Taken together, the findings of the SELECT trial provided the basis for the FDA approval of lenvatinib for the treatment of progressive RAI-refractory thyroid cancer. Another study investigated 59 patients with unresectable progressive MTC [41]. They received lenvatinib (24 mg daily, 28-day cycles) until disease progression, unmanageable toxicity, withdrawal, or death. Lenvatinib had a high objective response rate, a high disease control rate, and a short median time to response. Toxicities were managed with dose modifications and medications. Most hypertension and proteinuria events were grade 1 or 2 and managed with standard drug therapy. The median duration of treatment was 264 days (range, 13–547 days). Withdrawal from therapy due to hypertension: one patient (2%).

Hypertension is a condition that affects the cardiovascular system in the body. It is an abstract condition because not all patients have the same tolerance of raised pressure. Some may experience serious cardiovascular outcomes, while others can live their whole life with hypertension without showing any symptoms. Studies show that hypertension is associated with many different cardiovascular diseases, where almost all have a high mortality rate, such as intracerebral hemorrhage, subarachnoid hemorrhage, stable angina pectoris, myocardial infarction, aorta aneurysm and heart failure [42].

Most patients with hypertension as an adverse effect continue the treatment of cancer. In a lenvatinib-versus-placebo study, only 1.1% of patients had to stop the treatment because of hypertensive effects and 19.9% were given a lower dose because of a rise in BP [31]. Nearly all patients with this form of hypertension manage the symptoms with normal anti-hypertensive drugs. Patients with unmanageable hypertension and signs of organ damage, renal dysfunction or cardiovascular diseases need intervention in the form of either a lower dose or a total stop of treatment [38]. Hypertension, and other AE from TKI-treatment are currently under investigation. Several clinical trials are investigating this concern, both examining the direct AE of lenvatinib, but also whether lower doses of the drug can show the same effect while giving fewer unwanted effects. An overview of recent studies is given in Table 2.

Table 2. Overview of recent clinical trials studying lenvatinib, sunitinib, and sorafenib, website used on 7 March 2017 [43].

Title	Design	Objective	Status
A phase I trial of lenvatinib (multi-kinase inhibitor) and capecitabine (Antimetabolite) in patients with advanced malignancies. NCT02915172	Interventional open label	This phase I study aims to find the highest tolerable dose of lenvatinib and Capecitabine that can be given to patients with advanced cancer.	Not yet recruiting
Post-marketing surveillance of lenvatinib mesylate in patients with unresectable thyroid cancer. NCT02430714	Observational cohort prospective	The objective of this study is to find unknown adverse reactions, adverse drug reactions, efficacy, safety and effectiveness factors, incidence of hypertension, hemorrhagic, and thromboembolic effects and liver disorder.	Recruiting
A multi center, randomized, double-blind phase II trial of lenvatinib (E7080) in subjects with iodine-131 refractory differentiated thyroid cancer (RR-DTC) to evaluate whether an oral starting dose of 20 mg or 14 mg daily will provide comparable efficacy to a 24 mg starting dose, but have a better safety profile. NCT02702388	Interventional double blind randomized	This randomized double-blinded study aims to investigate whether a lower starting dose of lenvatinib can provide comparable efficacy whilst showing a better safety profile for the patients.	Active Not recruiting

Table 2. *Cont.*

Title	Design	Objective	Status
An open label phase I dose escalation study of E7080 administered to patients with solid tumors. NCT00280397	Interventional open label	This study investigates the maximum tolerable dose and the related effects of E7080 (lenvatinib) given to patients with solid tumors with no successful treatment.	Completed
A phase II study of E7080 in subjects with advanced thyroid cancer. NCT01728623	Interventional open label	This study was performed to evaluate the safety, efficacy and pharmacokinetics of E7080 (lenvatinib), taken orally daily in patients with advanced thyroid cancer.	Completed
An open label phase I dose escalation study of E7080. NCT00121719	Interventional open label	This study aims to find the maximum tolerated dose of lenvatinib in patients with solid tumors or lymphomas.	Active Not recruiting
Phase II, multi-center, open-label, single arm trial to evaluate the safety and efficacy of oral E7080 in medullary and iodine-131 refractory, unresectable differentiated thyroid cancers, stratified by histology. NCT00784303	Interventional open label non-randomized	This is a phase II study that aimed to investigate the safety and efficacy of oral E7080 (lenvatinib) in medullary and iodine-131 refractory, unresectable differentiated thyroid cancer.	Completed
Phase II study assessing the efficacy and safety of lenvatinib for anaplastic thyroid cancer (HOPE). NCT02726503	Interventional open label	This phase II study aims to investigate the efficacy and safety of lenvatinib for unresectable anaplastic thyroid cancer.	Recruiting
A multi-center, randomised, double-blind, placebo-controlled, phase III trial of lenvatinib (E7080) in I-131-refractory differentiated thyroid cancer in China. NCT02966093	Interventional double blind randomized	This phase III study primarily aims to compare progression-free survival of participants with radioiodine refractory differentiated thyroid cancer treated with lenvatinib or placebo, and secondarily to investigate adverse events.	Not yet recruiting
Post-marketing surveillance of Lenvima in Korean patients. NCT02764554	Observational prospective	This study aims to observe the safety profile of lenvatinib (Lenvima) in normal clinical practice.	Recruiting
Prospective, non-interventional, post-authorization safety study that includes all patients diagnosed as unresectable differentiated thyroid carcinoma and treated with sorafenib (JPMS-DTC). NCT02185560	Observational	Safety study that includes all patients diagnosed as unresectable differentiated thyroid carcinoma (DTC) and treated with sorafenib within a certain period.	Recruiting
Safety and efficacy of sorafenib in patients with advanced thyroid cancer: a Phase II clinical study. NCT02084732	Interventional	Describe the clinical activity and safety profile of sorafenib in the treatment of patients with advanced thyroid cancer (metastatic or recurrent) among a selected group of patients refractory to or ineligible to radioactive iodine (RAI) therapy.	Recruiting
Prospective, non-interventional, post-authorization safety study that includes all patients diagnosed as unresectable differentiated thyroid carcinoma and treated with sorafenib (JPMS-DTC). NCT02185560	Observational	This is a non-interventional, multi center post-authorization safety study that includes all patients diagnosed as unresectable differentiated thyroid carcinoma (DTC) and treated with sorafenib within a certain period.	Recruiting

Table 2. *Cont.*

Title	Design	Objective	Status
Thyroid cancer and sunitinib (THYSU). NCT00510640	Interventional	The objective of the trial is to determine the objective tumor response rate (efficacy) in patients with locally advanced or metastatic anaplastic, differentiated or medullary thyroid carcinoma treated with sunitinib; a secondary objective is to evaluate the safety of sunitinib in these patients	Completed
Sutent adjunctive treatment of differentiated thyroid cancer (IIT Sutent). NCT00668811	Interventional	The primary objective is to assess clinical benefit rate, defined as complete response, partial response, or stable disease per RECIST criteria. The secondary objective will be to assess the safety of Sutent in this patient population.	Completed

The study NCT02915172 includes both a phase I and a phase II study. The phase I study aims to decide the highest tolerable dose of lenvatinib and capecitabine in patients with advanced cancer. The initial dose for the first included group is 10 mg orally taken lenvatinib in a 21-day cycle. The phase II study wishes to determine whether the maximal tolerated dose found in the phase I study can be used for treating patients with advanced cancer including thyroid cancer. NCT02915172 takes place at the MD Anderson Cancer Center, Texas, USA, which is also the responsible party. The study will include 46 participants over the age of 18. The study was set to start December 2016 and the final data collection date is set to December 2022.

NCT02430714 and NCT02764554 are post-marketing surveillance studies of lenvatinib in patients with unresectable thyroid cancer. In NCT02430714, a dose of 24 mg orally taken once daily is administered to the patients and, in NCT02764554, the participants are in groups with 4 mg and 10 mg orally taken lenvatinib. The primary outcome is to decide the number of AE in the following year. NCT02430714 is conducted by the Japanese pharmaceutical company Eisai Co., Ltd., Japan, recruiting 400 participants including children, adults and seniors from centers in Osaka and Tokyo, Japan. The study was set to start May 2015 and the final data collection date is March 2025. NCT02764554 is performed by the Korean pharmaceutical company Eisai Korea Inc. The estimated number of enrolled patients is 3000 including children, adults and seniors recruiting from Seoul, Republic of Korea. The start date of the surveillance study was September 2016 with a final data collection date of July 2021.

The study NCT02702388 investigates whether a lower dose, than the currently approved dose of 24 mg orally taken lenvatinib once daily, shows a comparable effect but a better safety profile in patients with RAI-refractory DTC. The two experimental starting doses are 20 mg and 14 mg reducing down to a dose of 4 mg. NCT02702388 includes 41 participants over the age of 18. The start date was set to March 2016 and the final data collection date is May 2019. The 78 centers of the study are located in Australia, Austria, Belgium, Denmark, France, Germany, Italy, Philippines, Poland, Portugal, Romania, Spain, Sweden, United Kingdom and United States of America. The responsible party of the study is Eisai Inc.

The aim of the studies NCT00280397 (solid tumors) and NCT00121719 (solid tumors) is to investigate the maximal tolerated dose of lenvatinib and furthermore, to provide a summary of AE and the pharmacokinetic properties of lenvatinib. NCT00280397 started in January 2006 and had its final data collection date in September 2008. The study consisted of 27 participants with an age span from 20 to 75 years. The study was performed by Eisai Inc. with a center in Tokyo, Japan. The study found the maximal tolerated dose of lenvatinib to be 13 mg, when orally taken two times daily two-weeks-on/one-week-off [44]. NCT00121719 began in July 2005 with a final data collection date in June 2009. Eighty-seven participants over the age of 18 were enrolled in the study. The study was conducted by Eisai Inc. with centers in Amsterdam, Netherlands and Glasgow, United Kingdom. The study found the maximal tolerated dose of lenvatinib to be 25 mg orally taken once daily.

The studies NCT01728623, NCT02726503, NCT00784303, and NCT02966093 intend to determine the overall- and progression-free survival rates of participants treated with 24 mg lenvatinib orally taken once daily, moreover the studies aim to investigate the AE of the drug. NCT02726503 also studies the efficacy on ATC (Phase II) and NCT00784303 furthermore investigates the tumor response rate and the pharmacokinetic profile of lenvatinib. NCT01728623 started September 2012 and had its last data collection date in July 2015. In this study 37 participants over the age of 20 years were enrolled from three centers in Japan. The responsible party of the study was Eisai Inc. NCT02726503 started in January 2016 and has its last data collection date in July 2018. The study consists of 39 participants over the age of 20 years recruiting from 12 Japanese hospitals. The responsible party of this study is the Translational Research Informatics Center, Kobe, Hyogo, Japan. NCT00784303 started in August 2009 and had its last data collection date in April 2011. One hundred sixteen participants over the age of 18 were enrolled from 51 centers in the United States, Australia, France, Germany, Italy, Poland and the United Kingdom. The responsible party is Eisai Inc. NCT02966093 is set to start in January 2017 and has its final data collection date in January 2020. The estimated number of participants is 150 over the age of 18. Recruiting from 13 centers in China. Eisai Co., Ltd. is the responsible party of this study.

Study NCT02185560 aims to analyze patients of all ages and sexes, suffering from unresectable differentiated thyroid carcinoma and receiving sorafenib treatment. Standard follow-up length will be 9 months (or until lost to follow-up). For patients with a possible follow-up of 24 months, additional data, such as survival status and keratoacanthoma and/or squamous cell cancer development will be collected. Primary end points are the number of participants with adverse and serious adverse drug reactions as well as the number of participants with serious AE as a measure of safety and tolerability. This study is conducted by Bayer Clinical Trials.

The Phase 2 study NCT02084732, conducted by the Instituto Nacional de Cancerologia, Columbia, is also directed towards an assessment of the safety and efficacy of sorafenib in the treatment of patients with advanced thyroid cancer. It is currently in the recruiting stage and is open for adult patients of both sexes. The follow-up period is 24 months, and besides its primary objective of determining the clinical activity and safety profile of sorafenib in the treatment of patients with advanced thyroid cancer, its secondary aims are the measurement of PFS and the description of AE associated with sorafenib used in advanced thyroid cancer.

Study NCT02185560 is similar in design and scope to NCT02185560 discussed above, just substituting sorafenib with sunitinib. Bayer Clinical Trials is the responsible party.

Study NCT00510640 has been conducted by the University Hospital Bordeaux, France under collaboration with Pfizer. Patients suffering from locally advanced or metastatic anaplastic, differentiated or medullary thyroid carcinoma receive 50 mg/day sunitinib for four weeks followed by a rest-period of two weeks. Cycles are repeated until disease progression or severe toxicity. Based on tolerability, doses can be reduced to 12.5 mg/day. Adult patients of both sexes are eligible. The primary outcome measure of this study is the objective response rate (ORR) (the proportion of patients with confirmed complete (CR) or partial response (PR) according to the RECIST), secondary outcome measures are the safety of sunitinib in patients with thyroid carcinoma and time to disease progression, response, and duration of response. The study is completed, and no results are published so far.

In the study NCT00668811 a total of 23 patients with advanced differentiated thyroid cancer received 37.5 mg/day sunitinib for treatment cycles of 28 days. Adult patients with advanced DTC were eligible and of the 23 subjects, 74% (17) were male. The primary objective is to assess clinical benefit rate, defined as complete response, partial response, or stable disease and the secondary objective is to assess the safety of sunitinib in this patient population. Overall, this study found, that sunitinib was relatively well tolerated and had good anti-tumor activity for this kind of thyroid cancer [45]. The responsible party for this study was the Washington Hospital Center in collaboration with the company Pfizer.

For an overview of adverse effects observed in larger clinical trials with lenvatinib, sofrafenib, and sunitinib, see Table 3.

Table 3. Overview of serious AE observed in clinical trials investigating TKI treatment; website used on 7 March 2017 [43].

Clinical Trial Title ID	Dose (mg/day)	# of Patients	Most Frequent Serious Adverse Effects	
SELECT: A multi center, randomized, double-blind, placebo-controlled, phase 3 trial of Lenvatinib (E7080) in ^{131}I-refractory differentiated thyroid cancer. NCT01321554 [31,40]	24 Per os (PO)	261	4%	Pneumonia
			3%	Hypertension
			3%	Dehydration
			2%	Physical health deterioration
			2%	Renal failure
			2%	Pulmonary embolism
			2%	Sepsis
An open label phase I dose escalation study of E7080 (solid tumors). NCT00121719	0.1–32 PO	93	5%	Abdominal pain
			5%	Vomiting
			4%	Hypertension
			3%	Physical health deterioration
			3%	Pyrexia
Sorafenib in treating patients with advanced anaplastic thyroid cancer. NCT00126568 [46]	2 × 400 PO	20	15%	Disease progression
			10%	Death
			10%	Dyspnea
			5%	Thrombosis
			5%	Pulmonary disorders
Nexavar® versus placebo in locally advanced/metastatic RAI-refractory differentiated thyroid cancer. NCT00984282 [24–26]	2 × 400 PO	207	5%	Secondary malignancy
			4%	Dyspnea
			4%	Musculoskeletal disorders
			3%	Pleural effusion
			2%	Fever
A continuation study using sunitinib malate for patients leaving treatment on a previous sunitinib study. NCT00428220	37.5 PO	223	5%	Disease progression
			4%	Abdominal pain
			3%	Vomiting
			3%	Diarrhea
			2%	Physical health deterioration
			2%	Pyrexia
			2%	Anemia
Sutent adjunctive treatment of differentiated thyroid cancer (IIT Sutent). NCT00668811 [45]	37.5 PO	23	13%	Hypertension
			13%	Leukopenia
			9%	Hand-foot syndrome
			9%	Anorexia
			4%	Neutropenia
			4%	Lymphopenia
			4%	Thrombocytopenia
			4%	Nausea
			4%	Gastrointestinal bleeding

2.5. Management of Multi-Kinase Inhibitor-Caused Hypertension

It is important to measure baseline BP in order to better determine cardiovascular risk factors after administering treatment. If the BP < 120/80 mmHg, TKI therapy can begin [47]. BP monitoring under TKI therapy should be performed every week for the first eight weeks, and before any infusion or cycle. When patients show prehypertension values (120 < SBP < 140 mmHg and 80 < DBP < 90 mmHg), it is important to search for cardiovascular risk factors. If no risk factors are present, TKI treatment can begin and BP should be controlled accordingly. In cases when cardiovascular risk factors are detectable, an antihypertensive therapy, for example, calcium channel blockers (CCB), should be started 3–7 days before the start of TKI therapy.

Nebivolol is a β-adrenoceptor antagonist (BAA) whose antihypertensive effect is mainly related to a reduction in peripheral resistance; therefore, this is a good candidate for treating the TKI-induced rise in BP [48]. However, it is important to consider each patient individually, as there is no golden standard for treatment of TKI-induced hypertension and because every patient is diverse in his or her anamnesis. Some patients may take medicine that affects the metabolism of the antihypertensive or the anti-angiogenic drug and an individual plan must be made for each patient. For example, the multi-kinase inhibitor sunitinib affects the CYP3A4 enzyme, which is of high importance in metabolizing many drugs. This property of sunitinib is important to keep in mind when planning an antihypertensive strategy [49].

Hypertension should be monitored and treated according to standard medical practice, i.e., it should be managed as in patients without cancer. For a patient without comorbidities, the target blood pressure should be <140/90 mmHg. For patients with chronic kidney disease, the target BP is <135/85 mmHg, and for patients with proteinuria, drugs inhibiting the renin-angiotensin system should be applied. The anti-angiogenic therapy should be continued without dose reduction unless severe or persistent hypertension is present [47].

It is difficult to determine which is the best antihypertensive drug because there has been a lack of controlled clinical studies until now. In general, there are several classes of antihypertensive drugs, which can be applied. When a patient exerts signs of proteinuria, angiotensin converting enzyme inhibitors (ACEi) or angiotensin 2 receptor antagonists can be given. In addition, calcium channel blockers such as amlodipine can be considered. Non-dihydropyridine drugs such as verapamil or diltiazem should be avoided because of CYP3A4 metabolism. They are contraindicated in combination with oral angiogenesis inhibitors. In addition, medicaments increasing NO, like nitrates, or the BAA nebivolol can be applied. Common treatments of hypertension involve BAA, CCB, diuretics, ACEi, angiotensin-II receptor antagonists (AIIA), and NO donors [49]. The current treatment of TKI-induced hypertension uses therapeutic options that are already established and is summarized in Table 4. Studies have shown that dihydropyridine CCBs, such as amlodipine, have a great effect on relaxing smooth muscle cells and thereby lowering the total periphery resistance and hence the blood pressure. ACEi and AIIA have also shown good results in the antihypertensive treatment and concurrent renal-protecting effects in these treated cancer patients [39,50].

Finally, based on the proposed mechanisms regarding the lack of NO production, therapeutic considerations may include NO derivatives. Agents that increase NO bioavailability, such as long-acting nitrates are interesting. By giving long lasting precursors of NO, the desired effect is to neutralize or minimize the multi-kinase inhibitor-induced hypertension. In general, NO donors are known to be well tolerated and cause relatively harmless adverse effects such as headache, flushing, nausea and hypotension. Therefore, long lasting NO donors are suggested as a first line treatment [51].

An alternative and newer way to manage TKI-induced hypertension is the application of endothelin receptor antagonists, because it is assumed that ET-1 concentration increases in patients treated with multi-kinase inhibitors. This kind of hypertensive treatment is a novel regimen to handle the AE. Studies have shown that the combination of two or three of the suggested treatments may have a beneficial effect on TKI-induced hypertension [39,52].

Table 4. Different antihypertensive drugs for the management of TKI-induced hypertension.

Class	Drug	Dose	Recommendation
CCB Dihydropyridines	Amlodipine	2.5–10 mg/day	Great potency for reducing arterial smooth muscle cell contractility [39], effective therapy [49].
ACEi	Enalapril	Start with 5–20 mg/12–24 h, then max 40 mg/12–24 h	Particularly indicated in the setting of proteinuria [39], effective [49].
	Ramipril	Start with 2.5 mg/day, then 5 mg/day after 2 weeks, after another 2 weeks max 10 mg/day	
ARB	Losartan	50–100 mg/day	Particularly indicated in the setting of proteinuria [39], effective [49].
	Valsartan	80–320 mg/day	
	Irbesartan	150–300 mg/day	
BBA	Nebivolol	2.5–5 mg/day	Indicated for DTC; begin therapy of hypertension with a BBA [53].
Diuretics/Thiazides	Hydrochlorothiazide	Start with 12.5–25 mg/day, then 12.5 mg/day	Less-effective than CCB, ACEi or ARB [39], but often used [54].
Nitrate derivates	Long-acting nitrates: Isosorbide dinitrate (ISDN) or Isosorbide mononitrate (ISMN)	40–60 mg/day	Adequate response in hypertension refractory to ACEi and CCB [51].
α-blockers	Prazosin	2–20 mg/day	Used as additional therapy if BP is not sufficiently controlled.

CCB, calcium channel blockers; ACEi, Angiotensin converting enzyme inhibitors; ARB, angiotensin II receptor blockers; BBA, β-adrenoceptor antagonists; d, day.

2.6. Discussion

Hypertension as an AE of multi-kinase inhibitor treatment is an important and frequent concern over TKI treatment. As summarized in the previous sections, numerous AE are occurring with multi-kinase inhibitor treatment of cancer. Grade 1 and grade 2 hypertension, experienced by patients treated with TKI, can frequently be managed without the need for dose reduction or discontinuation of treatment. Hypertension is linked with several cardiovascular and renal diseases, and higher mortality, which make it crucial to investigate [54–59].

Not all patients who meet the requirements of the definition of hypertension are affected. Some will never feel any change in their state of health or experience any pathogenic outcomes as a result of hypertension. Meanwhile, other groups of patients suffer from the impact of hypertension consisting of either cardiovascular events or even death. In particular, patients already suffering from risk factors, such as diabetes mellitus, cardiovascular conditions, chronic renal disease or are weakened in other ways are affected [60].

The patients treated with multi-kinase inhibitors, such as lenvatinib and sorafenib, are already in a late stage of their cancer illness and it must be assumed that this group of patients is, in some way, weakened or have sequelae from either the illness itself or from previous treatment. When considering whether hypertension as an AE is critical, it is important to be aware of the individual patient's anamnesis and general state of health. If the patient shows no previous signs of, for example, cardiovascular events, hypertension usually takes some time to affect the well-being of the patient or to induce life-threatening conditions.

Currently, treatment with multi-kinase inhibitors prolongs progression-free survival by about 12 to 16 months, from a starting point of a couple of months to half a year. Whether this time is enough to develop life-threatening conditions arising from unwanted effects such as hypertension is questionable, but must be considered when evaluating AE. Some above-mentioned studies [31,32] showed that patients either had to stop treatment entirely or lower the dose because of an alarming rise in BP. Other studies suggested the presence of hypertension as a biomarker for a good response to the anti-angiogenic treatment and therefore that hypertension may be necessary, but must also be treated [1,20,32,61]. Similarly, the sunitinib-associated rise in BP, as well as neutropenia, is discussed as biomarkers in metastatic renal cell carcinoma patients: both side effects were associated with longer progression-free survival and a higher overall-survival rate [20]. At the moment, various possible anti-angiogenic biomarkers are under examination, such as hypertension, altered VEGF plasma levels, interleukin (IL)-8 polymorphisms, or a change in tumor microvessel density. Today, promising candidates are detected, but important challenges limit their translation into practice.

Since there is still no generally accepted explanation for the cause of hypertension as an AE in multi-kinase inhibition, a specific therapy to manage this form of hypertension is not given. Various strategies of treating hypertension are suggested, some with good results, but these treatments are only symptomatic and also have accompanying side effects. Therefore, it is important to perform more studies investigating the mechanisms of hypertension induced by TKI, as well as to know the risk factors and the frequency of hypertension induced by anti-angiogenic drugs in different cancer types.

3. Conclusions

The multi-kinase inhibition by targeted therapy offers new strategies of treating cancer diseases that otherwise are considered untreatable. Unfortunately, many AE follow this kind of treatment, especially hypertension.

Hypertension is a well-known systemic AE of treatment with VEGF-inhibitors. Treatment induced-hypertension has been associated with sunitinib therapy for different forms of cancer. The TKIs are included in international clinical guidelines as first-line and second-line therapy in metastatic renal cell carcinoma (mRCC). Hypertension is an adverse effect of these drugs and the degree of hypertension associates with the anti-tumor effect in mRCC [62]. More recent phase II trials have shown a significant risk of treatment-induced hypertension with sunitinib in patients suffering from pancreatic neuroendocrine tumors [63] and endometrial carcinoma [64].

In addition, the incidence of treatment-associated hypertension with sorafenib was increased in patients with hepatocellular carcinoma [65] and non-small cell lung cancer [50]. Besides to locally recurrent or metastatic, progressive, RAI-refractory differentiated thyroid cancer, lenvatinib is indicated for the treatment of patients with advanced renal cell carcinoma in combination with everolimus following prior anti-angiogenic therapy. The frequency of hypertension, which is a known class effect of VEGF-targeting agents, was increased in both treatment groups in which lenvatinib was administered [66].

Therapy with multi-kinase inhibitors leads to a relatively large increase in the progression-free survival of patients with late stage metastatic thyroid cancer and not all patients are directly affected by the rise in BP. For some groups of patients with diseases other than thyroid cancer, either simultaneously or previously, hypertension is a great concern that can be dangerous for the individual. Currently, biomarkers for the prediction of the effectiveness of anti-angiogenic therapy are being investigated [67] and in this course, it might be helpful to analyze possible predictors of AE such as hypertension, too. Management of hypertension is somewhat possible, but not with a curative result, and is thereby only symptomatic with the antihypertensive drugs listed in Table 4. Hence, it is important to focus on different regimens for managing high blood pressure, or even to use other strategies to specifically target the cancer cells without causing systemic adverse events.

In summary, the relationship between angiogenic inhibitors and a rise in BP has now been established: angiogenic inhibitors used to treat cancer may exacerbate cardiac risk factors. Introduction or even prophylactic use of antihypertensive drugs can allow maintenance of therapy despite the onset of hypertension. In addition to cancer therapy, the reduction of hypertension risk factors should be addressed.

4. Outlook

Several studies have already shown hypertension and other unwanted events in response to multi-kinase inhibition treatment. Future and on-going studies have adjusted their aims to consider the AE of this type of treatment, since many studies with these aims are now registered. It is of great interest and import to examine how to reduce or eliminate AE by dose-finding studies or by determining the mechanisms inducing hypertension and thus being able to properly treat this effect.

5. Materials and Methods

Literature and information used for this review can be found using online databases, such as Pubmed, Scopus and clinicaltrials.gov by using the search terms "multi-kinase inhibitors", "antiangiogenesis", "multi-kinase inhibitors and hypertension", "thyroid cancer" and others.

Acknowledgments: This review was the bachelor thesis of Ole Vincent Ancker. The authors would like to thank Petra Wise, USC, Los Angeles, CA, USA and Proof-Reading-Service.com, Devonshire Business Center, Works Road, Letchworth Garden City, SG6 1GJ, Hertfordshire, United Kingdom for reviewing the article for language and grammatical errors.

Author Contributions: Ole Vincent Ancker, Markus Wehland and Daniela Grimm wrote the review; Johann Bauer searched and checked the references; Manfred Infanger and Johann Bauer supported the paper; and Markus Wehland, and Ole Vincent Ancker contributed with drafting the figures and tables.

Conflicts of Interest: The authors have declared no conflicts of interest.

Abbreviations

AIIA	angiotensin II receptor antagonists
ACEi	angiotensin converting enzyme inhibitors
AE	adverse effects
AKT	protein kinase B
ATC	anaplastic thyroid cancer
CCB	calcium channel blockers
cGMP	cyclic guanosine monophosphate
DTC	differentiated thyroid cancer
EGF	endothelial growth factor
eNOS	endothelial nitric oxide synthase
ERK	extracellular signal regulated kinase
ET-1	endothelin-1
ET_A	endothelin receptor type A
FGF	fibroblast growth factor
FGFR	fibroblast growth factor receptor
FTC	follicular thyroid cancer
MAPKK	mitogen activated protein kinase kinase
MTC	medullary thyroid cancer
NO	nitric oxide
PDGF	platelet derived growth factor
PDGFR	platelet derived growth factor receptor
PI3K	phosphatidylinositol-3-kinase
PDTC	poorly differentiated thyroid cancer
PO	per os
PTC	papillary thyroid cancer
RAI	radioactive iodine
RAS	rat sarcoma
RET	ret proto-oncogene

sGC	soluble guanylyl cyclase
TKI	tyrosine kinase inhibitor
VEGF	vascular endothelial growth factor
VEGFR	vascular endothelial growth factor receptor

References

1. Costa, R.; Carneiro, B.A.; Chandra, S.; Pai, S.G.; Chae, Y.K.; Kaplan, J.B.; Garrett, H.B.; Agulnik, M.; Kopp, P.A.; Giles1, F.J. Spotlight on lenvatinib in the treatment of thyroid cancer: Patient selection and perspectives. *Drug Des. Dev. Ther.* **2016**, *10*, 873–884. [CrossRef] [PubMed]
2. O'Neill, C.J.; Oucharek, J.; Learoyd, D.; Sidhu, S.B. Standard and emerging therapies for metastatic differentiated thyroid cancer. *Oncologist* **2010**, *15*, 146–156. [CrossRef] [PubMed]
3. Cabanillas, M.E.; Habra, M.A. Lenvatinib: Role in thyroid cancer and other solid tumors. *Cancer Treat. Rev.* **2016**, *42*, 47–55. [CrossRef] [PubMed]
4. Mancia, G.; Fagard, R.; Narkiewicz, K.; Redon, J.; Zanchetti, A.; Bohm, M.; Christiaens, T.; Cifkova, R.; De Backer, G.; Dominiczak, A.; et al. 2013 ESH/ESC guidelines for the management of arterial hypertension: The Task Force for the Management of Arterial Hypertension of the European Society of Hypertension (ESH) and of the European Society of Cardiology (ESC). *Eur. Heart J.* **2013**, *34*, 2159–2219. [CrossRef] [PubMed]
5. He, J.; Whelton, P.K. Elevated systolic blood pressure and risk of cardiovascular and renal disease: Overview of evidence from observational epidemiologic studies and randomized controlled trials. *Am. Heart J.* **1999**, *138*, 211–219. [CrossRef]
6. Thyroid Cancer 2016 Updated 31 March 2016. Available online: http://www.cancer.org/cancer/thyroidcancer/detailedguide/thyroid-cancer-what-is-thyroid-cancer (accessed on 27 December 2016).
7. Tiedje, V.; Schmid, K.W.; Weber, F.; Bockisch, A.; Fuhrer, D. Differentiated thyroid cancer. *Internist* **2015**, *56*, 153–166. [CrossRef] [PubMed]
8. Patel, K.N.; Shaha, A.R. Poorly differentiated thyroid cancer. *Curr. Opin. Otolaryngol. Head Neck Surg.* **2014**, *22*, 121–126. [CrossRef] [PubMed]
9. Roy, M.; Chen, H.; Sippel, R.S. Current understanding and management of medullary thyroid cancer. *Oncologist* **2013**, *18*, 1093–1100. [CrossRef] [PubMed]
10. Krook, K.A.; Fedewa, S.A.; Chen, A.Y. Prognostic indicators in well-differentiated thyroid carcinoma when controlling for stage and treatment. *Laryngoscope* **2015**, *125*, 1021–1027. [CrossRef] [PubMed]
11. Laursen, R.; Wehland, M.; Kopp, S.; Pietsch, J.; Infanger, M.; Grosse, J.; Grimm, D. Effects and Role of Multi-kinase Inhibitors in Thyroid Cancer. *Curr. Pharm. Des.* **2016**, *22*, 5915–5926. [CrossRef] [PubMed]
12. Ferlay, J.; Steliarova-Foucher, E.; Lortet-Tieulent, J.; Rosso, S.; Coebergh, J.W.; Comber, H.; Forman, D.; Bray, F. Cancer incidence and mortality patterns in Europe: Estimates for 40 countries in 2012. *Eur. J. Cancer* **2013**, *49*, 1374–1403. [CrossRef] [PubMed]
13. GLOBOCAN 2012: Estimated Cancer Incidence, Mortality and Prevalence Worldwide in 2012. Available online: http://globocan.iarc.fr/Pages/fact_sheets_population.aspx (accessed on 31 December 2016).
14. Anderson, R.T.; Linnehan, J.E.; Tongbram, V.; Keating, K.; Wirth, L.J. Clinical, safety, and economic evidence in radioactive iodine-refractory differentiated thyroid cancer: A systematic literature review. *Thyroid* **2013**, *23*, 392–407. [CrossRef] [PubMed]
15. Dvorak, H.F. Angiogenesis: Update 2005. *J. Thromb. Haemost.* **2005**, *3*, 1835–1842. [CrossRef] [PubMed]
16. Infanger, M.; Shakibaei, M.; Kossmehl, P.; Hollenberg, S.M.; Grosse, J.; Faramarzi, S.; Schulze-Tanzil, G.; Paul, M.; Grimm, D. Intraluminal application of vascular endothelial growth factor enhances healing of microvascular anastomosis in a rat model. *J. Vasc. Res.* **2005**, *42*, 202–213. [CrossRef] [PubMed]
17. Infanger, M.; Kossmehl, P.; Shakibaei, M.; Baatout, S.; Witzing, A.; Grosse, J.; Bauer, J.; Cogoli, A.; Faramarzi, S.; Derradji, H.; et al. Induction of three-dimensional assembly and increase in apoptosis of human endothelial cells by simulated microgravity: Impact of vascular endothelial growth factor. *Apoptosis* **2006**, *11*, 749–764. [CrossRef] [PubMed]
18. Infanger, M.; Grosse, J.; Westphal, K.; Leder, A.; Ulbrich, C.; Paul, M.; Grimm, D. Vascular endothelial growth factor induces extracellular matrix proteins and osteopontin in the umbilical artery. *Ann. Vasc. Surg.* **2008**, *22*, 273–284. [CrossRef] [PubMed]

19. Ferrara, N. Vascular endothelial growth factor: Basic science and clinical progress. *Endocr. Rev.* **2004**, *25*, 581–611. [CrossRef] [PubMed]

20. Frandsen, S.; Kopp, S.; Wehland, M.; Pietsch, J.; Infanger, M.; Grimm, D. Latest Results for Anti-Angiogenic Drugs in Cancer Treatment. *Curr. Pharm. Des.* **2016**, *22*, 5927–5942. [CrossRef] [PubMed]

21. Kowanetz, M.; Ferrara, N. Vascular endothelial growth factor signaling pathways: Therapeutic perspective. *Clin. Cancer Res.* **2006**, *12*, 5018–5022. [CrossRef] [PubMed]

22. Stjepanovic, N.; Capdevila, J. Multi-kinase inhibitors in the treatment of thyroid cancer: Specific role of lenvatinib. *Biologics* **2014**, *8*, 129–139. [PubMed]

23. Hayman, S.R.; Leung, N.; Grande, J.P.; Garovic, V.D. VEGF inhibition, hypertension, and renal toxicity. *Curr. Oncol. Rep.* **2012**, *14*, 285–294. [CrossRef] [PubMed]

24. Worden, F.; Fassnacht, M.; Shi, Y.; Hadjieva, T.; Bonichon, F.; Gao, M.; Fugazzola, L.; Ando, Y.; Hasegawa, Y.; Park do, J.; et al. Safety and tolerability of sorafenib in patients with radioiodine-refractory thyroid cancer. *Endocr. Relat. Cancer* **2015**, *22*, 877–887. [CrossRef] [PubMed]

25. Brose, M.S.; Nutting, C.M.; Jarzab, B.; Elisei, R.; Siena, S.; Bastholt, L.; de la Fouchardiere, C.; Pacini, F.; Paschke, R.; Shong, Y.K.; et al. Sorafenib in radioactive iodine-refractory, locally advanced or metastatic differentiated thyroid cancer: A randomised, double-blind, phase 3 trial. *Lancet* **2014**, *384*, 319–328. [CrossRef]

26. Brose, M.S.; Nutting, C.M.; Sherman, S.I.; Shong, Y.K.; Smit, J.W.; Reike, G.; Chung, J.; Kalmus, J.; Kappeler, C.; Schlumberger, M. Rationale and design of decision: A double-blind, randomized, placebo-controlled phase III trial evaluating the efficacy and safety of sorafenib in patients with locally advanced or metastatic radioactive iodine (RAI)-refractory, differentiated thyroid cancer. *BMC Cancer* **2011**, *11*, 349.

27. Ma, X.; Wehland, M.; Schulz, H.; Saar, K.; Hübner, N.; Infanger, M.; Bauer, J.; Grimm, D. Genomic approach to identify factors that drive the formation of three-dimensional structures by EA.hy926 endothelial cells. *PLoS ONE* **2013**, *8*, e64402. [CrossRef] [PubMed]

28. Kristensen, T.B.; Knutsson, M.L.; Wehland, M.; Laursen, B.E.; Grimm, D.; Warnke, E.; Magnusson, N.E. Anti-vascular endothelial growth factor therapy in breast cancer. *Int. J. Mol. Sci.* **2014**, *15*, 23024–23041. [CrossRef] [PubMed]

29. Gupta-Abramson, V.; Troxel, A.B.; Nellore, A.; Puttaswamy, K.; Redlinger, M.; Ransone, K.; Mandel, S.J.; Flaherty, K.T.; Loevner, L.A.; O'Dwyer, P.J.; et al. Phase II trial of sorafenib in advanced thyroid cancer. *J. Clin. Oncol.* **2008**, *26*, 4714–4719. [CrossRef] [PubMed]

30. Nexavar (Sorafenib). *Tablets Prescribing Information*; Bayer Health Care Pharmaceuticals, Inc.: Wayne, NJ, USA, 2012.

31. Schlumberger, M.; Tahara, M.; Wirth, L.J.; Robinson, B.; Brose, M.S.; Elisei, R.; Habra, M.A.; Newbold, K.; Shah, M.H.; Hoff, A.O.; et al. Lenvatinib versus placebo in radioiodine-refractory thyroid cancer. *N. Engl. J. Med.* **2015**, *372*, 621–630. [CrossRef] [PubMed]

32. Cabanillas, M.E.; Schlumberger, M.; Jarzab, B.; Martins, R.G.; Pacini, F.; Robinson, B.; McCaffrey, J.C.; Shah, M.H.; Bodenner, D.L.; Topliss, D.; et al. A phase 2 trial of lenvatinib (E7080) in advanced, progressive, radioiodine-refractory, differentiated thyroid cancer: A clinical outcomes and biomarker assessment. *Cancer* **2015**, *121*, 2749–2756. [CrossRef] [PubMed]

33. Bair, S.M.; Choueiri, T.K.; Moslehi, J. Cardiovascular complications associated with novel angiogenesis inhibitors: Emerging evidence and evolving pers Phase I dose-escalation study and biomarker analysis of E7080 in patients with advanced solid tumors pectives. *Trends. Cardiovasc. Med.* **2013**, *23*, 104–113. [CrossRef] [PubMed]

34. Ollero, M.; Sahali, D. Inhibition of the VEGF signalling pathway and glomerular disorders. *Nephrol. Dial. Transplant.* **2015**, *30*, 1449–1455. [CrossRef] [PubMed]

35. Kappers, M.H.; de Beer, V.J.; Zhou, Z.; Danser, A.H.; Sleijfer, S.; Duncker, D.J.; van den Meiracker, A.H.; Merkus, D. Sunitinib-induced systemic vasoconstriction in swine is endothelin mediated and does not involve nitric oxide or oxidative stress. *Hypertension* **2012**, *59*, 151–157. [CrossRef] [PubMed]

36. Kappers, M.H.; van Esch, J.H.; Sluiter, W.; Sleijfer, S.; Danser, A.H.; van den Meiracker, A.H. Hypertension induced by the tyrosine kinase inhibitor sunitinib is associated with increased circulating endothelin-1 levels. *Hypertension* **2010**, *56*, 675–681. [CrossRef] [PubMed]

37. Steeghs, N.; Gelderblom, H.; Roodt, J.O.; Christensen, O.; Rajagopalan, P.; Hovens, M.; Putter, H.; Rabelink, T.J.; de Koning, E. Hypertension and rarefaction during treatment with telatinib, a small molecule angiogenesis inhibitor. *Clin. Cancer Res.* **2008**, *14*, 3470–3476. [CrossRef] [PubMed]

38. Eskens, F.A.; Verweij, J. The clinical toxicity profile of vascular endothelial growth factor (VEGF) and vascular endothelial growth factor receptor (VEGFR) targeting angiogenesis inhibitors: A review. *Eur. J. Cancer* **2006**, *42*, 3127–3139. [CrossRef] [PubMed]

39. De Jesus-Gonzalez, N.; Robinson, E.; Moslehi, J.; Humphreys, B.D. Management of antiangiogenic therapy-induced hypertension. *Hypertension* **2012**, *60*, 607–615. [CrossRef] [PubMed]

40. Kiyota, N.; Schlumberger, M.; Muro, K.; Ando, Y.; Takahashi, S.; Kawai, Y.; Wirth, L.; Robinson, B.; Sherman, S.; Suzuki, T.; et al. Subgroup analysis of Japanese patients in a phase 3 study of lenvatinib in radioiodine-refractory differentiated thyroid cancer. *Cancer Sci.* **2015**, *106*, 1714–1721. [CrossRef] [PubMed]

41. Schlumberger, M.; Jarzab, B.; Cabanillas, M.E.; Robinson, B.; Pacini, F.; Ball, D.W.; McCaffrey, J.; Newbold, K.; Allison, R.; Martins, R.G.; et al. A Phase II Trial of the Multitargeted Tyrosine Kinase Inhibitor Lenvatinib (E7080) in Advanced Medullary Thyroid Cancer. *Clin Cancer Res.* **2016**, *22*, 44–53. [CrossRef] [PubMed]

42. Rapsomaniki, E.; Timmis, A.; George, J.; Pujades-Rodriguez, M.; Shah, A.D.; Denaxas, S.; White, I.R.; Caulfield, M.J.; Deanfield, J.E.; et al. Blood pressure and incidence of twelve cardiovascular diseases: Lifetime risks, healthy life-years lost, and age-specific associations in 1.25 million people. *Lancet* **2014**, *383*, 1899–1911. [CrossRef]

43. ClinicalTrials.gov. Available online: http://www.clinicaltrials.gov (accessed on 7 March 2017).

44. Yamada, K.; Yamamoto, N.; Yamada, Y.; Nokihara, H.; Fujiwara, Y.; Hirata, T.; Koizumi, F.; Nishio, K.; Koyama, N.; Tamura, T. Phase I dose-escalation study and biomarker analysis of E7080 in patients with advanced solid tumors. *Clin. Cancer Res.* **2011**, *17*, 2528–2537. [CrossRef] [PubMed]

45. Bikas, A.; Kundra, P.; Desale, S.; Mete, M.; O'Keefe, K.; Clark, B.G.; Gandhi, R.; Barett, C.; Jelinek, J.S.; Wexler, J.A.; et al. Phase 2 clinical trial of sunitinib as adjunctive treatment in patients with advanced differentiated thyroid cancer. *Eur. J. Endocrinol.* **2016**, *174*, 373–380. [CrossRef] [PubMed]

46. Savvides, P.; Nagaiah, G.; Lavertu, P.; Fu, P.; Wright, J.J.; Chapman, R.; Wasman, J.; Dowlati, A.; Remick, S.C. Phase II trial of sorafenib in patients with advanced anaplastic carcinoma of the thyroid. *Thyroid* **2013**, *23*, 600–604. [CrossRef] [PubMed]

47. Izzedine, H.; Ederhy, S.; Goldwasser, F.; Soria, J.C.; Milano, G.; Cohen, A.; Khayat, D.; Spano, J.P. Management of hypertension in angiogenesis inhibitor-treated patients. *Ann. Oncol.* **2009**, *20*, 807–815. [CrossRef] [PubMed]

48. Porta, C.; Paglino, C.; Imarisio, I.; Bonomi, L. Uncovering Pandora's vase: The growing problem of new toxicities from novel anticancer agents. The case of sorafenib and sunitinib. *Clin. Exp. Med.* **2007**, *7*, 127–134. [CrossRef] [PubMed]

49. Leon-Mateos, L.; Mosquera, J.; Anton Aparicio, L. Treatment of sunitinib-induced hypertension in solid tumor by nitric oxide donors. *Redox. Biol.* **2015**, *6*, 421–425. [CrossRef] [PubMed]

50. Wasserstrum, Y.; Kornowski, R.; Raanani, P.; Leader, A.; Pasvolsky, O.; Iakobishvili, Z. Hypertension in cancer patients treated with anti-angiogenic based regimens. *Cardio-Oncology* **2015**, *1*, 6. [CrossRef]

51. Kruzliak, P.; Novak, J.; Novak, M. Vascular endothelial growth factor inhibitor-induced hypertension: From pathophysiology to prevention and treatment based on long-acting nitric oxide donors. *Am. J. Hypertens* **2014**, *27*, 3–13. [CrossRef] [PubMed]

52. Laffin, L.J.; Bakris, G.L. Endothelin Antagonism and Hypertension: An Evolving Target. *Semin. Nephrol.* **2015**, *35*, 168–175. [CrossRef] [PubMed]

53. Brose, M.S.; Frenette, C.T.; Keefe, S.M.; Stein, S.M. Management of sorafenib-related adverse events: A clinician's perspective. *Semin. Oncol.* **2014**, *41*, S1–S16. [CrossRef] [PubMed]

54. Walko, C.M.; Grande, C. Management of common adverse events in patients treated with sorafenib: Nurse and pharmacist perspective. *Semin. Oncol.* **2014**, *41*, S17–S28. [CrossRef] [PubMed]

55. Sim, J.J.; Bhandari, S.K.; Shi, J.; Reynolds, K.; Calhoun, D.A.; Kalantar-Zadeh, K.; Jacobsen, S.J. Comparative risk of renal, cardiovascular, and mortality outcomes in controlled, uncontrolled resistant, and nonresistant hypertension. *Kidney Int.* **2015**, *88*, 622–632. [CrossRef] [PubMed]

56. Wehland, M.; Grosse, J.; Simonsen, U.; Infanger, M.; Bauer, J.; Grimm, D. The Effects of Newer β-Adrenoceptor Antagonists on Vascular Function in Cardiovascular Disease. *Curr. Vasc. Pharmacol.* **2012**, *10*, 378–390. [CrossRef] [PubMed]

57. Fisker, F.Y.; Grimm, D.; Wehland, M. Third-generation β-adrenoceptor antagonists in the treatment of hypertension and heart failure. *Basic Clin. Pharmacol. Toxicol.* **2015**, *117*, 5–14. [CrossRef] [PubMed]

58. Andersen, M.B.; Simonsen, U.; Wehland, M.; Pietsch, J.; Grimm, D. LCZ696 (Valsartan/Sacubitril)—A possible new treatment for hypertension and heart failure. *Basic Clin. Pharmacol. Toxicol.* **2016**, *118*, 14–22. [CrossRef] [PubMed]

59. Semeniuk-Wojtaś, A.; Lubas, A.; Stec, R.; Szczylik, C.; Niemczyk, S. Influence of Tyrosine Kinase Inhibitors on Hypertension and Nephrotoxicity in Metastatic Renal Cell Cancer Patients. *Int. J. Mol. Sci.* **2016**, *17*, 2073. [CrossRef] [PubMed]

60. Sim, J.J.; Shi, J.; Kovesdy, C.P.; Kalantar-Zadeh, K.; Jacobsen, S.J. Impact of achieved blood pressures on mortality risk and end-stage renal disease among a large, diverse hypertension population. *J. Am. Coll. Cardiol.* **2014**, *64*, 588–597. [CrossRef] [PubMed]

61. Small, H.Y.; Montezano, A.C.; Rios, F.J.; Savoia, C.; Touyz, R.M. Hypertension due to antiangiogenic cancer therapy with vascular endothelial growth factor inhibitors: Understanding and managing a new syndrome. *Can. J. Cardiol.* **2014**, *30*, 534–543. [CrossRef] [PubMed]

62. Randrup Hansen, C.; Grimm, D.; Bauer, J.; Wehland, M.; Magnusson, N.E. Effects and side effects of using sorafenib and sunitinib in the treatment of metastatic renal cell carcinoma. *Int. J. Mol. Sci.* **2017**, *18*, 461. [CrossRef] [PubMed]

63. Raymond, E.; Dahan, L.; Raoul, J.L.; Bang, Y.J.; Borbath, I.; Lombard-Bohas, C.; Valle, J.; Metrakos, P.; Smith, D.; Vinik, A.; et al. Sunitinib malate for the treatment of pancreatic neuroendocrine tumors. *N. Engl. J. Med.* **2011**, *364*, 501–513. [CrossRef] [PubMed]

64. Castonguay, V.; Lheureux, S.; Welch, S.; Mackay, H.J.; Hirte, H.; Fleming, G.; Morgan, R.; Wang, L.; Blattler, C.; Ivy, P.S.; et al. A phase II trial of sunitinib in women with metastatic or recurrent endometrial carcinoma: A study of the Princess Margaret, Chicago and California Consortia. *Gynecol. Oncol.* **2014**, *134*, 274–280. [CrossRef] [PubMed]

65. Granito, A.; Marinelli, S.; Negrini, G.; Menetti, S.; Benevento, F.; Bolondi, L. Prognostic significance of adverse events in patients with hepatocellular carcinoma treated with sorafenib. *Therap. Adv. Gastroenterol.* **2016**, *9*, 240–249. [CrossRef] [PubMed]

66. Motzer, R.J.; Hutson, T.E.; Glen, H.; Michaelson, M.D.; Molina, A.; Eisen, T.; Jassem, J.; Zolnierek, J. Lenvatinib, everolimus, and the combination in patients with metastatic renal cell carcinoma: A randomised, phase 2, open-label, multicentre trial. *Lancet Oncol.* **2015**, *16*, 1473–1482. [CrossRef]

67. Wehland, M.; Bauer, J.; Magnusson, N.E.; Infanger, M.; Grimm, D. Biomarkers for anti-angiogenic therapy in cancer. *Int. J. Mol. Sci.* **2013**, *14*, 9338–9364. [CrossRef] [PubMed]

International Journal of
Molecular Sciences

|MDPI|

Review

New Insights in Thyroid Cancer and p53 Family Proteins

Livia Manzella [1,*], Stefania Stella [1], Maria Stella Pennisi [1], Elena Tirrò [1], Michele Massimino [1], Chiara Romano [1], Adriana Puma [1], Martina Tavarelli [2] and Paolo Vigneri [1]

[1] Department of Clinical and Experimental Medicine, University of Catania, 95124 Catania, Italy; stefania.stel@gmail.com (S.S.); perny76@gmail.com (M.S.P.); ele_tir@yahoo.it (E.T.); michedot@yahoo.it (M.M.); chiararomano83@libero.it (C.R.); adry.p88@hotmail.it (A.P.); pvigneri@libero.it (P.V.)

[2] Endocrinology, Department of Clinical and Experimental Medicine, Garibaldi Nesima Medical Center, University of Catania, 95122 Catania, Italy; martinatava@hotmail.it

* Correspondence: manzella@unict.it; Tel.: +39-095-312-389

Received: 10 April 2017; Accepted: 17 June 2017; Published: 21 June 2017

Abstract: Thyroid cancers are common endocrine malignancies that comprise tumors with different clinical and histological features. Indeed, papillary and follicular thyroid cancers are slow-growing, well-differentiated tumors, whereas anaplastic thyroid cancers are undifferentiated neoplasias that behave much more aggressively. Well-differentiated thyroid carcinomas are efficiently cured by surgery and radioiodine, unlike undifferentiated tumors that fail to uptake radioactive iodine and are usually resistant to chemotherapy. Therefore, novel and more effective therapies for these aggressive neoplasias are urgently needed. Whereas most genetic events underlying the pathogenesis of well-differentiated thyroid cancers have been identified, the molecular mechanisms that generate undifferentiated thyroid carcinomas are still unclear. To date, one of the best-characterized genetic alterations leading to the development of poorly differentiated thyroid tumors is the loss of the p53 tumor suppressor gene. In addition, the existence of a complex network among p53 family members (p63 and p73) and their interactions with other factors that promote thyroid cancer progression has been well documented. In this review, we provide an update on the current knowledge of the role of p53 family proteins in thyroid cancer and their possible use as a therapeutic target for the treatment of the most aggressive variants of this disease.

Keywords: thyroid cancer; p53; p63; p73; genetic alterations; p53 inhibition mechanisms; target therapies

1. Introduction

Thyroid cancer is the most common endocrine neoplasm, accounting for about 1.7% of total cancer diagnoses each year in the USA [1]. Worldwide, disease incidence has increased three-fold over the past 30 years because of both a higher prevalence in thyroid gland screening and environmental and life-style changes [2]. The risk of developing a thyroid tumor depends on genetic factors, age, gender, histological type, radiation exposure, and geographical region [3]. Neoplastic transformation of the follicular epithelium generates well-differentiated thyroid cancers (WDTC), including papillary (PTC) and follicular (FTC) thyroid carcinomas, and undifferentiated tumors represented by poorly differentiated (PDTC) and anaplastic thyroid carcinomas (ATC). PTCs and FTCs account for 85% and 10% of all thyroid carcinomas, respectively, with the remaining tumors unevenly distributed between PDTCs and ATCs [4].

The treatment for WDTC is based on surgery, radioactive iodine, and thyroid hormone therapy [5]. On the other hand, PDTCs or ATCs are refractory to hormone therapy and lack expression of

Int. J. Mol. Sci. **2017**, *18*, 1325

the Sodium/Iodide Symporter (NIS) required for radioactive iodine uptake. Moreover, although metastases are observed in 10 to 20% of patients with ATC, most of them succumb to locally invasive, inoperable disease [6]. Furthermore, ATC is usually refractory to conventional chemotherapy [6] with more than 50% of affected individuals dying within one year from diagnosis [7].

Multiple studies have improved our understanding of the mechanisms underlying thyroid carcinogenesis. Presently, we know that several forms of thyroid cancer are characterized by mutually exclusive somatic mutations and/or gene rearrangements, which trigger increased cell proliferation and survival [8–10]. PTCs are often characterized by activating mutations of the BRAF (V600E substitution) that have been reported in up to 70% of cases. Rat Sarcoma (*RAS*) point mutations in codons 12, 13, and 61 and rearrangements of the Rearrangement During Transfection (*RET*) gene (generating *RET/PTC* chimeric oncogenes) represent other frequent genetic alterations (Figure 1). These events share a common downstream signaling pathway as they lead to the improper activation of the mitogen-activated protein kinase (MAPK) [11].

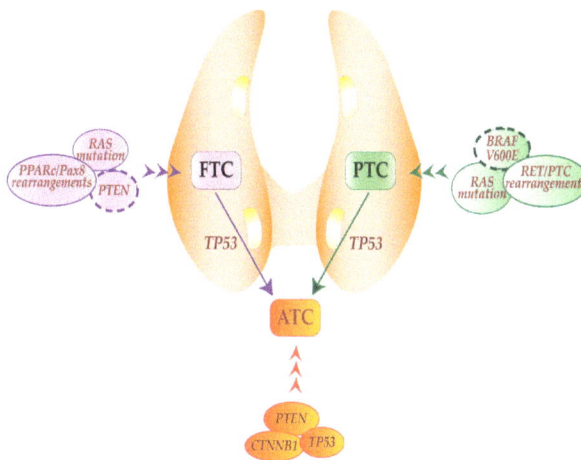

Figure 1. Schematic representation of thyroid cancer hystotypes and their causative genetic events. Papillary (PTC), Follicular (FTC) and Anaplastic Thyroid carcinomas (ATC) originate from thyroid follicular cells. PTCs display BRAF (V600E substitution) and/or Rat Sarcoma (*RAS*) mutations as well as Rearrangement During Transfection (*RET*)/*PTC* rearrangements. FTCs present *PPARc/Pax8* rearrangements, *RAS*, mutations and *PTEN* inactivating mutations or deletions. ATCs are characterized by *PTEN* and *CTNNB1* mutations and p53 inactivation. Furthermore, ATCs may arise from FTCs and PTCs as a result of p53 loss of function, dysregulation of the PTEN/PI3K/AKT pathway or additional genetic alterations. Dashed lines indicate genes and mechanisms involved in progression to ATC.

FTCs display a different genetic profile (Figure 1) characterized by the presence of peroxisome proliferator-activated receptor c/paired box 8 (*PPARc/Pax8*) genetic rearrangements (30%), *RAS* activating point mutations (20–40%) and *PTEN* inactivation by point mutations, deletions, or promoter methylation (<10%) [12].

On the other hand, ATCs can be initiated by the coexistence of activating mutations in *CTNNB1* (β-catenin encoding gene) and inactivation of both *p53* and *PTEN* [13]. In addition, ATCs may arise from WDTC when *PTEN* inactivation (in the case of FTC) or *BRAF* activation (in the case of PTC) are associated with *p53* loss of function (Figure 1). Finally, ATC may display mutations in the catalytic subunit alpha of the phosphatidyl-inositol-3-kinase (PIK3CA) that, in addition to PTEN inactivation, cause constitutive activation of the AKT pathway resulting in a more aggressive phenotype [14].

These alterations represent prognostic markers associated with tumor progression and potential targets that could become therapeutically actionable in the near future. In this review we intend to clarify the role of the p53 protein family in thyroid tumorigenesis. Furthermore we provide an update on the new therapeutic approaches involving p53 for the treatment of the most aggressive variants of the disease.

2. The p53 Protein Family

p53, *p63* and *p73* are tumor suppressor genes encoding for proteins with high homology. The common protein domains consist of an amino-terminal transactivation domain (TAD), a central DNA binding domain (DBD), a carboxy-terminal oligomerization domain (OD), and a proline-rich sequence recognition domain (PRD) [15] (Figure 2A).

The TAD is the binding site for several transcription regulators [16]. The DBD mediates binding to target DNA sequences [17] while the OD induces oligomers formation that influence DNA-binding and transcriptional activity [18]. In addition to the aforementioned domains, p63 and p73 present a sterile α motif domain (SAMD) that modulates protein-protein interactions [19,20] and a transactivation inhibitory domain (TID) that regulates their transcription activity [21] (Figure 2A).

p53, *p63* and *p73* express different internal promoters located at the amino-terminus of each gene. The alternative activation of each promoter leads to the expression of several isoforms containing a complete N-terminal transactivation domain (TA-isoform) and/or an N-terminally truncated isoform lacking the transactivation domain (Figure 2B). In addition, several COOH-terminus transcripts, originated by alternative splicing, have been identified in p53 family members: α, β, and γ in the case of p53 and p63; α, β, γ, δ, ε, ζ, η, and ϕ in the case of p73 [15] (Figure 2B).

Since p53, p63, and p73 play a pivotal role in DNA damage response, cell differentiation, proliferation, and death, they currently represent very appealing targets in anticancer drug development [22,23]. Early evidence suggested significant redundancy in the biologic role of the p53 family members. However, additional findings have also indicated non p53-redundant functions for TAp63 and TAp73 as they regulate genes that are not p53 transcriptional targets [24]. Furthermore, while p53 is a tumor suppressor inactivated in half of human cancers, p63 and p73 rarely display mutations in their sequence. It was later demonstrated that TAp63 and TAp73 transactivate distinct but functionally overlapping subsets of known p53-regulated genes involved in cell-cycle arrest and apoptosis [25,26]. It should also be noted that although TAp63 and TAp73 show tumor suppressor properties in different tumors [27,28], several reports suggest an oncogenic function for these proteins as they may inhibit p53 DNA binding activity [29–31]. Likewise, the p63 and p73 ΔN variants were considered dominant-negative inhibitors of their respective TA isoforms and of the p53 tumor-suppressor. Indeed, their overexpression in a wide range of tumors is usually associated with an inferior prognosis [17]. However, later experiments also revealed a physiologic role for ΔN isoforms as expression modulators for different p53 family members. For example, p53 and TAp73 are both transcriptional inducers of ΔNp73, thus defining a negative regulatory loop in which p53 and TAp73 activation leads to their functional inhibition by ΔNp73 [26].

In summary, the need to account for a plethora of biological parameters—such as the TA/ΔN p63/p73 ratios, p63/p73 protein-protein interactions, and p63/p73 binding to the promoter of different p53-target genes—has yet to define an unequivocal role for p63 and p73 in human tumorigenesis [16].

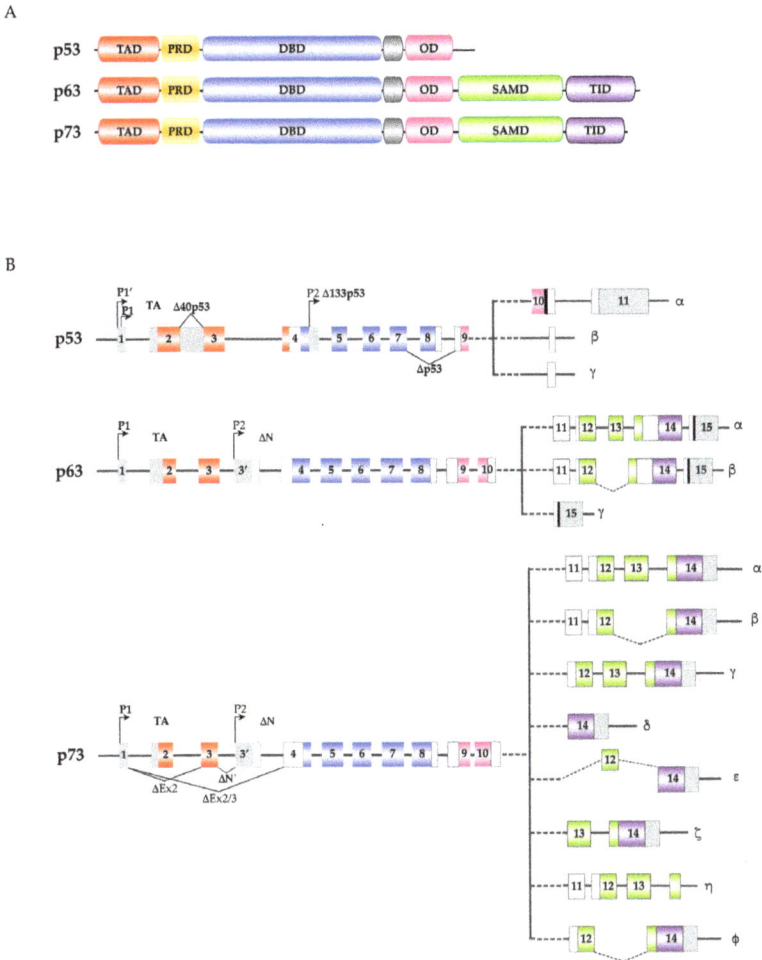

Figure 2. p53 family proteins and isoforms. (**A**) Functional domains of the p53 family members. Red, Transactivation Domain (TAD); yellow, Proline-Rich sequence Domain (PRD); blue, DNA Binding Domain (DBD); pink, Oligomerization Domain (OD); green, Sterile α Motif Domain (SAMD); purple, Transactivation Inhibitory Domain (TID); (**B**) *p53*, *p63* and *p73* intron/exon structure. Introns are depicted in gray, while exon coloring reflects the corresponding functional domains. All three genes express multiple splice variants and contain different internal promoters. p53 includes TAp53, Δ40p53 (generated by an alternative splicing of intron 2), Δp53 (produced by alternative splicing of exons 7/9), and Δ133p53 (generated using an internal promoter in intron 4). The alternative splicing of intron 9 gives rise to α, β, and γ isoforms. In *p63*, the proximal P1 promoter yields the TA isoforms, while the distal P2 promoter in intron 3′, gives rise to ΔNp63 truncated variants. In addition, the COOH-terminal splicing leads to p63 α, β and γ isoforms for both the TA and ΔN variants. As for p73 the P1 promoter generates the TA isoforms, while the P2 distal promoter in intron 3′, gives rise to ΔNp73 truncated variants. Moreover, p73 can use an additional NH2-terminal splicing site, within exon 2, that produces ΔN like proteins Ex2p73, Ex2/3p73 and ΔN′p73. The COOH-terminal splicing leads to p73 α, β, γ, δ, ε, ζ, η, and φ isoforms for both TA and ΔN variants.

2.1. p53 and Thyroid Cancer

It is well documented that thyroid carcinoma initiation and progression occurs through the gradual accumulation of multiple genetic alterations. One of the pivotal molecular alterations discriminating ATCs from WDTCs is the inactivation of the *p53* tumor suppressor gene. p53 mutations are common in undifferentiated thyroid tumors (50–80% in ATCs) [32,33]. However, several findings indicate that alterations in the *p53* sequence may also play a role in the early stages of thyroid cancerogenesis. Indeed, p53 mutations were recently found in up to 40% of PTCs and in 22% of oncocytic FTCs [34,35]. Usually, *p53* point mutations are located in the region between exons 5 and 8 [36].

DNA damaging agents activate p53, leading to its binding of specific DNA responsive elements that transcriptionally regulate selected target genes. In thyroid cancer cells, p53 tumor suppressor activity is inhibited by three different mechanisms hindering p53 transcriptional activity, protein stability, and downstream signaling.

Frasca and colleagues reported that overexpression of high mobility group A factors (HMGA1a, HMGA1b, and HMGA2) functionally disables all p53 family members [15,37] possibly by reducing their DNA-binding activity [37–39] (Figure 3A). The POZ/BTB and AT hook containing zinc finger protein (PATZ1) is down-regulated in PDTCs and ATCs. As PATZ1 facilitates p53 binding to its responsive elements. Its down-regulation in thyroid cancer cells reduces p53 biological activity favoring both epithelial-mesenchymal transition and cell migration [40] (Figure 3A).

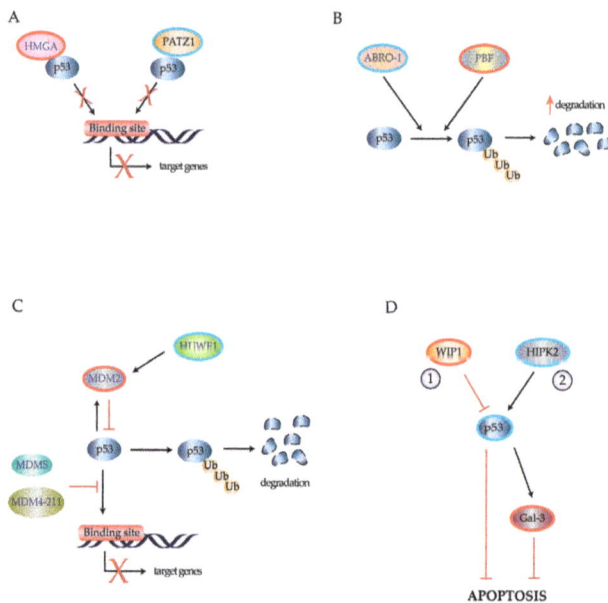

Figure 3. p53 inactivating mechanisms in thyroid cancer. (**A**) HGMA1 over-expression (red) and PATZ1 down-regulation (blue) inhibit p53 ability to bind to its DNA responsive elements; (**B**) ABRO-1 down-regulation (blue) and PBF over-expression (red) decrease p53 stability through an increase in ubiquitin-mediated p53 degradation; (**C**) MDM-S, MDM4-211, and MDM-2 over-expression (red), following ubiquitin E3 ligase HUWE1 down-regulation (blue), inhibit p53 transactivation activity through p53 poly-ubiquitination and by blocking p53 interaction with the DNA binding sites of its target genes; (**D**) Mechanisms leading to reduced apoptotic sensibility in thyroid cancer cells include (1) WIP1 over-expression (red) and (2) decreased HIPK2 expression (blue) causing a reduction of p53 levels (blue) and subsequent increases of Gal-3 (red).

p53 stability is regulated through a process of ubiquitin-dependent protein degradation. In thyroid cancer cells, the over-expression or down-regulation of different p53 regulatory proteins heavily influence this mechanism.

Zhang et al. reported that Abraxas brother 1 (ABRO1), a component of the BRISC multiprotein complex that specifically cleaves "Lys-63"-linked ubiquitin, is frequently downregulated in thyroid, breast, liver, and kidney tumors [41]. The authors showed that ABRO1 stabilizes p53 by facilitating its interaction with deubiquitinase USP7. Depletion of ABRO1 in thyroid cancer reduces this interaction causing p53 poly-ubiquitination and enhancing cellular transformation of thyroid neoplastic clones [41] (Figure 3B). Likewise, the proto-oncogene PTTG1-Binding Factor (PBF) is highly expressed in WDTC [42], where it interacts with p53 thereby enhancing its poly-ubiquitination, which depends on the E3 ligase activity of MDM2 [43] (Figure 3B).

Murine double minute (MDM) family members are key regulators of p53 expression and function. Both MDM2 and MDM4 negatively regulate p53 [44,45] (Figure 3C). Furthermore, Prodosmo et al. found that MDM-S and MDM4-211—shorter MDM4 spliced variants expressed in thyroid tumors—were strong in vitro p53 inhibitors [46] (Figure 3C). The HECT, UBA, and WWE domain-containing protein 1 (HUWE1) is a ubiquitin E3 ligase for MDM2 that shows deregulated expression in several human cancers. Recently, Ma and colleagues demonstrated that HUWE1 is down-regulated in human thyroid carcinomas thus increasing MDM2 expression and reducing p53 protein stability [47] (Figure 3C).

Alterations in p53 signaling can also contribute to thyroid carcinogenesis by determining checkpoint defects, genomic instability, and inhibition of apoptosis.

Wild type p53 induced phosphatase 1 (WIP1) is a member of the PP2C family of evolutionarily conserved protein phosphatases and is considered a novel proto-oncogene [48,49]. Originally described as a p53-regulated gene, it is overexpressed in several tumors including PTCs where it inhibits p53 as well as p38MAPK and p16 [50] (Figure 3D). Galectin-3 (Gal-3) is an anti-apoptotic molecule that is down-regulated by wild-type p53 [51]. Thyroid tumors usually display low levels of transcription factor Homeodomain Interacting Protein Kinase 2 (HIPK2) that reduces p53 activation resulting in Gal-3 overexpression thus promoting neoplastic transformation [52] (Figure 3D). Zou et al. have recently described an additional mechanism contributing to p53 inactivation in a murine thyroid cancer model. They demonstrated that mice with thyroid-specific expression of the BRAFV600E mutation rapidly developed thyroid carcinomas. In this model, high levels of thyroid-stimulating hormone upregulated the PI3K/AKT pathway thus reducing p53 expression and suppressing BRAFV600E-induced senescence [53]. A further mechanism disregulating p53 downstream signaling involves *FOXE1*, a single-exon coding gene belonging to the forkhead/winged helix-domain protein family. FOXE1 is essential for thyroid gland development and has also been implicated in thyroid cancerogenesis following the discovery of inactivating mutations detected in neoplastic thyrocytes [54–56]. Recent evidence suggests that myosin 9 (MYH9) is a binding partner of the papillary thyroid cancer susceptibility candidate 2 (*PTCSC2*) long noncoding RNA. The interaction between MYH9 and PTCSC2 inhibits the *FOXE1* promoter. In turn, lack of FOXE1 results in the transcriptional repression of *THBS1* and *IGFBP3*, pivotal members of the p53 signaling pathway [57].

2.2. p63 and Thyroid Cancer

The *p63* gene is infrequently mutated in human cancer. p63 overexpression has been reported in basal and squamous cell carcinomas of the head and neck, in thymomas, basal-like breast cancer, adenocarcinoma of the prostate, and poorly differentiated cervical tumors. p63 may also be aberrantly expressed in thyroid cancer [27,58–60].

Several evidences support the tumor suppressor role of TAp63 after interaction with ΔNp63 or other p53 family members. Furthermore, TAp63 induces cell death and suppress metastasis formation by decreasing mobility and invasion [27].

Different studies reported a possible role for p63 in thyrocyte neoplastic transformation [15,30]. Malaguarnera et al. demonstrated that, unlike normal thyroid cells and benign adenomas, most thyroid neoplasias express TAp63α. They also showed that endogenous TAp63α does not play a p53-like role as it fails to induce p21, BAX, and MDM2 and therefore does not cause cell growth arrest and apoptosis. Furthermore, they reported that TAp63α antagonizes p53 by interfering with the binding of more transcriptionally active p63 homologues (TAp63β and TAp63γ) and that of TAp73 [30]. As both TAp63α and ΔNp63α successfully inhibit p53-dependent suppression of colony formation, these findings imply that in thyroid cancer cells both TAp63α and ΔNp63α display a tumor-promoting role [15]. In this respect, Lazzari and colleagues have reported that HIPK2 induces ΔNp63α phosphorylation causing its proteasomal degradation. This mechanism removes the dominant negative effect of ΔNp63α on p53 restoring its pro-apoptotic activity [61].

2.3. p73 and Thyroid Cancer

Different studies have reported p73 expression in human thyroid tumors, even if the role of p73 in thyroid cancer progression is still controversial [15,62,63]. Both TAp73α and ΔNp73 were detected by RQ-PCR and immunoblots in the large majority of thyroid cancer cell lines isolated from different histotypes (papillary, follicular, and anaplastic), but not in normal cultured thyrocytes [15,64]. Likewise, TAp73 and ΔNp73 transcripts were found in a consistent number of human thyroid carcinomas, although no correlation was found with their clinical and pathological characteristics [65]. Puppin and colleagues published evidence suggesting that ΔNp73α transcriptionally stimulates periostin gene expression in papillary, follicular, and undifferentiated thyroid cancer cells [66] (Figure 4A).

As periostin is associated with accelerated tumor growth, increased neoangiogenesis and higher metastatic potential [67], ΔNp73α expression results in a more aggressive cancer phenotype [68,69]. Vella et al. reported that ΔNp73α represses the *PTEN* promoter [63]. *PTEN* down-regulation increases PIP3 half-life reducing its conversion in PIP2. High levels of PIP3 increase AKT phosphorylation causing MDM2-mediated ubiquitination and p53 degradation [70–72] (Figure 4A). *PTEN* repression by ΔNp73α requires binding to a DNA region outside of the canonical p53 site, confirming that p53 and its family members may recognize different sequences on the same promoter. In this scenario, the functional interactions among p53 family members may be either synergistic or antagonistic with respect to tumor suppression. This is in line with previous reports showing that, in different thyroid cancer models, co-expression of p53 and TAp73α may alternatively result in TAp73α down-regulation [62] or stronger p53 tumor suppressor activity [73]. In the latter case, TAp73α inhibits MDM2-mediated p53 protein degradation by reducing MDM2 binding to p53 and by directly antagonizing p53 activity on the *MDM2* promoter (Figure 4B).

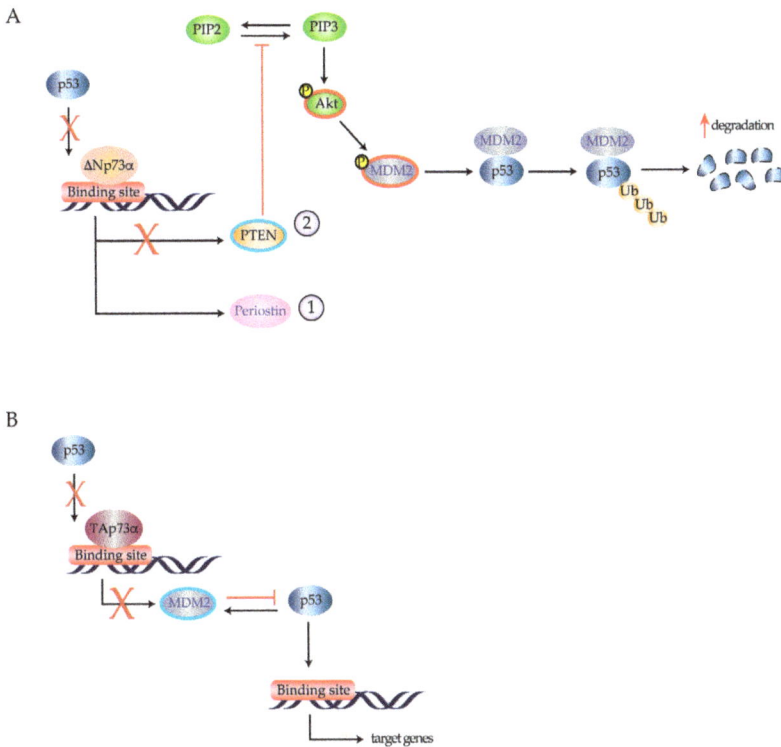

Figure 4. p73 pathway activation in thyroid cancer. (**A**) ΔNp73α binding on the p53 promoter causes the activation of *periostin* (1) and a reduction in *PTEN* expression (blue) (2). Transcriptional repression of the *PTEN* promoter determines an activation of the PI3K-Akt pathway resulting in MDM2 phosphorylation that enhances MDM2-dependent degradation of p53; (**B**) TAp73α blocks p53-dependent transcription on the MDM2 promoter. In turn, this leads to reduced MDM2 expression and increased p53 stability.

3. Novel Therapeutic Approaches for Undifferentiated Thyroid Cancer

The aforementioned advances in understanding the contribution of p53 family members to the pathogenesis of thyroid cancer have provided several opportunities for the use of molecular targeted therapies.

Multiple clinical trials using various multi-kinase inhibitors (MKIs) have led to the FDA and EMA approval of Sorafenib [74] and Lenvatinib [75] for undifferentiated thyroid cancer. Grassi and colleagues have also investigated SP600125, a reversible ATP-competitive MKI with anticancer properties against undifferentiated thyroid cancer. The compound reduces cell migration while killing human thyrocytes through the activation of mutant p53 and the concomitant inhibition of the ROCK/HDAC6 pathway [76] (Figure 5).

Besides MKIs, additional strategies for thyroid cancer treatment are currently being investigated that employ strategies that modulate epigenetic changes in thyroid cancer DNA, restore the transcriptional activity of mutant p53, and block signal transduction downstream of different p53 family members.

Studies on histone post-translational modifications that support tumor development and progression have focused on DNA methyltransferases (DNMT) inhibitors such as 5′-azadeoxycytidine with the aim of enabling thyroid cancer differentiation, thus restoring its sensitivity to radioactive iodine treatment [77]. This effect has also been investigated with the MEK1/MEK2 inhibitor

Selumetinib [78]. Thailandepsin A (TDP-A) is a novel class I histone deacetylase inhibitor that induces a dose- and time-dependent anti-proliferative effect on human thyroid cancer cells mainly attributed to activation of the extrinsic apoptotic pathway coupled with cell cycle arrest [79]. Proteasome inhibitors such as Bortezomib have also been tested on thyroid cancer cells leading to significant inhibition of neoplastic thyrocyte proliferation [80,81].

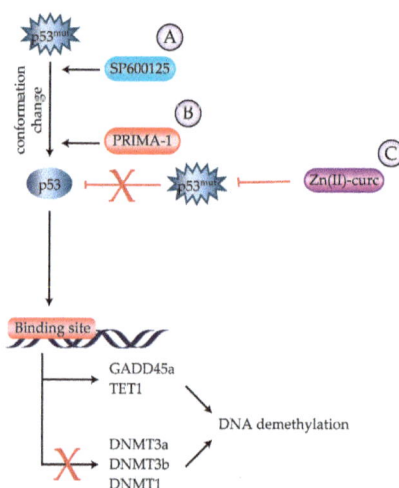

Figure 5. Strategies to reactivate p53 in thyroid cancer. (**A**) SP600125 and (**B**) PRIMA-1 reactivate mutant p53 via conformational changes. In thyroid cancer, modifications caused by PRIMA-1 induce global DNA demethylation through the up-regulation of GADD45a and TET1 and down-regulation of DNMT1, 3a, and 3b; (**C**) Zn(II)-curc reduces mutant p53 expression by restoring wild-type p53-DNA binding activity to target gene promoters.

p53 reactivation and induction of massive apoptosis (PRIMA-1) is a compound that reactivates the DNA binding ability of mutant p53 by restoring its conformation through two different mechanisms: (i) it induces Hsp90 over-expression thus enhancing its binding to mutant p53 and (ii) it covalently binds to mutant p53 [82]. Qiang and colleagues have recently defined the mechanism responsible for PRIMA-1-dependent rescue of mutant p53 function in thyroid cancer cells [83]. They showed that PRIMA-1 causes global DNA demethylation in cancer cells expressing mutant p53 mainly through inhibition of DNMT 1, 3a, and 3b, and upregulation of GADD45a. PRIMA-1 also increased the expression of the ten-eleven translocation (TET) family of 5mC-hydroxylases, particularly TET1, further contributing to DNA demethylation. Messina et al. studied the effect of PRIMA-1 in thyroid cancer cell lines with both wild-type and D259Y and K286E p53 mutants. They found that PRIMA-1 successfully killed undifferentiated thyroid carcinoma cells carrying mutant p53 that were refractory to chemotherapy [84] (Figure 5).

A curcumin-based zinc compound [Zn(II)-curc] also reduced mutant p53 expression in thyroid cancer cells, reactivating p53 tumor suppressor activity and resulting in increased response to anti-cancer therapies [85] (Figure 5).

Finally, the mammalian target of rapamycin (mTOR) pathway inhibitor Everolimus has been studied in a phase II trial of 40 patients with locally advanced or metastatic thyroid cancer [86] producing stable disease in 29 (76%) patients and a median Progression Free Survival of 47 weeks (95% CI 14.9–78.5).

4. Conclusions

The concept of precision medicine postulates that effective treatment strategies should be tailored to the individual variability of both the patient and her/his disease. While alterations in p53 sequence, stability, and downstream signaling are heavily involved in thyroid carcinogenesis, the role of the remaining p53 family members in the development and progression of thyroid cancer has yet to be fully elucidated. Nevertheless, increasing evidence indicates that p53 family members contribute to the development of multiple thyroid cancer variants and an ever-increasing number of therapeutic molecules targeting these proteins may soon be available in the clinical setting.

Conflicts of Interest: The authors declare no conflict of interest.

Abbreviations

WDTC	Well Differentiated Thyroid Cancer
PDTC	Poorly Differentiated Thyroid Cancer
ATC	Anaplastic Thyroid Cancer
RET	Rearranged During Transfection
PTC	Papillary Thyroid Cancer
FTC	Follicular Thyroid cancer
MAPK	Mitogen-Activated Protein Kinase
PPARc	Peroxisome Proliferator-Activated Receptor C
PAX8	Paired Box 8
RAS	Rat Sarcoma
PI3K	Phosphatidyl-Inositol-3-Kinase
PTEN	Phosphatase and Tensin Homolog
CTNNB1	Catenin beta 1
TAD	Transactivation Domain
DBD	DNA Binding Domain
OD	Oligomerization Domain
PRD	Prolin-Rich Sequence recognition Domain
SAMD	Steril Alpha Motif Domain
TID	Transactivation Inhibitory Domain
HMGA	High Mobility Group A
PATZ1	POZ7BTB and AT Hook containing Zinc-finger Protein
ABRO1	Abraxas Brother 1
PBF	PTTG1-Binding Factor
MDM	Murine Double Minute
HUWE1	HECT, UBA, and WWE Domain-Containing Protein 1
WIP1	Wild Type p53 Induced Phosphatases 1
Gal-3	Galectin 3
HIPK2	Homeodomain Interacting Protein Kinase 2
FOXE1	Forkhead7Winged Helix-Domain Protein
MYH9	Myosin-9
PTCSC2	Papillary Thyroid Cancer Susceptibility Candidate 2
THBS1	Thrombospondin 1
IGFBP3	Insulin-Like Growth Factor Binding Protein 3
MKIs	Multi-Kinase Inhibitors
HDAC	Histone Deacetylase

ROCK	Rho Kinase
DNMT	DNA Methyltransferase
MEK	Mitogen-Activated protein kinase kinase
TDP-A	Thailandepsin A
PRIMA-1	p53 Reactivation and Induction of Massive Apoptosis
GADD45A	Growth Arrest and DNA Damage Inducible Alpha
TET	Ten-Eleven Translocation
mTOR	Mammalian Target of Rapamycine

References

1. Siegel, R.; Ma, J.; Zou, Z.; Jemal, A. Cancer statistics, 2014. *CA Cancer J. Clin.* **2014**, *64*, 9–29. [CrossRef] [PubMed]
2. Chen, A.Y.; Jemal, A.; Ward, E.M. Increasing incidence of differentiated thyroid cancer in the United States, 1988–2005. *Cancer* **2009**, *115*, 3801–3807. [CrossRef] [PubMed]
3. Pellegriti, G.; Frasca, F.; Regalbuto, C.; Squatrito, S.; Vigneri, R. Worldwide increasing incidence of thyroid cancer: Update on epidemiology and risk factors. *J. Cancer Epidemiol.* **2013**, *2013*, 965212. [CrossRef] [PubMed]
4. Nikiforov, Y.E.; Steward, D.L.; Robinson-Smith, T.M.; Haugen, B.R.; Klopper, J.P.; Zhu, Z.; Fagin, J.A.; Falciglia, M.; Weber, K.; Nikiforova, M.N. Molecular testing for mutations in improving the fine-needle aspiration diagnosis of thyroid nodules. *J. Clin. Endocrinol. Metab.* **2009**, *94*, 2092–2098. [CrossRef] [PubMed]
5. Sherman, S.I. Thyroid carcinoma. *Lancet* **2003**, *361*, 501–511. [CrossRef]
6. McIver, B.; Hay, I.D.; Giuffrida, D.F.; Dvorak, C.E.; Grant, C.S.; Thompson, G.B.; van Heerden, J.A.; Goellner, J.R. Anaplastic thyroid carcinoma: A 50-year experience at a single institution. *Surgery* **2001**, *130*, 1028–1034. [CrossRef] [PubMed]
7. Hadar, T.; Mor, C.; Shvero, J.; Levy, R.; Segal, K. Anaplastic carcinoma of the thyroid. *Eur. J. Surg. Oncol. J. Eur. Soc. Surg. Oncol. Br. Assoc. Surg. Oncol.* **1993**, *19*, 511–516.
8. Massimino, M.; Vigneri, P.; Fallica, M.; Fidilio, A.; Aloisi, A.; Frasca, F.; Manzella, L. IRF5 promotes the proliferation of human thyroid cancer cells. *Mol. Cancer* **2012**, *11*, 21. [CrossRef] [PubMed]
9. Soares, P.; Trovisco, V.; Rocha, A.S.; Lima, J.; Castro, P.; Preto, A.; Maximo, V.; Botelho, T.; Seruca, R.; Sobrinho-Simoes, M. BRAF mutations and RET/PTC rearrangements are alternative events in the etiopathogenesis of PTC. *Oncogene* **2003**, *22*, 4578–4580. [CrossRef] [PubMed]
10. Xing, M. Molecular pathogenesis and mechanisms of thyroid cancer. *Nat. Rev. Cancer* **2013**, *13*, 184–199. [CrossRef] [PubMed]
11. Kimura, E.T.; Nikiforova, M.N.; Zhu, Z.; Knauf, J.A.; Nikiforov, Y.E.; Fagin, J.A. High prevalence of BRAF mutations in thyroid cancer: Genetic evidence for constitutive activation of the RET/PTC-RAS-BRAF signaling pathway in papillary thyroid carcinoma. *Cancer Res.* **2003**, *63*, 1454–1457. [PubMed]
12. Kroll, T.G.; Sarraf, P.; Pecciarini, L.; Chen, C.J.; Mueller, E.; Spiegelman, B.M.; Fletcher, J.A. PAX8-PPARgamma1 fusion oncogene in human thyroid carcinoma [corrected]. *Science* **2000**, *289*, 1357–1360. [CrossRef] [PubMed]
13. Omur, O.; Baran, Y. An update on molecular biology of thyroid cancers. *Crit. Rev. Oncol. Hematol.* **2014**, *90*, 233–252. [CrossRef] [PubMed]
14. Xing, M. Genetic alterations in the phosphatidylinositol-3 kinase/Akt pathway in thyroid cancer. *Thyroid Off. J. Am. Thyroid Assoc.* **2010**, *20*, 697–706. [CrossRef] [PubMed]
15. Malaguarnera, R.; Vella, V.; Vigneri, R.; Frasca, F. p53 family proteins in thyroid cancer. *Endocr. Relat. Cancer* **2007**, *14*, 43–60. [CrossRef] [PubMed]
16. Ferraiuolo, M.; Di Agostino, S.; Blandino, G.; Strano, S. Oncogenic Intra-p53 Family Member Interactions in Human Cancers. *Front. Oncol.* **2016**, *6*, 77. [CrossRef] [PubMed]
17. Collavin, L.; Lunardi, A.; Del Sal, G. p53-family proteins and their regulators: Hubs and spokes in tumor suppression. *Cell Death Differ.* **2010**, *17*, 901–911. [CrossRef] [PubMed]
18. Sauer, M.; Bretz, A.C.; Beinoraviciute-Kellner, R.; Beitzinger, M.; Burek, C.; Rosenwald, A.; Harms, G.S.; Stiewe, T. C-terminal diversity within the p53 family accounts for differences in DNA binding and transcriptional activity. *Nucleic Acids Res.* **2008**, *36*, 1900–1912. [CrossRef] [PubMed]

19. Bernard, H.; Garmy-Susini, B.; Ainaoui, N.; Van Den Berghe, L.; Peurichard, A.; Javerzat, S.; Bikfalvi, A.; Lane, D.P.; Bourdon, J.C.; Prats, A.C. The p53 isoform, Δ133p53α, stimulates angiogenesis and tumour progression. *Oncogene* **2013**, *32*, 2150–2160. [CrossRef] [PubMed]
20. Melino, G.; Lu, X.; Gasco, M.; Crook, T.; Knight, R.A. Functional regulation of p73 and p63: Development and cancer. *Trends Biochem. Sci.* **2003**, *28*, 663–670. [CrossRef] [PubMed]
21. Ghioni, P.; Bolognese, F.; Duijf, P.H.; Van Bokhoven, H.; Mantovani, R.; Guerrini, L. Complex transcriptional effects of p63 isoforms: Identification of novel activation and repression domains. *Mol. Cell. Biol.* **2002**, *22*, 8659–8668. [CrossRef] [PubMed]
22. Candi, E.; Agostini, M.; Melino, G.; Bernassola, F. How the TP53 family proteins TP63 and TP73 contribute to tumorigenesis: Regulators and effectors. *Hum. Mutat.* **2014**, *35*, 702–714. [CrossRef] [PubMed]
23. Gomes, S.; Leao, M.; Raimundo, L.; Ramos, H.; Soares, J.; Saraiva, L. p53 family interactions and yeast: Together in anticancer therapy. *Drug Discov. Today* **2016**, *21*, 616–624. [CrossRef] [PubMed]
24. Barbieri, C.E.; Pietenpol, J.A. p63 and epithelial biology. *Exp. Cell Res.* **2006**, *312*, 695–706. [CrossRef] [PubMed]
25. Deyoung, M.P.; Ellisen, L.W. p63 and p73 in human cancer: Defining the network. *Oncogene* **2007**, *26*, 5169–5183. [CrossRef] [PubMed]
26. Melino, G.; De Laurenzi, V.; Vousden, K.H. p73: Friend or foe in tumorigenesis. *Nat. Rev. Cancer* **2002**, *2*, 605–615. [CrossRef] [PubMed]
27. Melino, G. p63 is a suppressor of tumorigenesis and metastasis interacting with mutant p53. *Cell Death Differ.* **2011**, *18*, 1487–1499. [CrossRef] [PubMed]
28. Su, X.; Chakravarti, D.; Flores, E.R. p63 steps into the limelight: Crucial roles in the suppression of tumorigenesis and metastasis. *Nat. Rev. Cancer* **2013**, *13*, 136–143. [CrossRef] [PubMed]
29. Freebern, W.J.; Smith, J.L.; Chaudhry, S.S.; Haggerty, C.M.; Gardner, K. Novel cell-specific and dominant negative anti-apoptotic roles of p73 in transformed leukemia cells. *J. Biol. Chem.* **2003**, *278*, 2249–2255. [CrossRef] [PubMed]
30. Malaguarnera, R.; Mandarino, A.; Mazzon, E.; Vella, V.; Gangemi, P.; Vancheri, C.; Vigneri, P.; Aloisi, A.; Vigneri, R.; Frasca, F. The p53-homologue p63 may promote thyroid cancer progression. *Endocr. Relat. Cancer* **2005**, *12*, 953–971. [CrossRef] [PubMed]
31. Vikhanskaya, F.; D'Incalci, M.; Broggini, M. p73 competes with p53 and attenuates its response in a human ovarian cancer cell line. *Nucleic Acids Res.* **2000**, *28*, 513–519. [CrossRef] [PubMed]
32. Quiros, R.M.; Ding, H.G.; Gattuso, P.; Prinz, R.A.; Xu, X. Evidence that one subset of anaplastic thyroid carcinomas are derived from papillary carcinomas due to BRAF and p53 mutations. *Cancer* **2005**, *103*, 2261–2268. [CrossRef] [PubMed]
33. Salvatore, D.; Celetti, A.; Fabien, N.; Paulin, C.; Martelli, M.L.; Battaglia, C.; Califano, D.; Monaco, C.; Viglietto, G.; Santoro, M.; et al. Low frequency of p53 mutations in human thyroid tumours; p53 and Ras mutation in two out of fifty-six thyroid tumours. *Eur. J. Endocrinol.* **1996**, *134*, 177–183. [CrossRef] [PubMed]
34. Cancer Genome Atlas Research Network. Integrated genomic characterization of papillary thyroid carcinoma. *Cell* **2014**, *159*, 676–690.
35. Dralle, H.; Machens, A.; Basa, J.; Fatourechi, V.; Franceschi, S.; Hay, I.D.; Nikiforov, Y.E.; Pacini, F.; Pasieka, J.L.; Sherman, S.I. Follicular cell-derived thyroid cancer. *Nat. Rev. Dis. Primers* **2015**, *1*, 15077. [CrossRef] [PubMed]
36. Farid, N.R. P53 mutations in thyroid carcinoma: Tidings from an old foe. *J. Endocrinol. Investig.* **2001**, *24*, 536–545. [CrossRef] [PubMed]
37. Frasca, F.; Rustighi, A.; Malaguarnera, R.; Altamura, S.; Vigneri, P.; Del Sal, G.; Giancotti, V.; Pezzino, V.; Vigneri, R.; Manfioletti, G. HMGA1 inhibits the function of p53 family members in thyroid cancer cells. *Cancer Res.* **2006**, *66*, 2980–2989. [CrossRef] [PubMed]
38. Reeves, R. Molecular biology of HMGA proteins: Hubs of nuclear function. *Gene* **2001**, *277*, 63–81. [CrossRef]
39. Sgarra, R.; Rustighi, A.; Tessari, M.A.; Di Bernardo, J.; Altamura, S.; Fusco, A.; Manfioletti, G.; Giancotti, V. Nuclear phosphoproteins HMGA and their relationship with chromatin structure and cancer. *FEBS Lett.* **2004**, *574*, 1–8. [CrossRef] [PubMed]

40. Chiappetta, G.; Valentino, T.; Vitiello, M.; Pasquinelli, R.; Monaco, M.; Palma, G.; Sepe, R.; Luciano, A.; Pallante, P.; Palmieri, D.; et al. PATZ1 acts as a tumor suppressor in thyroid cancer via targeting p53-dependent genes involved in EMT and cell migration. *Oncotarget* **2015**, *6*, 5310–5323. [CrossRef] [PubMed]

41. Zhang, J.; Cao, M.; Dong, J.; Li, C.; Xu, W.; Zhan, Y.; Wang, X.; Yu, M.; Ge, C.; Ge, Z.; et al. ABRO1 suppresses tumourigenesis and regulates the DNA damage response by stabilizing p53. *Nat. Commun.* **2014**, *5*, 5059. [CrossRef] [PubMed]

42. Stratford, A.L.; Boelaert, K.; Tannahill, L.A.; Kim, D.S.; Warfield, A.; Eggo, M.C.; Gittoes, N.J.; Young, L.S.; Franklyn, J.A.; McCabe, C.J. Pituitary tumor transforming gene binding factor: A novel transforming gene in thyroid tumorigenesis. *J. Clin. Endocrinol. Metab.* **2005**, *90*, 4341–4349. [CrossRef] [PubMed]

43. Read, M.L.; Seed, R.I.; Fong, J.C.; Modasia, B.; Ryan, G.A.; Watkins, R.J.; Gagliano, T.; Smith, V.E.; Stratford, A.L.; Kwan, P.K.; et al. The PTTG1-binding factor (PBF/PTTG1IP) regulates p53 activity in thyroid cells. *Endocrinology* **2014**, *155*, 1222–1234. [CrossRef] [PubMed]

44. Marine, J.C.; Dyer, M.A.; Jochemsen, A.G. MDMX: From bench to bedside. *J. Cell Sci.* **2007**, *120*, 371–378. [CrossRef] [PubMed]

45. Marine, J.C.; Francoz, S.; Maetens, M.; Wahl, G.; Toledo, F.; Lozano, G. Keeping p53 in check: Essential and synergistic functions of MDM2 and MDM4. *Cell Death Differ.* **2006**, *13*, 927–934. [CrossRef] [PubMed]

46. Prodosmo, A.; Giglio, S.; Moretti, S.; Mancini, F.; Barbi, F.; Avenia, N.; Di Conza, G.; Schunemann, H.J.; Pistola, L.; Ludovini, V.; et al. Analysis of human MDM4 variants in papillary thyroid carcinomas reveals new potential markers of cancer properties. *J. Mol. Med.* **2008**, *86*, 585–596. [CrossRef] [PubMed]

47. Ma, W.; Zhao, P.; Zang, L.; Zhang, K.; Liao, H.; Hu, Z. Tumour suppressive function of HUWE1 in thyroid cancer. *J. Biosci.* **2016**, *41*, 395–405. [CrossRef] [PubMed]

48. Bulavin, D.V.; Demidov, O.N.; Saito, S.; Kauraniemi, P.; Phillips, C.; Amundson, S.A.; Ambrosino, C.; Sauter, G.; Nebreda, A.R.; Anderson, C.W.; et al. Amplification of PPM1D in human tumors abrogates p53 tumor-suppressor activity. *Nat. Genet.* **2002**, *31*, 210–215. [CrossRef] [PubMed]

49. Li, J.; Yang, Y.; Peng, Y.; Austin, R.J.; van Eyndhoven, W.G.; Nguyen, K.C.; Gabriele, T.; McCurrach, M.E.; Marks, J.R.; Hoey, T.; et al. Oncogenic properties of PPM1D located within a breast cancer amplification epicenter at 17q23. *Nat. Genet.* **2002**, *31*, 133–134. [CrossRef] [PubMed]

50. Yang, D.; Zhang, H.; Hu, X.; Xin, S.; Duan, Z. Abnormality of p16/p38MAPK/p53/Wip1 pathway in papillary thyroid cancer. *Gland Surg.* **2012**, *1*, 33–38. [PubMed]

51. Lavra, L.; Ulivieri, A.; Rinaldo, C.; Dominici, R.; Volante, M.; Luciani, E.; Bartolazzi, A.; Frasca, F.; Soddu, S.; Sciacchitano, S. Gal-3 is stimulated by gain-of-function p53 mutations and modulates chemoresistance in anaplastic thyroid carcinomas. *J. Pathol.* **2009**, *218*, 66–75. [CrossRef] [PubMed]

52. Lavra, L.; Rinaldo, C.; Ulivieri, A.; Luciani, E.; Fidanza, P.; Giacomelli, L.; Bellotti, C.; Ricci, A.; Trovato, M.; Soddu, S.; et al. The loss of the p53 activator HIPK2 is responsible for galectin-3 overexpression in well differentiated thyroid carcinomas. *PLoS ONE* **2011**, *6*, e20665. [CrossRef] [PubMed]

53. Zou, M.; Baitei, E.Y.; Al-Rijjal, R.A.; Parhar, R.S.; Al-Mohanna, F.A.; Kimura, S.; Pritchard, C.; Binessa, H.A.; Alzahrani, A.S.; Al-Khalaf, H.H.; et al. TSH overcomes Braf(V600E)-induced senescence to promote tumor progression via downregulation of p53 expression in papillary thyroid cancer. *Oncogene* **2016**, *35*, 1909–1918. [CrossRef] [PubMed]

54. He, H.; Li, W.; Liyanarachchi, S.; Jendrzejewski, J.; Srinivas, M.; Davuluri, R.V.; Nagy, R.; de la Chapelle, A. Genetic predisposition to papillary thyroid carcinoma: Involvement of FOXE1, TSHR, and a novel lincRNA gene, PTCSC2. *J. Clin. Endocrinol. Metab.* **2015**, *100*, E164–E172. [CrossRef] [PubMed]

55. Landa, I.; Ruiz-Llorente, S.; Montero-Conde, C.; Inglada-Perez, L.; Schiavi, F.; Leskela, S.; Pita, G.; Milne, R.; Maravall, J.; Ramos, I.; et al. The variant rs1867277 in FOXE1 gene confers thyroid cancer susceptibility through the recruitment of USF1/USF2 transcription factors. *PLoS Genet.* **2009**, *5*, e1000637. [CrossRef] [PubMed]

56. Mond, M.; Bullock, M.; Yao, Y.; Clifton-Bligh, R.J.; Gilfillan, C.; Fuller, P.J. Somatic Mutations of FOXE1 in Papillary Thyroid Cancer. *Thyroid Off. J. Am. Thyroid Assoc.* **2015**, *25*, 904–910. [CrossRef] [PubMed]

57. Wang, Y.; He, H.; Li, W.; Phay, J.; Shen, R.; Yu, L.; Hancioglu, B.; de la Chapelle, A. MYH9 binds to lncRNA gene PTCSC2 and regulates FOXE1 in the 9q22 thyroid cancer risk locus. *Proc. Natl. Acad. Sci. USA* **2017**, *114*, 474–479. [CrossRef] [PubMed]

58. Bonzanini, M.; Amadori, P.L.; Sagramoso, C.; Dalla Palma, P. Expression of cytokeratin 19 and protein p63 in fine needle aspiration biopsy of papillary thyroid carcinoma. *Acta Cytol.* **2008**, *52*, 541–548. [CrossRef] [PubMed]

59. Candi, E.; Dinsdale, D.; Rufini, A.; Salomoni, P.; Knight, R.A.; Mueller, M.; Krammer, P.H.; Melino, G. TAp63 and DeltaNp63 in cancer and epidermal development. *Cell Cycle* **2007**, *6*, 274–285. [CrossRef] [PubMed]

60. Candi, E.; Rufini, A.; Terrinoni, A.; Giamboi-Miraglia, A.; Lena, A.M.; Mantovani, R.; Knight, R.; Melino, G. DeltaNp63 regulates thymic development through enhanced expression of FgfR2 and Jag2. *Proc. Natl. Acad. Sci. USA* **2007**, *104*, 11999–12004. [CrossRef] [PubMed]

61. Lazzari, C.; Prodosmo, A.; Siepi, F.; Rinaldo, C.; Galli, F.; Gentileschi, M.; Bartolazzi, A.; Costanzo, A.; Sacchi, A.; Guerrini, L.; et al. HIPK2 phosphorylates DeltaNp63alpha and promotes its degradation in response to DNA damage. *Oncogene* **2011**, *30*, 4802–4813. [CrossRef] [PubMed]

62. Frasca, F.; Vella, V.; Aloisi, A.; Mandarino, A.; Mazzon, E.; Vigneri, R.; Vigneri, P. p73 tumor-suppressor activity is impaired in human thyroid cancer. *Cancer Res.* **2003**, *63*, 5829–5837. [PubMed]

63. Vella, V.; Puppin, C.; Damante, G.; Vigneri, R.; Sanfilippo, M.; Vigneri, P.; Tell, G.; Frasca, F. DeltaNp73alpha inhibits PTEN expression in thyroid cancer cells. *Int. J. Cancer* **2009**, *124*, 2539–2548. [CrossRef] [PubMed]

64. Dominguez, G.; Garcia, J.M.; Pena, C.; Silva, J.; Garcia, V.; Martinez, L.; Maximiano, C.; Gomez, M.E.; Rivera, J.A.; Garcia-Andrade, C.; et al. DeltaTAp73 upregulation correlates with poor prognosis in human tumors: Putative in vivo network involving p73 isoforms, p53, and E2F-1. *J. Clin. Oncol. Off. J. Am. Soc. Clin. Oncol.* **2006**, *24*, 805–815. [CrossRef] [PubMed]

65. Ferru, A.; Denis, S.; Guilhot, J.; Gibelin, H.; Tourani, J.M.; Kraimps, J.L.; Larsen, C.J.; Karayan-Tapon, L. Expression of TAp73 and DeltaNp73 isoform transcripts in thyroid tumours. *Eur. J. Surg. Oncol. J. Eur. Soc. Surg. Oncol. Br. Assoc. Surg. Oncol.* **2006**, *32*, 228–230.

66. Puppin, C.; Passon, N.; Frasca, F.; Vigneri, R.; Tomay, F.; Tomaciello, S.; Damante, G. In thyroid cancer cell lines expression of periostin gene is controlled by p73 and is not related to epigenetic marks of active transcription. *Cell. Oncol.* **2011**, *34*, 131–140. [CrossRef] [PubMed]

67. Ruan, K.; Bao, S.; Ouyang, G. The multifaceted role of periostin in tumorigenesis. *Cell. Mol. Life Sci. CMLS* **2009**, *66*, 2219–2230. [CrossRef] [PubMed]

68. Fluge, O.; Bruland, O.; Akslen, L.A.; Lillehaug, J.R.; Varhaug, J.E. Gene expression in poorly differentiated papillary thyroid carcinomas. *Thyroid Off. J. Am. Thyroid Assoc.* **2006**, *16*, 161–175. [CrossRef] [PubMed]

69. Puppin, C.; Fabbro, D.; Dima, M.; Di Loreto, C.; Puxeddu, E.; Filetti, S.; Russo, D.; Damante, G. High periostin expression correlates with aggressiveness in papillary thyroid carcinomas. *J. Endocrinol.* **2008**, *197*, 401–408. [CrossRef] [PubMed]

70. Mayo, L.D.; Dixon, J.E.; Durden, D.L.; Tonks, N.K.; Donner, D.B. PTEN protects p53 from MDM2 and sensitizes cancer cells to chemotherapy. *J. Biol. Chem.* **2002**, *277*, 5484–5489. [CrossRef] [PubMed]

71. Mayo, L.D.; Donner, D.B. A phosphatidylinositol 3-kinase/Akt pathway promotes translocation of MDM2 from the cytoplasm to the nucleus. *Proc. Natl. Acad. Sci. USA* **2001**, *98*, 11598–11603. [CrossRef] [PubMed]

72. Ogawara, Y.; Kishishita, S.; Obata, T.; Isazawa, Y.; Suzuki, T.; Tanaka, K.; Masuyama, N.; Gotoh, Y. Akt enhances MDM2-mediated ubiquitination and degradation of p53. *J. Biol. Chem.* **2002**, *277*, 21843–21850. [CrossRef] [PubMed]

73. Malaguarnera, R.; Vella, V.; Pandini, G.; Sanfilippo, M.; Pezzino, V.; Vigneri, R.; Frasca, F. TAp73 alpha increases p53 tumor suppressor activity in thyroid cancer cells via the inhibition of MDM2-mediated degradation. *Mol. Cancer Res. MCR* **2008**, *6*, 64–77. [CrossRef] [PubMed]

74. Brose, M.S.; Nutting, C.M.; Jarzab, B.; Elisei, R.; Siena, S.; Bastholt, L.; de la Fouchardiere, C.; Pacini, F.; Paschke, R.; Shong, Y.K.; et al. Sorafenib in radioactive iodine-refractory, locally advanced or metastatic differentiated thyroid cancer: A randomised, double-blind, phase 3 trial. *Lancet* **2014**, *384*, 319–328. [CrossRef]

75. Schlumberger, M.; Tahara, M.; Wirth, L.J.; Robinson, B.; Brose, M.S.; Elisei, R.; Habra, M.A.; Newbold, K.; Shah, M.H.; Hoff, A.O.; et al. Lenvatinib versus placebo in radioiodine-refractory thyroid cancer. *N. Engl. J. Med.* **2015**, *372*, 621–630. [CrossRef] [PubMed]

76. Grassi, E.S.; Vezzoli, V.; Negri, I.; Labadi, A.; Fugazzola, L.; Vitale, G.; Persani, L. SP600125 has a remarkable anticancer potential against undifferentiated thyroid cancer through selective action on ROCK and p53 pathways. *Oncotarget* **2015**, *6*, 36383–36399. [PubMed]

77. Riesco-Eizaguirre, G.; Santisteban, P. New insights in thyroid follicular cell biology and its impact in thyroid cancer therapy. *Endocr. Relat. Cancer* **2007**, *14*, 957–977. [CrossRef] [PubMed]

78. Ho, A.L.; Grewal, R.K.; Leboeuf, R.; Sherman, E.J.; Pfister, D.G.; Deandreis, D.; Pentlow, K.S.; Zanzonico, P.B.; Haque, S.; Gavane, S.; et al. Selumetinib-enhanced radioiodine uptake in advanced thyroid cancer. *N. Engl. J. Med.* **2013**, *368*, 623–632. [CrossRef] [PubMed]

79. Weinlander, E.; Somnay, Y.; Harrison, A.D.; Wang, C.; Cheng, Y.Q.; Jaskula-Sztul, R.; Yu, X.M.; Chen, H. The novel histone deacetylase inhibitor thailandepsin A inhibits anaplastic thyroid cancer growth. *J. Surg. Res.* **2014**, *190*, 191–197. [CrossRef] [PubMed]

80. Altmann, A.; Markert, A.; Askoxylakis, V.; Schoning, T.; Jesenofsky, R.; Eisenhut, M.; Haberkorn, U. Antitumor effects of proteasome inhibition in anaplastic thyroid carcinoma. *J. Nucl. Med.* **2012**, *53*, 1764–1771. [CrossRef] [PubMed]

81. Putzer, D.; Gabriel, M.; Kroiss, A.; Madleitner, R.; Eisterer, W.; Kendler, D.; Uprimny, C.; Bale, R.J.; Gastl, G.; Virgolini, I.J. First experience with proteasome inhibitor treatment of radioiodine nonavid thyroid cancer using bortezomib. *Clin. Nucl. Med.* **2012**, *37*, 539–544. [CrossRef] [PubMed]

82. Rehman, A.; Chahal, M.S.; Tang, X.; Bruce, J.E.; Pommier, Y.; Daoud, S.S. Proteomic identification of heat shock protein 90 as a candidate target for p53 mutation reactivation by PRIMA-1 in breast cancer cells. *Breast Cancer Res.* **2005**, *7*, R765–R774. [CrossRef] [PubMed]

83. Qiang, W.; Jin, T.; Yang, Q.; Liu, W.; Liu, S.; Ji, M.; He, N.; Chen, C.; Shi, B.; Hou, P. PRIMA-1 selectively induces global DNA demethylation in p53 mutant-type thyroid cancer cells. *J. Biomed. Nanotechnol.* **2014**, *10*, 1249–1258. [CrossRef] [PubMed]

84. Messina, R.L.; Sanfilippo, M.; Vella, V.; Pandini, G.; Vigneri, P.; Nicolosi, M.L.; Giani, F.; Vigneri, R.; Frasca, F. Reactivation of p53 mutants by prima-1 [corrected] in thyroid cancer cells. *Int. J. Cancer* **2012**, *130*, 2259–2270. [CrossRef] [PubMed]

85. Garufi, A.; D'Orazi, V.; Crispini, A.; D'Orazi, G. Zn(II)-curc targets p53 in thyroid cancer cells. *Int. J. Oncol.* **2015**, *47*, 1241–1248. [CrossRef] [PubMed]

86. Lim, S.M.; Chang, H.; Yoon, M.J.; Hong, Y.K.; Kim, H.; Chung, W.Y.; Park, C.S.; Nam, K.H.; Kang, S.W.; Kim, M.K.; et al. A multicenter, phase II trial of everolimus in locally advanced or metastatic thyroid cancer of all histologic subtypes. *Ann. Oncol.* **2013**, *24*, 3089–3094. [CrossRef] [PubMed]

International Journal of
Molecular Sciences

MDPI

Review

Impact of Gravity on Thyroid Cells

Elisabetta Albi [1], Marcus Krüger [2], Ruth Hemmersbach [3], Andrea Lazzarini [4], Samuela Cataldi [1], Michela Codini [1], Tommaso Beccari [1], Francesco Saverio Ambesi-Impiombato [5] and Francesco Curcio [5,*]

1   Department of Pharmaceutical Science, University of Perugia, San Costanzo, via Romana, 06121 Perugia, Italy; elisabetta.albi@unipg.it (E.A.); samuelacataldi@libero.it (S.C.); michela.codini@unipg.it (M.C.); tommaso.beccari@unipg.it (T.B.)
2   Clinic and Policlinic for Plastic, Aesthetic and Hand Surgery, Otto-von-Guericke-University, Leipziger Str. 44, 39120 Magdeburg, Germany; marcus.krueger@med.ovgu.de
3   German Aerospace Center (DLR), Institute of Aerospace Medicine, Gravitational Biology, Linder Höhe, 51147 Cologne, Germany; ruth.hemmersbach@dlr.de
4   Laboratory of Nuclear Lipid BioPathology, CRABiON, Perugia, via Ponchielli 4, 06073 Perugia, Italy; andrylazza@gmail.it
5   Dipartimento di Area Medica (DAME), University of Udine, p.le M. Kolbe 4, 33100 Udine, Italy; ambesi@me.com
*   Correspondence: francesco.curcio@uniud.it; Tel.: +39-0432-559201

Academic Editor: Daniela Gabriele Grimm
Received: 30 March 2017; Accepted: 28 April 2017; Published: 4 May 2017

Abstract: Physical and mental health requires a correct functioning of the thyroid gland, which controls cardiovascular, musculoskeletal, nervous, and immune systems, and affects behavior and cognitive functions. Microgravity, as occurs during space missions, induces morphological and functional changes within the thyroid gland. Here, we review relevant experiments exposing cell cultures (normal and cancer thyroid cells) to simulated and real microgravity, as well as wild-type and transgenic mice to hypergravity and spaceflight conditions. Well-known mechanisms of damage are presented and new ones, such as changes of gene expression for extracellular matrix and cytoskeleton proteins, thyrocyte phenotype, sensitivity of thyrocytes to thyrotropin due to thyrotropin receptor modification, parafollicular cells and calcitonin production, sphingomyelin metabolism, and the expression and movement of cancer molecules from thyrocytes to colloids are highlighted. The identification of new mechanisms of thyroid injury is essential for the development of countermeasures, both on the ground and in space, against thyroid cancer. We also address the question whether normal and cancer cells show a different sensitivity concerning changes of environmental conditions.

Keywords: thyroid gland; thyroid cancer; microgravity; hypergravity; space environment

1. Introduction

The effects of the space environment are mainly due to cosmic radiation, microgravity, and a confined habitat. Physical and mental equilibrium is very likely to be subject to significant perturbations during prolonged space missions. Astronauts are subjected to several physiological variations, such as cardio-circulatory, musculoskeletal, and immune disorders, together with changes of mental conditions, mood, and personality during and after re-entry from space missions [1]. The space research in the field is currently attracting more and more attention among researchers since, in the near future, an increasing number of astronauts will visit the International Space Station (ISS) and beyond for prolonged times and space is presently considered as the "next frontier" for humankind [1].

The mammalian thyroid gland consists of two lobes structurally composed by follicles and interfollicular spaces. Follicles are surrounded by thyrocytes which synthesize triiodothyronine (T3) and tetraiodothyronine (T4), while C cells in interfollicular spaces secrete calcitonin [2].

The diversity of cell types and the complexity of its hormone functions render the thyroid gland particularly relevant in the functioning of the entire organism. Therefore, maintaining good physical and mental health requires a correctly functioning thyroid gland.

2. How Gravity Influences Thyroid Function

All living organisms have evolved under the influence of the constant gravity force on Earth, which maintains the architecture and function of each kind of cell. The human organism possesses a series of adaptations to gravity changes not only at systemic, but also at the cellular level, such as the regulation of the circadian rhythm, the activation of mechanotransduction pathways, and inducing modifications in immune response, metabolism, and cell proliferation [3–5]. Additionally, genes that modify chromatin structure and methylation have been identified, suggesting that long-term adaptation to gravity may be mediated by epigenetic modifications [5].

2.1. Simulated Microgravity

Usually the impact of microgravity is simulated on the ground by different approaches: prolonged head-down tilt bed rest in humans induces comparable physiological changes with respect to fluid distribution, muscle, and bone loss as in space missions. Small animals, plants, and cell cultures are studied in ground-based facilities under conditions of a randomized influence of gravity. In the case that the exposed system does no longer perceive gravity as a stimulus, and the method applied does not induce non-gravitational effects, the term "simulated microgravity" is justified. Our assumption bears several prerequisites. The exposed system should possess a sensor for the detection of gravity and the threshold of the system should be less than the residual acceleration induced by the experimental conditions. In case of thyroid cells, a gravisensor has not yet been identified and, consequently, a threshold of graviperception is unknown. However, this type of cell reveals a large number of physiological changes under altered gravity conditions which might bring us closer to the identification of a general cellular gravisensory mechanism, which has also been postulated for other cell systems [6].

Microgravity is characterized by the prevention of sedimentation. On the ground this is achieved on so-called clinostats, in which samples are rotated around one axis positioned perpendicular to the direction of the gravity vector. By placing the sample container in the center of rotation and keeping its diameter small (in the range of a few mm) the induced residual accelerations can be kept as minimal as possible. A clinostat with one rotation axis (2D clinostat) is operated constantly in one direction which results in a static change of the gravity vector and in turn—at appropriate speed—prevention of sedimentation.

3D clinostats and random positioning machines (RPM) are further machines used for ground-based microgravity experiments aiming to simulate microgravity conditions [7,8]. Here, two rotation axes are mounted in a gimbal mount and their movement with respect to speed and direction is controlled by an algorithm [9]. While 3D clinostats are commonly rotating continuously, but with changed speed of the two axes at random, in a random positioning machine not only the velocity, but also the direction of rotation is altered in a real random mode. Comparative studies between the different experimental approaches are necessary to identify, experimentally, the induced non-gravitational effects, such as shear forces [8,10] and to avoid misinterpretations. Finally, simulations have to be verified in real microgravity.

Normal thyroid cells (HTU-5 strain) cultured on a 3D clinostat (with a rotation of 60°/s) show apoptotic signs highlighted by electron microscopy analysis, activation of caspase-3, increase in Fas and Bax, and elevation of 85-kDa apoptosis-related cleavage fragments resulting from enhanced poly (ADP-ribose) polymerase activity [11]. Within 12 h of 3D clinorotation the monolayer of human follicular thyroid carcinoma cells (ML-1 line) turns spontaneously into three-dimensional multicellular tumor spheroids with an increase of extracellular matrix proteins and of TGF-β1, while thyroglobulin, fT3 and fT4 secretion are reduced [12] (Figure 1). Within 24–48 h of exposure on a random positioning machine (RPM), ML-1 cells show an upregulation of intermediate filaments, cell adhesion molecules (vimentin and vinculin), extracellular matrix proteins (collagen I and III, laminin, fibronectin, chondroitin sulfate), and Fas protein, while Bcl-2 is downregulated [13]. Laser scanning confocal microscopy of HTU-5 and ML-1 cells immuno-stained with anti-cytokeratin demonstrate that cytokeratin filaments extend from the center, are thickened, coalesce, and shortened, while the vimentin network forms a coiled aggregate, more closely associated with the nucleus as compared to control cells [14]. Changes of enzymes involved in carbohydrate metabolism, protein synthesis, and degradation are present in HTU-5 normal thyroid cells, in FTC-133, and in CGTH W-1 thyroid cancer cells [15]. During long-term exposure on a RPM (7–14 days) thyroid cancer cells form larger and more spheroids than normal thyroid cells, what can be related to an earlier production of the cell adhesion molecule osteopontin [16].

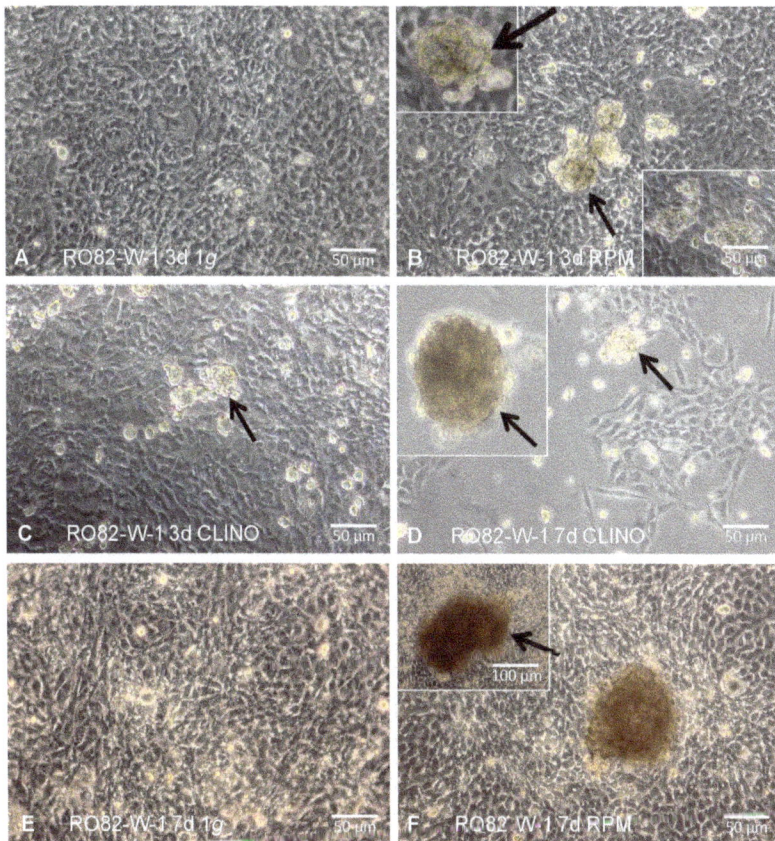

Figure 1. Formation of 3D spheroids: RO82-W-1 cells cultured for 3/7 days at static $1 \times g$ results in a 2D monolayer, while incubation on the RPM or on a clinostat shows 3D aggregates (multicellular spheroids; MCS). (**A**) RO82-W-1 cells cultured for 3 d at static $1 \times g$; (**B**) MCS (arrows) formed on the RPM after 3 days; (**C**) MCS (arrow) formed on the clinostat after 3 days; (**D**) MCS (arrows) of RO82-W-1 cells cultured for 7 days on the clinostat; (**E**) RO82-W-1 cells cultured for 7 days at static $1 \times g$; (**F**) Adherent cells and large MCS (arrow) formed on the RPM after 7 days. For more details see [17].

2.2. Real Microgravity

Short-term effects of real microgravity—in the order of seconds—can be studied with research platforms such as drop towers or parabolic flights (during which microgravity lasts a few seconds), and by sounding rockets (microgravity condition lasts minutes). During the TEXUS-44 mission (launched on 7 February 2008 from Kiruna, in Northern Sweden), FRTL-5 cells were treated with TSH (Thyroid-stimulating hormone) at the onset of microgravity and were fixed after 6 min and 19 s, just at the end of the microgravity period [18]. The study has clearly shown that FRTL-5 cells undergo relevant changes in real microgravity conditions: they aggregate, chromatin condensates, and TSH receptors are shed in the culture medium. This caused impaired production of cAMP and of proteins involved in cell signaling, such as PKC-ζ, PPAR-γ, and SMase and, consequently, failed to respond to TSH stimulation [18]. Live-cell imaging experiments with the fluorescence microscope FLUMIAS during a parabolic flight campaign and the TEXUS-52 mission, showed cytoskeletal changes of FTC-133 cells in real-time. These changes occur rapidly after entrance into microgravity. Under the microscope disturbance of F-actin bundles were detected. Another important finding was the formation of filopodia- and lamellipodia-like structures [19].

2.3. Hypergravity

During the initial launch phase of parabolic flights and spaceflight, hypergravity forces due to rocket acceleration are accompanied by launch vibration. In longer space missions hypergravity is not such an important issue, however, during parabolic flights each of the approximately 30 parabolas normally include about 22 s of microgravity intercalated by periods of normal and hypergravity, which may influence structural and functional changes in plant, animal, and human cells. In ML-1 thyroid cancer cells hypergravity ($1.8\times$ g) does not change *ACTB*, *KRT80*, or *COL4A5* mRNA for extracellular matrix proteins, but upregulates the mRNAs for the metastasis suppressor protein 1 and for the cytoskeleton-associated LIM domain and actin binding protein 1 [20]. Experiments with Nthy-3-1-ori cells at $1.8\times$ g showed that *ITGA10* expression is not gravity-dependent in normal thyroid cells [21]. Hypergravity is also studied on the ground by slow-rotating centrifuges [22]. The exposition of TSH-stimulated FRTL5 cells to hypergravity ($5\times$ g and $9\times$ g) increases cAMP production [23]. In ML-1 cells in culture, centrifugation at $1.8\times$ g induces significant variations of growth factor mRNAs. A PKCα-independent mechanism of *IL6* gene activation seems to be very sensitive to physical forces [24]. In rats, hypergravity increases TSH response and T3 production [25]. Repeated five-day $2\times$ g treatment of rats influences cell structure, enzymes, such as SMase and SM-synthase, and hormone content in the thyroid gland, which differ in quantity or quality from the changes arisen under the primary five days at $2\times$ g, pointing out the animal capability for "memorizing the change of gravity level" [26]. The $2\times$ g exposure upregulates TSHR (thyroid stimulating hormone receptor) surface proteins in mouse thyroids, but the response to TSH treatment remains unchanged without variations of cAMP, because hypergravity delocalizes TSHR [27]. In fact, immunofluorescence analysis of TSHR demonstrates that in control mice maintained in the vivarium the receptor is present on the surface of thyrocytes that surrounded follicles with a precise location, whereas in $2\times$ g samples the fluorescent signal is higher and spreads over the entire surface of the thyrocytes [27]. The cell membrane loses the β-subunit of the receptor in the culture medium together to cholesterol (CHO), whereas SM remains unchanged [28]. On the other hand, the SM metabolism also has no variations; in fact, SMase and SM-synthase 1 expression is not affected by $2\times$ g exposure [29], indicating that membrane CHO, and not SM, is critical for TSH–TSHR interaction. In addition, $2\times$ g exposure induces the loss of parafollicular cells and the reduction of calcitonin production [30].

3. Space Environment Drives Thyroid Cells in Culture toward the Change of Cell Phenotype

Proliferating and quiescent FRTL5 cells have been used in two different space missions performed by the European Space Agency (ESA): the Eneide mission with astronaut Roberto Vittori

(15–25 April 2005), and the Esperia mission with astronaut Paolo Nespoli (23 October–7 November 2007). In these missions, the behavior of thyrocytes in culture in the space environment has been analyzed [30]. During space missions, FRTL5 cells do not respond to TSH treatment; the space environment influences cell proliferation causing a pro-apoptotic condition in cells: this status is characterized by specific expression levels of STAT3, RNA polymerase II, and Bax, and by Smase- and SM-synthase-specific activities [31]. The SMase/SM-synthase activity ratio is very high in apoptotic cells, lower in pro-apoptotic cells, low in proliferating cells, and very low in quiescent cells. A comparison of the SMase/SM-synthase ratio value in space mission cells with that of proliferating, quiescent, pro-apoptotic, and apoptotic cells demonstrates that the space mission cells are in a pro-apoptotic state; thus, the SMase/SM-synthase ratio has been proposed as a biomarker for thyroid cell fate very useful in space missions [31]. FTC-133 analysis at the re-entry from the Shenzhou-8 mission (November 2011, 10 days) shows that the space environment induces a scaffold-free formation of extraordinarily large three-dimensional aggregates [32] and changes in the expression of genes and proteins involved in cancer cell proliferation, metastasis, and survival, shifting the cells toward a less tumor-aggressive phenotype [33]. Strong evidences from lipidomics studies, cancer preclinical models, and clinical trials have shown the key role of lipid molecular species in supporting cancer generation and progression. Such effects may result from fundamental changes in lipid raft composition, persistent ER stress which, through three distinct ERS sensor proteins: ATF6 (activating transcription factor 6), PERK (protein kinase RNA-like endoplasmic reticulum kinase) and IRE1 (inositol-requiring trans-membrane kinase/endonuclease 1), activate the unfolded protein response (UPR), and disruption of the lipid-mediated crosstalk between cancer and stromal cells. All of the processes mentioned above may be influenced by the space environment.

3.1. Thyroid Glands of Wild-Type and Transgenic Mice on Board the International Space Station

The effect of long-term exposure to the space environment on thyroid glands in vivo was shown for the first time by participating in the longest-duration spaceflight mission ever endured by any living animal within the "tissue sharing" team headed by Cancedda [1]. During the mission, six mice had been exposed to the space environment for 91 days (28 August–27 November 2009) on board the International Space Station (ISS), while kept inside the "mouse drawer system" (MDS). Both wild-type (WT) and transgenic mice (TG) with an overexpression of the pleiotropin (PTN) gene, which is under the control of the same specific human bone promoter of the osteocalcin gene, have participated in the mission. The thyroids of TG mice were analyzed with respect to bone metabolism in space and in a corresponding $1 \times g$ reference on the ground. Post-flight thyroid glands showed changes in follicular and parafollicular cells with different modifications in signal lipid and protein content [1,28,30].

3.2. Structural Changes

No significant changes in the proportions of major and minor axes of the glands between mice in space or on the ground were noted, but in the thyroid gland of WT ground control follicles showed variable size and spatial orientation. In contrast, spaceflight animals have a more homogenous thyroid tissue structure, with ordered follicles in which thyrocytes are thicker and the nuclear volume is increased; consequently, the thyroid epithelium vs. colloid volumetric ratio is higher in space than on the ground [1]. In addition, the interfollicular space is strongly reduced with the loss of C cells and defects of calcitonin production [30]. Since, on the ground, a spatial integration of follicular and parafollicular cells, and a functional coordination of both epithelial cells, are observed [34], the space environment evidently induces modifications of follicular cells responsible for C cell changes, causing defective bone homeostasis via thyroid disequilibrium [30]. The loss of thyrocytes arranged in a continuous rim around the colloid and irregularity in the parafollicular spaces with the reduction of calcitonin expression may, in part, be due to the confinement in MDS [35]. Overexpression of PTN on the ground does not protect the thyroid gland from the confinement-induced calcitonin reduction [35], whereas the space environment counteracts the effect of confinement, retains the thyroid C cells in shape, and strongly reduces the loss of C cells, thus exerting a protective action [30].

3.3. Modulation Ligand-Receptor: TSH–TSHR

In WT mice, the space environment induces both TSHR increased expression and different cell distribution, in comparison with the control mice maintained in the vivarium [1]. While in control mice thyroids, the TSHR is distributed uniformly on the thyrocyte surface, after spaceflight the receptor localizes instead in the intracellular junctions and cell membranes, while it is not found in the nuclei. Additionally, caveolin-1 is overexpressed in the space environment and its distribution is highly consistent with the TSHR localization [1]. It has been demonstrated that TSHR is a G protein-coupled receptor [36] associated with both no-raft and raft fractions of cell membranes [37]. In cancer cells, a wide range of signaling proteins and receptors regulating pro-oncogenic and apoptotic pathways during all stages of carcinogenesis reside in lipid rafts. Importantly, lipid rafts and their main component, cholesterol, are increased in the membranes of many types of cancer cells, as well as in the membranes of tumor-released exosomes. Important oncogenetic pathways and their aberrant activation correlate with increased lipid rafts. Lipid rafts may also be involved in cancer dissemination: they participate in cancer cell migration by regulating cytoskeletal reorganization and focal adhesion functions. Cholesterol-depleting agents inhibit the formation and activation of specific lipid raft entities called "clusters of apoptotic signaling molecule-enriched rafts" (CASMERs), which are co-aggregations of lipid rafts with death receptors and their downstream apoptotic molecules. They activate the apoptotic response independently of death receptor ligands.

Furthermore, lipid rafts are rich in SM, CHO [38], and caveolin-1 [39]. SM present in lipid rafts is rapidly metabolized with SMase and SM-synthase enzymes by changing the structure/function of rafts and generating the lipid second messengers involved in signal transduction [40]. The space environment increases the levels of SMase and SM-synthase, moves SMase from the nucleus to the cytoplasm and the cell membrane, and increases its activity [29]. Therefore, in the space environment there is a remodeling of cell membranes which become rich in lipid rafts containing TSHR, caveolin-1, and SMase. This induces the thyroid gland to respond to TSH treatment more intensively as compared to the thyroids of control mice. Thus, the cAMP release in spaceflight animals is higher than that of control animals [1]. Thyrotropin regulates the TSHR-rafts complexes: TSH activates cells through a specific signal transduction pathway and causes the disappearance of the raft-TSHR complexes because it stimulates their monomerization and rapid exit [39]. The production of cAMP by activation of the TSHR, continues in the pre-Golgi compartment and the spatial patterns of downstream signals are influenced by the location of the TSHR-cAMP signal [41]. Cytosolic protein kinase A (PKA) I and PKA II, which are mainly located in the Golgi complex, are activated by cAMP that spreads across the basolateral membrane; several other targets, situated in different cellular compartments, are also phosphorylated by PKAs [42,43]. The increase of TSHR in the thyroids of spaceflight animals is a compensatory mechanism, and explains the strong response to TSH after stimulation. The stress is, in part, due to the confined environment since, in the thyroids of mice maintained in MDS on the ground, intermediate changes between those of control spaceflight animals are observed [35]. During the confinement, TSHR and cAMP slightly increase without any displacement of the receptor within the cell, while no changes were found in SMase expression, localization and activity [35]. Therefore, confinement can only be a predisposing factor and the data in spaceflight animals are due to the combination of stress, confined environment, microgravity, and radiation [35].

3.4. How Space Environment Affects Cancer Proteins in Thyroid Glands

In the thyroid gland, Galectin 3 (Gal-3) plays a significant role in the pathogenesis of well-differentiated carcinoma, particularly in papillary carcinoma [44]. Gal-3, together with human bone marrow endothelial cell-1 (HBME-1) and Cytokeratin-19 (CK-19), are markers most commonly used to assist in distinguishing different thyroid lesions [45]. In addition, MIB-1 index is useful for evaluating proliferative activity and predicting the aggressiveness of thyroid carcinoma [46]. In the space environment, the MIB-1 proliferative index and CK-19 are negative in the thyroid tissue, whereas HBME-1 and Gal-3 show a higher expression in comparison with thyroid tissue in $1 \times g$ controls [47]. Gal-3, usually present in the cytoplasm, nucleus, and extracellular space, diffuses from thyrocytes in the colloid because of the remodeling of cell membrane proteins and lipids which occurs in microgravity. Alshenawy suggests that no single marker is completely sensitive and specific for the diagnosis of thyroid lesions, but only their combination [48] and Gal-3 + HBME-1 is considered highly significant for distinguishing benign from malignant lesions [49]. The possibility that HBME-1 and Gal-3 overexpression might indicate a thyroid tissue premalignant state cannot be excluded, considering that, in microgravity, follicle cells appear two times larger with darker colloids [1], similar to those of papillary carcinoma [50].

4. Cancer versus Non-Cancer Cells

When reviewing literature, the question becomes obvious whether normal cells differ in their sensitivity towards a change in the influence of gravity compared to cancer cells. Furthermore, we have to answer whether spaceflight induces carcinogenesis, understand the underlying mechanisms, and develop countermeasure or protection devices. Studies with normal and cancer thyroid cells might bring us closer to the answers. Table 1 reviews the current data.

Table 1. Research on thyroid cells under the effects of altered gravity (green: normal cells, yellow: cancer cells).

Cells	Exposure Device	Dur.	Analyses and Most Important Findings	Ref.
			HUMAN CELLS	
Primary thyrocytes	s-µg RCCS	14 d	• Morphology: spheroid formation • Protein content: thyroglobulin↑, KGF↑	[51]
HTU-5	s-µg RPM	1 d 2 d 3 d	• Immunofluorescence microscopy: cytokeratin filaments thickened and shortened, extended from poorly defined organizing centers; vimentin formed a coiled aggregate closely associated with the nucleus	[14]
		3 d	• STRING network analysis • Mass spectrometry: high quantities of glycolytic enzymes and marginal quantities of citric acid cycle enzymes	[15,52]
Nthy-3-1-ori	s-µg RPM	7 d 14 d	• Morphology: spheroid formation • Gene expression of genes involved in cytoskeleton forming (*ACTB*↑, *TUBB*↑, *PFN1*↑), growth (*OPN* ↗, *CPNE1*↑, *TGM2*↑→, *NGAL*↑, *COL1A1*↑, *VEGF*↓↑) and signaling (*IL6*↑, *IL8*↑, *IL17* ↗, *PBK*↑, *CASP9*↑, *ERK1/2*↑) • Protein content (MAP profiling): IL-8↑, BDNF↑, MMP-3↑, VEGF↓	[16]
	1.8× g SAHC	2 h	• Gene expression of genes involved in apoptosis (*TNFA*↓), extracellular matrix (*VCAM*↑), growth (*OPN*↓), cytoskeleton (*ABL2*↑, *ACTB*↑, *ITGA10*→), signaling (*CTGF*↑).	[21]
	Vibration Vibraplex		• Gene expression of genes involved in apoptosis (*ANXA2*↓, *TNFA*↓), extracellular matrix (*ADAM19*↓, *ITGB1*↓), cytoskeleton (*ACTB*↓, *VIM*↓), and signaling (*PRKAA1*↓,*PRKCA*↓)	

Table 1. *Cont.*

Cells	Exposure Device	Dur.	Analyses and Most Important Findings	Ref.
			HUMAN CELLS	
ML-1	r-µg PFC	22 s	Relationship between cytoskeleton and ECM under altered gravity • Morphology: F-actin/cytokeratin cytoskeleton altered, no signs of apoptosis or necrosis • Microarray: 2430 significantly regulated transcripts • Gene expression of genes involved in forming cytoskeleton (*ACTB*↑, *LIMA1* ↗), and extracellular matrix (*KER80*↑, *COL4A5*↓, *OPN*↑, *FN*↑). *MTSS1*↓	[20]
	s-µg RPM (CN)	12 h	• Morphology: spheroid formation • Protein content: ECM proteins↑, TGF-β↑ • Protein secretion: Tg↓, fT3↓, fT4↓,	[12]
		2 d	• Morphology: spheroid formation, signs of apoptosis • Protein content: apoptosis: Fas↑, p53↑, Bax↑, Casp3↑	[53]
		1 d 2 d	• Morphology: spheroid formation, induced apoptosis • Protein content: fT3↓, fT4↓, elevated intermediate filaments, cell adhesion molecules, and extracellular matrix proteins	[13]
		1 d 2 d 3 d	• Immunofluorescence microscopy: cytokeratin filaments coalesced and shortened, extended from poorly defined organizing centers; enormous elevation on vimentin filaments • Protein content: Talin↑, α-tubulin↑, β-tubulin↑, β1-integrin↑	[14]
		3 d 7 d	• Morphology: spheroid formation • Protein content: IL-6↑, MCP-1↑, integrin-β1↓ in spheroids	[17]
	s-µg RPM	7 d 11 d	Proteome analysis • Protein content: glutathione S-transferase P↑, nucleoside diphosphate kinase A↑, heat shock cognate 71 kDa protein↑ • Mass spectrometry: 202 different polypeptide chains identified compared to 1g controls (glycolytic enzymes, structural proteins, cytoplasmic actin, tubulin, various heat shock proteins), many proteins showed different Mascot scores	[54]
	1.8g SAHC MuSIC	22 s	• Gene expression of genes involved in forming cytoskeleton (*ACTB*→, *LIMA1*↑), and extracellular matrix (*KER80*→, *COL4A5*→). *MTSS1*↑	[20]
		2 h	• Gene expression of genes involved in cytoskeleton modulation (*EZR*↑, *RDX*↑, *MSN*→), growth factors (*EGF*→, *CTGF*↑), and signaling (*IL6*↓↑, *IL8*↓↑, *PRKAA1*↘↑, *PKC*→)	[24]
	Vibration Vibraplex	22 s	• Gene expression of genes involved in forming cytoskeleton (*ACTB*→, *LIMA1*→), and extracellular matrix (*KER80*→, *COL4A5*→). *MTSS1*→	[20]
		2 h	• Gene expression of genes involved in cytoskeleton forming (*MYO9B*→, *TUBB*→, *VIM*→) and cytoskeleton modulation (*EZR*→, *RDX*→, *MSN*→), growth factors (*EGF*→, *CTGF* ↗), and signaling (*IL6*↓, *IL8*↘, *PRKAA1*↘, *PKC*↘)	[24]
UCLA RO82-W-1	s-µg RPM CN	3 d 7 d	• Morphology: spheroid formation • Protein content: integrin-β1↓ in spheroids	[17]

Table 1. *Cont.*

Cells	Exposure Device	Dur.	Analyses and Most Important Findings	Ref.
			HUMAN CELLS	
CGTH W-1	s-µg RPM	3 d	• Morphology: collagen-chains found • Gene expression: *VIM*↓, *TUBB*↓, *ACTB*↓	[55]
			• STRING network analysis: Considerable number of candidates for gravi-sensitive proteins detected. Clusters of strongly interacting enzymes involved in carbohydrate metabolism, protein • Mass spectrometry: low quantities of glycolytic enzymes and marginal quantities of citrate cycle enzymes, abnormal LDH A-chains	[15,52]
	Vibration Vibraplex	2 h	• Gene expression of genes involved in cytoskeleton forming (*ACTB*↑, *MYO9B*→, *TUBB*↓, *VIM*→, *ITGB1*↑) and cytoskeleton modulation (*EZR*↗, *RDX*→, *MSN*↑), growth factors (*EGF*↗, *CTGF*↑), and signaling (*IL6*↑, *IL8*→, *PKC*↑)	[24]
FTC-133	r-µg Space	10 d	*Shenzhou-8/SIMBOX-mission* • Morphology: spheroid formation • Microarray: 2881 significantly regulated transcripts; genes involved in several biological processes: apoptosis, cytoskeleton, adhesion/extracellular matrix, proliferation, stress response, migration, angiogenesis, signal transduction, regulation of cancer cell proliferation and metastasis • Gene expression of genes involved in extracellular matrix (*OPN*↓), growth (*EGF*↑, *CTGF*↑, *VEGFA*↓, *VEGFD*↑), and signaling (*IL8*↓)	[32,33]
		12 d	*ISS/Cellbox 1-mission* • Morphology: no spheroid formation • Mass spectrometry: 180 different polypeptide chains identified • Protein content: enhanced production of proteins related to the extracellular matrix	[56]
	r-µg TEXUS-52	369 s	• Life-cell imaging (FLUMIAS) with FTC-133 cells expressing the Lifeact-GFP marker protein for the visualization of F-actin: significant alterations of the cytoskeleton	[19]
	r-µg PFC	~3 h	• Microarray: 63 significantly regulated transcripts • Gene expression during the PFC was often regulated in the opposite direction compared with the RPM or Space	[33]
			• Life-cell imaging • Gene expression of genes involved in cytoskeleton forming (*EZR*↑) and signaling (*SEPT11*↓)	[19]
	s-µg CN	4 h 1 d 3 d	• Morphology: spheroid formation • Protein content: decreased cytokine release	[10]
		4 h 1 d 3 d	Gene expression: *CAV1*↓ and *CTGF*↓ in spheroids • Protein content: increased cytokine release	
	s-µg RPM	1 d	Analysis of MCS formation • Morphology: spheroid formation, apoptosis enhanced • Microarray: 487 significantly regulated transcripts • Protein content: NF-κB p65↑ • Gene expression: AD: *IL6*↑, *IL8*↑, *CD44*↑, *OPN*↑, *ERK1/2*↓, *CAV2*↓, *TLN1*↓, *CTGF*↓; MCS: *ERK2*↓, *IL6*↓, *CAV2*↓, *TLN1*↓, *CTGF*↓	[57]
		3 d	Proteomic analysis with focus cytoskeletal and membrane-associated proteins to understand forming of larger MCS by FTC-133 cells • FF-IEF/SDS-PAGE/mass spectrometry: integrin α5 chains, myosin-10 and filamin B only found in protein solution of FTC-133 cells → possible role in binding fibronectin • Gene expression: *VIM*↑, *TUDD*↑, *ACTB*↓	[55]

Table 1. *Cont.*

Cells	Exposure Device	Dur.	Analyses and Most Important Findings	Ref.
			HUMAN CELLS	
	s-µg RPM	3 d	• STRING network analysis: considerable number of candidates for gravi-sensitive proteins detected. Clusters of strongly interacting enzymes involved in carbohydrate metabolism, protein formation, degradation, and cell shaping and proteins regulating cell growth. • Mass spectrometry: high quantities of glycolytic enzymes and moderate quantities of citric acid cycle enzymes, abnormal LDH B-chains	[15,52]
	s-µg RPM	10 d	• Morphology: spheroid formation • Microarray: 2881 significantly regulated transcripts • Gene expression: of genes involved in extracellular matrix (*OPN*↑), growth (*EGF*↑, *CTGF*↑, *VEGFA*↓, *VEGFD*↑), and signaling (*IL8*↓)	[32,33]
FTC-133	s-µg RPM	7 d 14 d	• Morphology: formation of larger and numerous spheroids than normal cells • Gene expression of genes involved in cytoskeleton forming (*ACTB*↑, *TUBB*↑, *PFN1*↑), growth (*OPN*↑, *CPNE1*↑, *TGM2*↑, *NGAL*↑, *COL1A1*↓, *VEGF*↓→) and signaling (*IL6*↑, *IL7*↑, *IL8*↑, *IL17*↗, *PBK*→, *CASP9*↑, *ERK1/2*↑, *FLT1*↑→, *FLK1*↑→) • Protein content (MAP profiling): IL-6↑, MIP-1α↑, IL-1β↑, IL-1ra↑, IL-12p40↑, IL-12p70↑, IL-15↑, IL-17↑, SCF↑, VEGF↓, NGAL↑	[16]
	1.8× *g* SAHC	2 h	• Life-cell imaging • Gene expression of genes involved in cytoskeleton forming (*ACTB*↑, *EZR*↑, *RDX*↑, *MSN*↑) and signaling (*LCP1*↓)	[19]
	Vibration Vibraplex	2 h	• Life-cell imaging • Gene expression of genes involved in cytoskeleton forming (*MSN*↓) and signaling	
			ANIMAL MODELS	
	s-µg CN	5–7 d	• Activity: less-responsive to TSH stimulation in terms of cAMP	[58]
FRTL-5 (rat)	5*g*/9*g* LSC	1 h	• cells functionally respond to the variable gravity force in a dose-dependent manner in terms of cAMP production following TSH-stimulation	[23]
	r-µg TEXUS-44	379 s	• Morphology: irregular shape, rearrangement of the cell membrane • Activity: no response to TSH, shedding of TSH-R in the supernatant • Protein content: Bax↑, sphingomyelin-synthase↑	[18]
	r-µg Space	91 d	• Morphology: increase in average follicle size • Protein content: sphingomyelinase↑, sphingomyelin-synthase↑	[29,30]
Thyroid gland (mouse)		3 mo	• Morphology: thyroid follicles appeared more organized • Protein content: caveolin-1↑, TSH-R↑	[1]
		90 d	• Protein content: HBME-1↑, galectin-3↑	[47]
	2× *g* centrifuge	90/91 d	• Protein content: TSHR↑, caveolin-1↑, STAT3↓ • cholesterol↓, cAMP→	[28,30]

AD, adherent cells; CN, clinostat; Dur., duration; d, day; h, hour; LSC, low speed centrifuge; MCS, multicellular spheroids; mo, month; MuSIC, multi-sample incubator centrifuge; PFC, parabolic flight campaign; r-µg, real microgravity; RPM, random positioning machine; s-µg, simulated microgravity; s, second; SAHC, short-arm human centrifuge; ↑, upregulation; ↓, downregulation; ↗, slight upregulation; ↘, slight downregulation; →, not regulated.

Table 1 shows that normal, as well as cancerous, thyroid cells react to altered gravity conditions. In addition to morphological changes (the induction of spheroid formation), primarily, the gene expression of those genes, which are involved in cytoskeleton formation, cytoskeleton modulation, and extracellular matrix, is altered. Up to now, it is difficult to say whether normal cells are more sensitive than cancer cells. Ivanova et al. found that long-term exposure to hypergravity stimulated cGMP efflux in cultured human melanocytes and non-metastatic melanoma cells, whereas highly-metastatic melanoma cells appeared to be insensitive to hypergravity, most probably due to an upregulated cGMP efflux at $1 \times g$ [59].

At the current status, a statement of a differential gravity response of normal versus cancerous thyroid cells is difficult. It cannot be excluded that experimental parameters, for example, radiation, hardware geometry, cell density, and experiment operations during spaceflight as well as in ground facilities influence the results. Thus, they have to be clearly defined and described. Further experiments are needed to address the gravity impact on thyroid function with respect to the risk of astronauts to develop thyroid cancer during long-term spaceflights.

5. Summary and Perspective

Life, as we know it in our planet, evolved not taking into account the effects of microgravity and space radiation. The space environment induces several changes in thyroid glands under the influence of the pituitary gland. These observations may be instrumental for developing protective measures and countermeasures, to be adopted for the health and safety of all individuals exposed on Earth to extreme living and/or working conditions, and to astronauts prior to exposing them to unpredictable and unsustainable risks during long-term space flight missions.

Acknowledgments: We wish to acknowledge the financial support from the University of Udine, Italy.

Conflicts of Interest: The authors have declared no conflicts of interest.

Abbreviations

2D	Two-dimensional
3D	Three-dimensional
fT3	Free triiodothyronine
fT4	Free thyroxine
g	Gravity, acceleration
MCS	Multicellular spheroids
MIB-1	Ki-67 equivalent antibodies
PKA	Protein kinase A
PKC	Protein kinase C
PPAR	Peroxisome proliferator-activated receptor
RPM	Random Positioning Machine
SM	Sphingomyelin
STAT3	Signal transducer and activator of transcription 3
TSH	Thyroid-stimulating hormone
TSHR	Thyroid-stimulating hormone receptor
μg	Microgravity

References

1. Masini, M.A.; Albi, E.; Barmo, C.; Bonfiglio, T.; Bruni, L.; Canesi, L.; Cataldi, S.; Curcio, F.; D'Amora, M.; Ferri, I.; et al. The Impact of Long-Term Exposure to Space Environment on Adult Mammalian Organisms: A Study on Mouse Thyroid and Testis. *PLoS ONE* **2012**, *7*, e35418. [CrossRef] [PubMed]
2. Capen, C.C.; Martin, S.L. The effects of xenobiotics on the structure and function of thyroid follicular and C-cells. *Toxicol. Pathol.* **1989**, *17*, 266–293. [CrossRef] [PubMed]
3. Hauschild, S.; Tauber, S.; Lauber, B.; Thiel, C.S.; Layer, L.E.; Ullrich, O. T cell regulation in microgravity—The current knowledge from in vitro experiments conducted in space, parabolic flights and ground-based facilities. *Acta Astronaut.* **2014**, *104*, 365–377. [CrossRef]
4. Li, N.; An, L.; Hang, H. Increased sensitivity of DNA damage response-deficient cells to stimulated microgravity-induced DNA lesions. *PLoS ONE* **2015**, *10*, e0125236. [CrossRef] [PubMed]

5. Najrana, T.; Sanchez-Esteban, J. Mechanotransduction as an Adaptation to Gravity. *Front.Pediatr.* **2016**, *4*, 140. [CrossRef] [PubMed]
6. Häder, D.-P.; Braun, M.; Grimm, D.; Hemmersbach, R. Gravireceptors in eukaryotes—A comparison of case studies on the cellular level. *Npj Nat. Microgravity* **2017**, *3*, 13. [CrossRef]
7. Brungs, S.; Egli, M.; Wuest, S.L.; Christianen, P.C.M.; W.A. van Loon, J.J.; Ngo Anh, T.J.; Hemmersbach, R. Facilities for Simulation of Microgravity in the ESA Ground-Based Facility Programme. *Microgravity Sci. Technol.* **2016**, *28*, 191–203. [CrossRef]
8. Hauslage, J.; Cevik, V.; Hemmersbach, R. *Pyrocystis noctiluca* represents an excellent bioassay for shear forces induced in ground-based microgravity simulators (Clinostat and Random Positioning Machine). *Npj Nat. Microgravity* **2017**, *3*, 12. [CrossRef]
9. Wuest, S.L.; Richard, S.; Kopp, S.; Grimm, D.; Egli, M. Simulated Microgravity: Critical Review on the Use of Random Positioning Machines for Mammalian Cell Culture. *BioMed Res. Int.* **2015**, *2015*, 8. [CrossRef] [PubMed]
10. Warnke, E.; Pietsch, J.; Wehland, M.; Bauer, J.; Infanger, M.; Gorog, M.; Hemmersbach, R.; Braun, M.; Ma, X.; Sahana, J.; et al. Spheroid formation of human thyroid cancer cells under simulated microgravity: A possible role of CTGF and CAV1. *Cell Commun. Signal* **2014**, *12*, 32. [CrossRef] [PubMed]
11. Kossmehl, P.; Shakibaei, M.; Cogoli, A.; Infanger, M.; Curcio, F.; Schonberger, J.; Eilles, C.; Bauer, J.; Pickenhahn, H.; Schulze-Tanzil, G.; et al. Weightlessness induced apoptosis in normal thyroid cells and papillary thyroid carcinoma cells via extrinsic and intrinsic pathways. *Endocrinology* **2003**, *144*, 4172–4179. [CrossRef] [PubMed]
12. Grimm, D.; Kossmehl, P.; Shakibaei, M.; Schulze-Tanzil, G.; Pickenhahn, H.; Bauer, J.; Paul, M.; Cogoli, A. Effects of simulated microgravity on thyroid carcinoma cells. *J. Gravit. Physiol.* **2002**, *9*, P253–P256. [PubMed]
13. Grimm, D.; Bauer, J.; Kossmehl, P.; Shakibaei, M.; Schoberger, J.; Pickenhahn, H.; Schulze-Tanzil, G.; Vetter, R.; Eilles, C.; Paul, M.; et al. Simulated microgravity alters differentiation and increases apoptosis in human follicular thyroid carcinoma cells. *Faseb J.* **2002**, *16*, 604–606. [CrossRef] [PubMed]
14. Infanger, M.; Kossmehl, P.; Shakibaei, M.; Schulze-Tanzil, G.; Cogoli, A.; Faramarzi, S.; Bauer, J.; Curcio, F.; Paul, M.; Grimm, D. Longterm conditions of mimicked weightlessness influences the cytoskeleton in thyroid cells. *J. Gravit. Physiol.* **2004**, *11*, P169–P172. [PubMed]
15. Pietsch, J.; Riwaldt, S.; Bauer, J.; Sickmann, A.; Weber, G.; Grosse, J.; Infanger, M.; Eilles, C.; Grimm, D. Interaction of proteins identified in human thyroid cells. *Int. J. Mol. Sci.* **2013**, *14*, 1164–1178. [CrossRef] [PubMed]
16. Kopp, S.; Warnke, E.; Wehland, M.; Aleshcheva, G.; Magnusson, N.E.; Hemmersbach, R.; Corydon, T.J.; Bauer, J.; Infanger, M.; Grimm, D. Mechanisms of three-dimensional growth of thyroid cells during long-term simulated microgravity. *Sci. Rep.* **2015**, *5*, 16691. [CrossRef] [PubMed]
17. Svejgaard, B.; Wehland, M.; Ma, X.; Kopp, S.; Sahana, J.; Warnke, E.; Aleshcheva, G.; Hemmersbach, R.; Hauslage, J.; Grosse, J.; et al. Common Effects on Cancer Cells Exerted by a Random Positioning Machine and a 2D Clinostat. *PLoS ONE* **2015**, *10*, e0135157. [CrossRef] [PubMed]
18. Albi, E.; Ambesi-Impiombato, F.S.; Peverini, M.; Damaskopoulou, E.; Fontanini, E.; Lazzarini, R.; Curcio, F.; Perrella, G. Thyrotropin receptor and membrane interactions in FRTL-5 thyroid cell strain in microgravity. *Astrobiology* **2011**, *11*, 57–64. [CrossRef] [PubMed]
19. Corydon, T.J.; Kopp, S.; Wehland, M.; Braun, M.; Schutte, A.; Mayer, T.; Hulsing, T.; Oltmann, H.; Schmitz, B.; Hemmersbach, R.; et al. Alterations of the cytoskeleton in human cells in space proved by life-cell imaging. *Sci. Rep.* **2016**, *6*, 20043. [CrossRef] [PubMed]
20. Ulbrich, C.; Pietsch, J.; Grosse, J.; Wehland, M.; Schulz, H.; Saar, K.; Hubner, N.; Hauslage, J.; Hemmersbach, R.; Braun, M.; et al. Differential gene regulation under altered gravity conditions in follicular thyroid cancer cells: Relationship between the extracellular matrix and the cytoskeleton. *Cell Physiol. Biochem.* **2011**, *28*, 185–198. [CrossRef] [PubMed]
21. Wehland, M.; Warnke, E.; Frett, T.; Hemmersbach, R.; Hauslage, J.; Ma, X.; Aleshcheva, G.; Pietsch, J.; Bauer, J.; Grimm, D. The Impact of Hypergravity and Vibration on Gene and Protein Expression of Thyroid Cells. *Microgravity Sci. Technol.* **2016**, *28*, 261–274. [CrossRef]
22. Frett, T.; Petrat, G.; van Loon, J.J.W.A.; Hemmersbach, R.; Anken, R. Hypergravity Facilities in the ESA Ground-Based Facility Program—Current Research Activities and Future Tasks. *Microgravity Sci. Technol.* **2016**, *28*, 205–214. [CrossRef]

23. Meli, A.; Perrella, G.; Curcio, F.; Hemmersbach, R.; Neubert, J.; Impiombato, F.A. Response to thyrotropin of normal thyroid follicular cell strain FRTL5 in hypergravity. *Biochimie* **1999**, *81*, 281–285. [CrossRef]

24. Ma, X.; Wehland, M.; Aleshcheva, G.; Hauslage, J.; Wasser, K.; Hemmersbach, R.; Infanger, M.; Bauer, J.; Grimm, D. Interleukin-6 expression under gravitational stress due to vibration and hypergravity in follicular thyroid cancer cells. *PLoS ONE* **2013**, *8*, e68140. [CrossRef] [PubMed]

25. Krasnov, I.B.; Alekseev, E.I.; Loginov, V.I. Role of the endocrine glands in divergence of plastic processes and energy metabolism in rats after extended exposure to hypergravity: Cytologic investigation. *Aviakosm. Ekolog. Med.* **2006**, *40*, 29–34. [PubMed]

26. Krasnov, I.B. Gravity induced postponed potentiation as a result of repeated 2 G influence on rats. *J. Gravit. Physiol.* **2002**, *9*, P41–P42. [PubMed]

27. Lazzarini, A.; Albi, E.; Floridi, A.; Lazzarini, R.; Loreti, E.; Ferri, I.; Curcio, F.; Ambesi-Impiombato, F. Hypergravity delocalizes thyrotropin receptor. *FASEB J.* **2014**, *28* (Suppl. S1), 650.3.

28. Albi, E.; Curcio, F.; Lazzarini, A.; Floridi, A.; Cataldi, S.; Lazzarini, R.; Loreti, E.; Ferri, I.; Ambesi-Impiombato, F.S. A firmer understanding of the effect of hypergravity on thyroid tissue: Cholesterol and thyrotropin receptor. *PLoS ONE* **2014**, *9*, e98250. [CrossRef] [PubMed]

29. Albi, E.; Curcio, F.; Spelat, R.; Lazzarini, A.; Lazzarini, R.; Loreti, E.; Ferri, I.; Ambesi-Impiombato, F.S. Observing the mouse thyroid sphingomyelin under space conditions: A case study from the MDS mission in comparison with hypergravity conditions. *Astrobiology* **2012**, *12*, 1035–1041. [CrossRef] [PubMed]

30. Albi, E.; Curcio, F.; Spelat, R.; Lazzarini, A.; Lazzarini, R.; Cataldi, S.; Loreti, E.; Ferri, I.; Ambesi-Impiombato, F.S. Loss of Parafollicular Cells during Gravitational Changes (Microgravity, Hypergravity) and the Secret Effect of Pleiotrophin. *PLoS ONE* **2012**, *7*, e48518. [CrossRef] [PubMed]

31. Albi, E.; Ambesi-Impiombato, S.; Villani, M.; de Pol, I.; Spelat, R.; Lazzarini, R.; Perrella, G. Thyroid cell growth: Sphingomyelin metabolism as non-invasive marker for cell damage acquired during spaceflight. *Astrobiology* **2010**, *10*, 811–820. [CrossRef] [PubMed]

32. Pietsch, J.; Ma, X.; Wehland, M.; Aleshcheva, G.; Schwarzwalder, A.; Segerer, J.; Birlem, M.; Horn, A.; Bauer, J.; Infanger, M.; et al. Spheroid formation of human thyroid cancer cells in an automated culturing system during the Shenzhou-8 Space mission. *Biomaterials* **2013**, *34*, 7694–7705. [CrossRef] [PubMed]

33. Ma, X.; Pietsch, J.; Wehland, M.; Schulz, H.; Saar, K.; Hubner, N.; Bauer, J.; Braun, M.; Schwarzwalder, A.; Segerer, J.; et al. Differential gene expression profile and altered cytokine secretion of thyroid cancer cells in space. *Faseb J.* **2014**, *28*, 813–835. [CrossRef] [PubMed]

34. Kalisnik, M.; Vraspir-Porenta, O.; Kham-Lindtner, T.; Logonder-Mlinsek, M.; Pajer, Z.; Stiblar-Martincic, D.; Zorc-Pleskovic, R.; Trobina, M. The interdependence of the follicular, parafollicular, and mast cells in the mammalian thyroid gland: A review and a synthesis. *Am. J. Anat.* **1988**, *183*, 148–157. [CrossRef] [PubMed]

35. Albi, E.; Ambesi-Impiombato, F.S.; Lazzarini, A.; Lazzarini, R.; Floridi, A.; Cataldi, S.; Loreti, E.; Ferri, I.; Curcio, F. Reinterpretation of mouse thyroid changes under space conditions: The contribution of confinement to damage. *Astrobiology* **2014**, *14*, 563–567. [CrossRef] [PubMed]

36. Graves, P.N.; Vlase, H.; Bobovnikova, Y.; Davies, T.F. Multimeric complex formation by the thyrotropin receptor in solubilized thyroid membranes. *Endocrinology* **1996**, *137*, 3915–3920. [CrossRef] [PubMed]

37. Latif, R.; Ando, T.; Davies, T.F. Lipid rafts are triage centers for multimeric and monomeric thyrotropin receptor regulation. *Endocrinology* **2007**, *148*, 3164–3175. [CrossRef] [PubMed]

38. Cascianelli, G.; Villani, M.; Tosti, M.; Marini, F.; Bartoccini, E.; Magni, M.V.; Albi, E. Lipid microdomains in cell nucleus. *Mol. Biol. Cell* **2008**, *19*, 5289–5295. [CrossRef] [PubMed]

39. Latif, R.; Ando, T.; Daniel, S.; Davies, T.F. Localization and regulation of thyrotropin receptors within lipid rafts. *Endocrinology* **2003**, *144*, 4725–4728. [CrossRef] [PubMed]

40. Albi, E.; Lazzarini, A.; Lazzarini, R.; Floridi, A.; Damaskopoulou, E.; Curcio, F.; Cataldi, S. Nuclear lipid microdomain as place of interaction between sphingomyelin and DNA during liver regeneration. *Int. J. Mol. Sci.* **2013**, *14*, 6529–6541. [CrossRef] [PubMed]

41. Calebiro, D.; Nikolaev, V.O.; Gagliani, M.C.; de Filippis, T.; Dees, C.; Tacchetti, C.; Persani, L.; Lohse, M.J. Persistent cAMP-signals triggered by internalized G-protein-coupled receptors. *PLoS Biol.* **2009**, *7*, e1000172. [CrossRef] [PubMed]

42. Calebiro, D.; de Filippis, T.; Lucchi, S.; Martinez, F.; Porazzi, P.; Trivellato, R.; Locati, M.; Beck-Peccoz, P.; Persani, L. Selective modulation of protein kinase A I and II reveals distinct roles in thyroid cell gene expression and growth. *Mol. Endocrinol.* **2006**, *20*, 3196–3211. [CrossRef] [PubMed]

43. Vassart, G.; Dumont, J.E. The thyrotropin receptor and the regulation of thyrocyte function and growth. *Endocr. Rev.* **1992**, *13*, 596–611. [PubMed]

44. Yoshii, T.; Inohara, H.; Takenaka, Y.; Honjo, Y.; Akahani, S.; Nomura, T.; Raz, A.; Kubo, T. Galectin-3 maintains the transformed phenotype of thyroid papillary carcinoma cells. *Int. J. Oncol.* **2001**, *18*, 787–792. [CrossRef] [PubMed]

45. De Matos, L.L.; del Giglio, A.B.; Matsubayashi, C.O.; de Lima Farah, M.; del Giglio, A.; da Silva Pinhal, M.A. Expression of CK-19, galectin-3 and HBME-1 in the differentiation of thyroid lesions: Systematic review and diagnostic meta-analysis. *Diagn. Pathol.* **2012**, *7*, 97. [CrossRef] [PubMed]

46. Kjellman, P.; Wallin, G.; Hoog, A.; Auer, G.; Larsson, C.; Zedenius, J. MIB-1 index in thyroid tumors: A predictor of the clinical course in papillary thyroid carcinoma. *Thyroid* **2003**, *13*, 371–380. [CrossRef] [PubMed]

47. Albi, E.; Curcio, F.; Lazzarini, A.; Floridi, A.; Cataldi, S.; Lazzarini, R.; Loreti, E.; Ferri, I.; Ambesi-Impiombato, F.S. How microgravity changes galectin-3 in thyroid follicles. *BioMed Res. Int.* **2014**, *2014*, 652863. [CrossRef] [PubMed]

48. Alshenawy, H.A. Utility of immunohistochemical markers in differential diagnosis of follicular cell-derived thyroid lesions. *JMAU* **2014**, *2*, 127–136. [CrossRef]

49. Saleh, H.A.; Feng, J.; Tabassum, F.; Al-Zohaili, O.; Husain, M.; Giorgadze, T. Differential expression of galectin-3, CK-19, HBME1, and Ret oncoprotein in the diagnosis of thyroid neoplasms by fine needle aspiration biopsy. *Cytojournal* **2009**, *6*, 18. [CrossRef] [PubMed]

50. Lloyd, R.V.; Buehler, D.; Khanafshar, E. Papillary thyroid carcinoma variants. *Head Neck Pathol.* **2011**, *5*, 51–56. [CrossRef] [PubMed]

51. Martin, A.; Zhou, A.; Gordon, R.E.; Henderson, S.C.; Schwartz, A.E.; Schwartz, A.E.; Friedman, E.W.; Davies, T.F. Thyroid organoid formation in simulated microgravity: Influence of keratinocyte growth factor. *Thyroid* **2000**, *10*, 481–487. [PubMed]

52. Pietsch, J.; Sickmann, A.; Weber, G.; Bauer, J.; Egli, M.; Wildgruber, R.; Infanger, M.; Grimm, D. Metabolic enzyme diversity in different human thyroid cell lines and their sensitivity to gravitational forces. *Proteomics* **2012**, *12*, 2539–2546. [CrossRef] [PubMed]

53. Kossmehl, P.; Shakibaei, M.; Cogoli, A.; Pickenhahn, H.; Paul, M.; Grimm, D. Simulated microgravity induces programmed cell death in human thyroid carcinoma cells. *J. Gravit. Physiol.* **2002**, *9*, P295–P296. [PubMed]

54. Pietsch, J.; Bauer, J.; Weber, G.; Nissum, M.; Westphal, K.; Egli, M.; Grosse, J.; Schönberger, J.; Eilles, C.; Infanger, M.; et al. Proteome Analysis of Thyroid Cancer Cells After Long-Term Exposure to a Random Positioning Machine. *Microgravity Sci. Technol.* **2011**, *23*, 381–390. [CrossRef]

55. Pietsch, J.; Sickmann, A.; Weber, G.; Bauer, J.; Egli, M.; Wildgruber, R.; Infanger, M.; Grimm, D. A proteomic approach to analysing spheroid formation of two human thyroid cell lines cultured on a random positioning machine. *Proteomics* **2011**, *11*, 2095–2104. [CrossRef] [PubMed]

56. Riwaldt, S.; Pietsch, J.; Sickmann, A.; Bauer, J.; Braun, M.; Segerer, J.; Schwarzwalder, A.; Aleshcheva, G.; Corydon, T.J.; Infanger, M.; et al. Identification of proteins involved in inhibition of spheroid formation under microgravity. *Proteomics* **2015**, *15*, 2945–2952. [CrossRef] [PubMed]

57. Grosse, J.; Wehland, M.; Pietsch, J.; Schulz, H.; Saar, K.; Hubner, N.; Eilles, C.; Bauer, J.; Abou-El-Ardat, K.; Baatout, S.; et al. Gravity-sensitive signaling drives 3-dimensional formation of multicellular thyroid cancer spheroids. *FASEB J.* **2012**, *26*, 5124–5140. [CrossRef] [PubMed]

58. Meli, A.; Perrella, G.; Curcio, F.; Ambesi-Impiombato, F.S. Response to hypogravity of normal in vitro cultured follicular cells from thyroid. *Acta Astronaut.* **1998**, *42*, 465–472. [CrossRef]

59. Ivanova, K.; Das, P.K.; Gerzer, R. Melanocytes: Interface of cell biology and pathobiology with a focus on nitric oxide and cGMP signaling. *BMC Pharmacol.* **2007**, *7*, S31. [CrossRef]

International Journal of
Molecular Sciences

MDPI

Communication

Proteome Analysis of Human Follicular Thyroid Cancer Cells Exposed to the Random Positioning Machine

Johann Bauer [1,*], Sascha Kopp [2], Elisabeth Maria Schlagberger [1], Jirka Grosse [3], Jayashree Sahana [4], Stefan Riwaldt [4], Markus Wehland [2], Ronald Luetzenberg [2], Manfred Infanger [2] and Daniela Grimm [2,4]

[1] Max-Planck-Institute for Biochemistry, Scientific Information Services, 82152 Martinsried, Germany; schlagberger@biochem.mpg.de
[2] Clinic and Policlinic for Plastic, Aesthetic and Hand Surgery, Otto-von-Guericke-University, 39120 Magdeburg, Germany; sascha.kopp@med.ovgu.de (S.K.); markus.wehland@med.ovgu.de (M.W.); ronald.luetzenberg@med.ovgu.de (R.L.); manfred.infanger@med.ovgu.de (M.I.); dgg@biomed.au.dk (D.G.)
[3] Department of Nuclear Medicine, University Hospital, University of Regensburg, 95053 Regensburg, Germany; jirka.grosse@klinik.uni-regensburg.de
[4] Department of Biomedicine, Aarhus University, 8000 Aarhus C, Denmark; jaysaha@biomed.au.dk (J.S.); sr@biomed.au.dk (S.R.)
* Correspondence: jbauer@biochem.mpg.de; Tel.: +49-89-85783803; Fax: +49-89-141-7931

Academic Editor: Anthony Lemarié
Received: 26 October 2016; Accepted: 27 February 2017; Published: 3 March 2017

Abstract: Several years ago, we detected the formation of multicellular spheroids in experiments with human thyroid cancer cells cultured on the Random Positioning Machine (RPM), a ground-based model to simulate microgravity by continuously changing the orientation of samples. Since then, we have studied cellular mechanisms triggering the cells to leave a monolayer and aggregate to spheroids. Our work focused on spheroid-related changes in gene expression patterns, in protein concentrations, and in factors secreted to the culture supernatant during the period when growth is altered. We detected that factors inducing angiogenesis, the composition of integrins, the density of the cell monolayer exposed to microgravity, the enhanced production of caveolin-1, and the nuclear factor kappa B p65 could play a role during spheroid formation in thyroid cancer cells. In this study, we performed a deep proteome analysis on FTC-133 thyroid cancer cells cultured under conditions designed to encourage or discourage spheroid formation. The experiments revealed more than 5900 proteins. Their evaluation confirmed and explained the observations mentioned above. In addition, we learned that FTC-133 cells growing in monolayers or in spheroids after RPM-exposure incorporate vinculin, paxillin, focal adhesion kinase 1, and adenine diphosphate (ADP) ribosylation factor 6 in different ways into the focal adhesion complex.

Keywords: cellular compartments; mass spectrometry; proteomics; pathway analysis; random positioning machine

---

1. Introduction

Multicellular spheroids (MCS) are interesting models of cancer [1,2]. They resemble natural tumors more than cell monolayers, but they are not as complex as those [3]. Spheroids may be generated in various ways [1]. One example is their exposure to real (r-μg) and simulated microgravity (s-μg) [3]. s-μg can be achieved by application of the Random Positioning Machine (RPM), a device created to simulate microgravity on Earth. During RPM-exposure, human cells are oriented randomly with respect to the gravity vector, so that cell sedimentation does not occur [3]. In addition, the RPM is an interesting tool for novel applications, such as three-dimensional cell culturing as well as tissue engineering [3].

Technical characteristics for microgravity-dependent spheroid formation include the following: matrices are not required, frictional forces are extremely low, and cell-cell contact is established by cell surface features. Spheroid formation under microgravity in vitro includes a cell leaving a two-dimensional monolayer and joining neighbor cells in a three-dimensional manner. This process is supposed to mirror, in part, the change of a cancer cell's kind of growth in vivo as it is observed during metastasis [4]. Hence, knowing the detailed cellular changes of spheroid formation may provide information about cellular events occurring during metastasis.

For more than a decade we have aimed to define the cellular mechanisms causing the transition from two- to three-dimensional growth in cells cultured under r-µg or s-µg. Meanwhile, we know that subconfluent cultures of FTC-133 thyroid cancer cells divide into two populations when they are exposed to r-µg or s-µg [5,6]. One continues to grow adherently on the bottom of the culture dish and the other one forms three-dimensional aggregates (spheroids), which detach from the surface of the culture dish and float in the culture medium. Interestingly, FTC-133 cells do not form spheroids when they are exposed to s-µg created by an RPM or r-µg during spaceflight after previously forming confluent monolayers [7]. Under these conditions, all cells continue to grow within a monolayer, independent of the gravitational force they are exposed to.

In order to gain more information about the molecular mechanisms causing this differential growth behavior, we studied FTC-133 cells after encouraging or discouraging spheroid formation in vitro [5–7]. The harvested samples were analyzed by applying microscopy, gene array technology, quantitative real time-PCR, and Multi-Analyte Profiling with the aim to examine cell morphologies, gene expression patterns, and proteins which could be associated with the transition from a two- to a three-dimensional growth pattern [5,6,8,9]. Taken together, these efforts revealed that exposure of FTC-133 cells to microgravity enhances apoptosis and promotes nuclear factor kappa B (NFκB) p65 activities, while the caveolin 1 (*CAV-1*) gene is down-regulated during spheroid formation [6,10–12]. Most interestingly, several factors triggering angiogenesis were found in supernatants of thyroid cells exposed to the RPM [8]. In addition, Western blotting and mass spectrometry (MS) experiments revealed increased concentrations of several proteins, including vinculin, during growth on the RPM, while caveolin-1 proteins were enriched in confluent FTC-133 cultures which do not form spheroids, even if incubated on a RPM [7,12].

In this study, we applied advanced MS [13] and analyzed cells harvested after growth under five distinct conditions, as shown in Table 1. The proteins detected were assigned to their original cell compartments and the canonical pathways, wherein they are active, were assigned using modern computer programs. Thereby, we recognized that different proteins regulating the structure of the focal adhesion complexes were detectable in spheroid forming cells compared to cells that continued an adherent growth under equal conditions. The results challenged the conclusion that variable structures of the cell adhesion complex determine whether the cell leaves or perseveres in a given cell monolayer.

Table 1. Cell preparation and number of proteins detected.

Culture Condition	1	2 and 3		4	5 and 6
Sample Number	#1	#2	#3	#4	#5
Pre-incubation	2 days	2 days		5 days	5 days
Following 3 days incubation under …	1g	s-µg		1g	s-µg
Kind of growth at time of harvest	Adherent	Spheroid	Adherent	Adherent	Adherent
Number of proteins detected	4419	4505	4544	4621	4961

2. Results and Discussion

2.1. Spheroid Formation

In the present study, three flasks completely filled with cell suspension were incubated under each of the different culture conditions shown in Table 1. At the time of harvest, cells grown according to culture conditions 1, 4, 5, and 6 had formed monolayers only, while cells incubated under culture conditions 2 and 3 were separated into two parts: one grew as spheroids; the other one within adherent monolayers (Table 1). Hence, by these in vitro experiments, we reproduced earlier studies where FTC-133 thyroid cancer cells formed monolayers in plastic flasks if incubated under normal gravity. When subconfluent monolayers of this cell type were exposed to a spaceflight or to an RPM, one part of their cells formed spheroids [5,6]. Interestingly, spheroid formation did not occur when confluent monolayers were grown on the RPM or in space [7]. After harvest, the cells of three flasks of the same condition were pooled. This way, we obtained five samples (Table 1) usable for mass spectrometry (MS). Sample 1 contained cells grown in a monolayer, cultured under culture condition 1. Sample 2 included cells that formed spheroids under culture conditions 2 and 3, sample 3 was comprised of cells grown adherently under culture conditions 2 and 3. Sample 4 contained cells cultured under culture condition 4 and grown in a monolayer, as well as sample 5, which comprised cells grown as a monolayer only, although the cells had been cultured on the RPM (see culture conditions 5 and 6). As no spheroids were found when confluent monolayers were exposed to the RPM, a sample 6 could not be collected. All the samples were flash frozen immediately after harvest and stored until their preparation for MS.

2.2. Quantitative Overview on Proteins Detected by Mass Spectrometry

In order to see possible differences in the protein expression of the FTC-133 cells cultured under different conditions, MS was performed on the cell preparations of each of the five samples indicated above. The analyses revealed a total of 5924 different human proteins in FTC-133 cells. Of those, 3841 proteins were detected in all cells, independent of their incubation history. A total of 4419, 4505, 4544, 4621, and 4961 proteins were found in samples 1, 2, 3, 4, and 5, respectively (Table 1). Hence, applying modern electrospray ionization (ESI) MS technology we found much more than the 821 different proteins which we had detected about eight years ago when we applied matrix-assisted laser desorption/ionization (MALDI) technology after protein pre-separation with the help of electrophoretic techniques [14,15]. Interestingly, we confirmed more than 95% of the proteins, which had been detected previously in the examined thyroid cell lines FTC-133, ML-1, CGTH-W1, and HTU-5. Of the 35 proteins detected previously but not in the current experiment, 22 were present in ML-1, HTU-5, or CTGH-W1 cells, but only 15 were detectable in FTC-133 eight years ago. The first result might be explained by differences in the protein patterns in different thyroid cell lines. The second result could be due to a shift of isoforms as carbonic anhydrase 12, secernin 1, and high mobility group protein 1-like 10 appeared now instead of carbonic anhydrase 13, secernin 2, and high mobility group protein 1-like 1. However, it could also be due to the pre-separation procedure performed earlier as, for example, most of these proteins had either a rather high or low isoelectric point [16].

In order to gain information about the biological roles of the proteins detected by MS, we assigned the detected FTC-133 proteins to their original cellular compartments using Elsevier Pathway Studio®software. The analysis revealed that the coverage of major cellular compartments by the detected proteins was very similar in all cells treated by the different incubation methods. For example, 100% of all proteins typical for large ribosomal subunits and 93% of proteins characteristic for small ribosomal proteins were found in all five samples, while on average 27.6%, 30.2%, and 29.6% of known human membrane, cytoplasmic, and nuclear proteins, respectively, were detected (Table 2). Furthermore, less than 21% of proteins normally secreted into the extracellular space were detected in each sample. This may be explained by the washing of the cell samples prior to protein analysis. However, proteins of other cellular components such as mitochondria, endoplasmic reticulum, and nucleolus were found at average rates of 48.8%, 36.8%, and 49.8%, respectively, and of the focal adhesion complexes at 65% (Table 2). Comparing the numbers of Tables 1 and 2, one may suggest that MS revealed around 40% of the proteins that can be found in the human cell proteome [13]. Of course, a protein not detected in our study does not directly imply that the protein is not expressed in FTC-133 cells.

Table 2. Percentage of proteins of cellular compartments covered by the detected proteins.

Cellular Compartment	Sample 1	Sample 2	Sample 3	Sample 4	Sample 5
Secreted proteins	19%	19%	19%	20%	21%
Membrane	27%	27%	27%	28%	29%
Nucleus	29%	29%	29%	30%	31%
Cytoplasm	29%	30%	30%	30%	32%
Golgi apparatus	31%	31%	32%	34%	35%
Cytoskeleton	33%	33%	34%	35%	37%
Endoplasmic reticulum membrane	35%	36%	37%	36%	39%
Endoplasmic reticulum	36%	36%	37%	36%	39%
Perinuclear region of cytoplasm	37%	38%	37%	38%	40%
Intracellular-membrane-bounded organelle	37%	37%	38%	39%	40%
Endosome	38%	38%	38%	40%	41%
Cytosol	38%	39%	39%	40%	42%
Extracellular exosome	41%	43%	42%	44%	45%
Nucleoplasm	42%	42%	42%	42%	45%
Mitochondrion	49%	48%	49%	48%	50%
Nucleolus	50%	50%	50%	49%	50%
Nuclear speck	51%	50%	50%	50%	57%
Mitochondrial inner membrane	60%	60%	61%	61%	62%
Focal adhesion	63%	64%	65%	65%	68%
Mitochondrial matrix	65%	62%	63%	65%	63%
Spliceosomal complex	68%	69%	66%	66%	71%
Ribonucleoprotein complex	72%	72%	71%	72%	73%
Mitochondrial small ribosomal subunit	93%	93%	93%	93%	93%
Large ribosomal subunit	100%	100%	100%	100%	100%

2.3. Coverage of Different Canonical Pathways by the Proteins Detected in Different Cell Samples

In further experiments, we sought information about the signaling behavior of the detected proteins by applying Ingenuity Pathway Analysis (IPA). This in silico method revealed 478 canonical pathways, in which the detected proteins covered at least 20% of the pathways' proteins. A total of 229 of the indicated pathways comprised more than 20 contributing proteins. First, we took a closer look at pathways including caveolin 1 (CAV-1), which according to earlier results appears to inhibit spheroid formation [3,7]. This protein was detected by Western blotting (Figure 1). In MS measurement, it showed Label-free Quantitation (LfQ) values of 6.47×10^9, 5.35×10^9, 9.25×10^9, 12.57×10^9, and 12.86×10^9 in samples 1, 2, 3, 4, and 5, respectively. The numbers clearly indicated that the spheroid forming cells (sample 2) produced less CAV-1 proteins than the controls (sample 1), while in samples 3–5 the cells produced more CAV-1. This result corresponds to the gene analyses described in [6,11]. It was partially confirmed by Western blot analysis shown in Figure 1C.

CAV-1 was found as a member of the following pathways: caveolar-mediated endocytosis signaling, endothelial nitric oxide synthase (eNOS) signaling, G beta gamma signaling, gap junction signaling, Gαi signaling, integrin signaling, nitric oxide signaling in the cardiovascular system, platelet-derived growth factor (PDGF) signaling, and virus entry via endocytosis. It is not known via which one of these pathways CAV-1 exerts its inhibition of spheroid formation. Therefore, based on earlier studies [17,18], we examined the proteins detected in samples 2 and 3 that are known to be members of the integrin signaling pathway first.

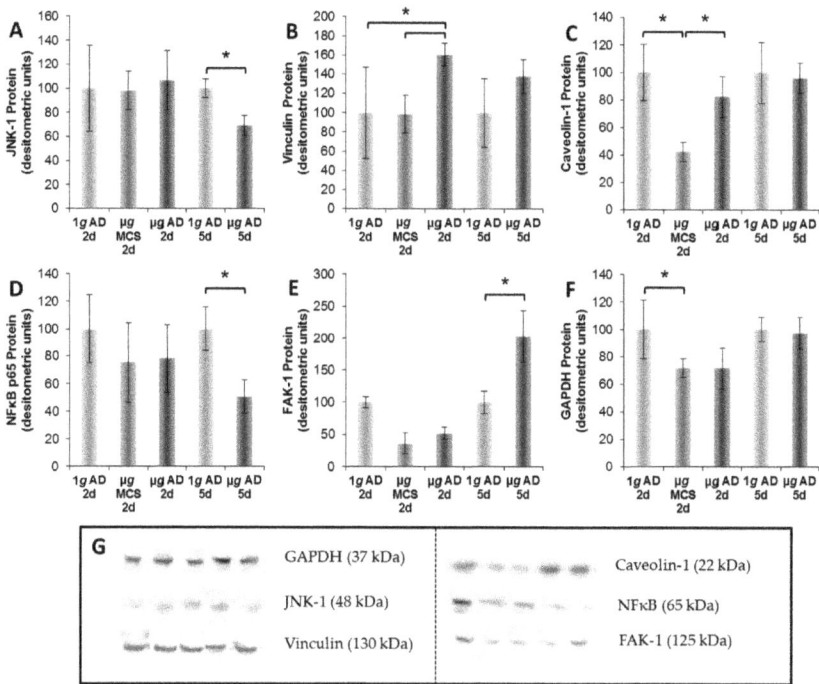

Figure 1. Western blot analyses and densitometric evaluation of FTC-133 cells exposed to culture conditions 1–5, normalized to the total protein content: (**A**) The c-Jun N-terminal kinase 1 (JNK-1) protein was not significantly altered when subconfluent monolayers were exposed to the Random Positioning Machine (RPM) for 3 days (d). JNK-1 was significantly reduced in adherent cells (AD) from culture condition 5 compared to that of the 1*g*-condition; (**B**) Vinculin was reduced in cells of multicellular spheroids (MCS) compared to that of AD (culture condition 3), and in AD 1*g* compared to that in AD (culture condition 3); (**C**) Caveolin 1 (Cav-1) protein was significantly reduced in MCS compared to that with 1*g* and AD (RPM); (**D**) Nuclear factor kappa B (NFκB) p65 protein was significantly decreased in the AD-RPM samples compared to that of the corresponding 1*g*-condition (culture condition 4); (**E**) Focal adhesion kinase 1 (FAK-1 also known as protein tyrosine kinase 2 or PTK2) protein was elevated in the AD of culture condition 5 compared to that in 1*g*; (**F**) Glycerinaldehyd-3-phosphat-Dehydrogenase (GAPDH) densitometric evaluation, normalized to the total protein content (Ponceau S red stain); (**G**) Western blot bands. * $p < 0.05$; $n = 5$ samples per group.

According to the IPA tool, the integrin signaling pathway comprises 207 proteins. We detected 98 and 102 of them in cells which grew in spheroids (sample 2) or monolayers (sample 3), respectively, although they had been subjected to equal conditions (see Table 1: culture condition 2 and 3) within the same flask. Amongst them were a considerable number of various types of membrane proteins including integrins, which had not been removed from the cells during the various wash steps. To our surprise, nuclear proteins linked to the integrin signaling pathways were not detected (Figure 2). Studying the 98 and 102 proteins mentioned above in detail, we found that 96 of the proteins detected in samples 2 and 3 and belonging to the integrin signaling pathway were identical. Only two of the 98 proteins of sample 2 and six of the 102 proteins of sample 3 were unique (Figure 2).

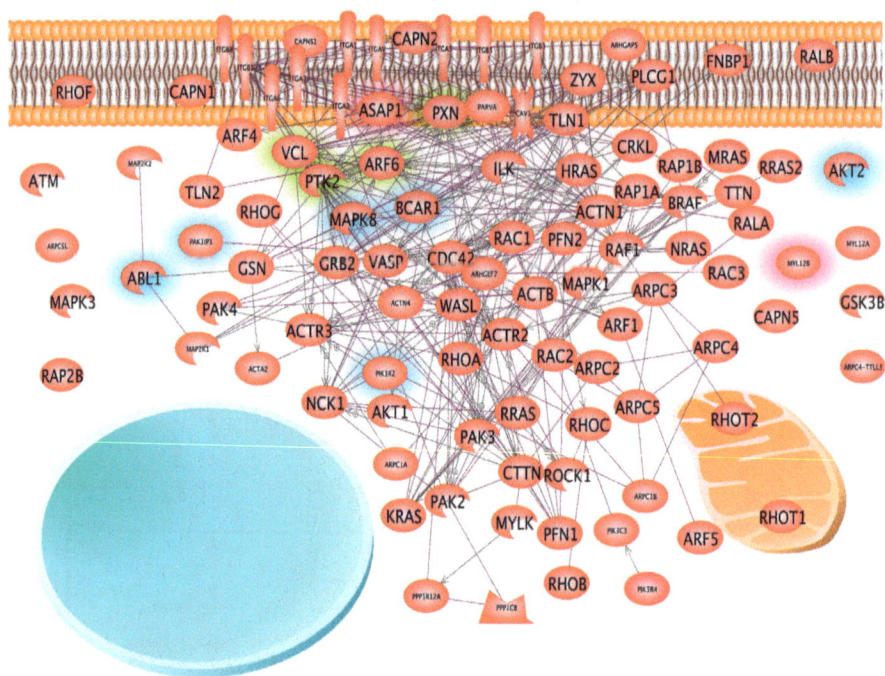

Figure 2. The interaction and localization of proteins belonging to the integrin signaling pathway and found in samples 2 and/or 3. The focal adhesion proteins paxillin (PXN), vinculin (VCL), PTK2 as well as ADB-ribosylation factor 6 (ARF6), marked by a green rim, may be influenced by the protein of ASAP1 (Arf-GAP with SH3 domain, ANK repeat and PH domain-containing protein 1), marked by a red rim, found only in sample 2 as well as by p130cas and Mitogen-activated protein kinase 8 (MAPK8) proteins, marked by a blue rim, found only in sample 3. Arrows indicate interaction; schemes of a membrane bilayer, a nucleus (circle) and a mitochondrion outline localization.

One of the proteins of the integrin pathway found only in sample 2 was the ASAP1 protein (Arf-GAP with SH3 domain, ANK repeat and PH domain-containing protein 1). As we demonstrated by Elsevier Pathway Studio® analysis, this protein interacts with several other proteins (Figure 2). Amongst them are vinculin (VCL), a rather abundant protein detectable by MS with an average LfQ of 11.82×10^9, and less abundant proteins such as paxillin (PXN) with an average LfQ value of 0.46×10^9, focal adhesion kinase 1 (PTK2, average LfQ = 0.72×10^9) as well as ADP-ribosylation factor 6 (ARF6, average LfQ = 0.6×10^9) (Figure 2). Their presence could be confirmed by Western blotting (Figure 1). ASAP1 regulates membrane traffic and cytoskeleton organization [19] and influences cell spreading and migration [20]. Directed by the Crk-like protein, it accumulates within the focal adhesion complex [21]. There, it co-localizes with PXN and VCL, which are able to bind to each other [19,22], and ASAP1 associates also with the focal adhesion kinase. Under unknown conditions, ASAP1 may cause a repositioning of PXN and focal adhesion kinase within the focal adhesion complex, which retards cell spreading [23]. ASAP1 has also been observed to bind to ARF6 [24]. This binding appears to facilitate the recruitment of ASAP1 to the focal adhesion complex [25]. Interestingly, the *ASAP1* gene was up-regulated in adherent FTC-133 cells after 24 h of incubation [11]. However, after 10 d in space, the *ASAP1* gene was 7-fold down-regulated in adherent cells, while it remained unregulated in spheroid cells [5].

The p130cas and MAPK8 (Mitogen-activated protein kinase 8) protein, which both are members of the integrin signaling pathway, were detected in sample 3 but not in sample 2. The p130cas protein (see BCAR1, breast cancer anti-estrogen resistance protein 1, in Figure 2) interacts also with PXN, VCL, and PTK2. Paxillin is a membrane protein that regulates cell-matrix interaction, and associates with p130cas [26]. When PXN-p130cas complexes are phosphorylated, they constitutively activate cell migration by inducing gene 5 proteins (RAC1) to abolish shear stress induced cell polarization [27]. Phosphorylation occurs by interaction with focal adhesion kinase after cell adhesion [28,29]. p130cas interacts also with VCL in focal adhesion complexes that mediate cell-matrix interactions in the presence of PXN [30,31]. MAPK8 proteins signal shear stress to focal adhesion sites in endothelial cells [32]. Its activation is controlled by ARF6 [33]. In human multiple myeloma cells, the expression of MAPK8 and interleukin-8 genes is suppressed simultaneously by azidothymidine [34]. Recently, we found that MCF-7 breast cancer cells enhance interleukin-8 gene expression only in adherent cells treated just as the cells of sample 3 [35]. In addition, when a population of FTC-133 was flown through the orbit on a Chinese rocket (Shenzhou-8 space mission), the *MAPK8* gene was up-regulated in the cells remaining adherent. This suggested that a strengthening of the *MAPK8* signal over extracellular shear forces could hinder the cells to aggregate to floating spheroids. The conclusion has still to be confirmed, as the c-Jun N-terminal kinase 1 (JNK-1), which is also called MAPK8, was, according to Figure 1A, similar in MCS and adherent cells, both harvested from culture condition 2 and 3.

Taken together, the results shown in Figure 2 suggest that composition and interaction of PXN, VCL, PTK2, and ARF6 are strongly influenced by ASAP1 when cells form spheroids (sample 2). When FTC-133 cells grow in monolayers, even under conditions of microgravity (sample 3), the same proteins are predominantly affected by p130cas and MAPK8. Enhanced ASAP1 appears to weaken the focal adhesion complexes, so that the removal of sedimentation forces together with minimal shear forces present in culture flasks on the RPM may trigger the cells to detach from the bottom of the culture flask but anchor to neighboring cells [6,36]. At the same time, a predominance of p130cas appears to strengthen lamellipodia [28] as well as cell-matrix interaction [30] via PXN [26], which facilitates continued adherent growth.

Furthermore, we looked at the angiopoietin pathway. It comprises 66 proteins, 22 of which were detected in sample 2. In sample 3, the 22 proteins of sample 2 were also found plus an additional seven (Figure 3). Nineteen of the 29 proteins detected in samples 2 and 3 belong to both the angiopoietin and the integrin signaling pathways. The finding of proteins of the angiopoietin pathway was not surprising; we had repeatedly found increased gene expression and protein secretion of vascular endothelial growth factor (VEGF) of the neutrophil gelatinase-associated lipocalin, osteopontin, and the interleukins 6 and 17 in thyroid cancer cells, suggesting that factors of the angiopoietin pathway may play a role in spheroid formation of FTC-133 cancer cells [5,8,37]. In this proteome study, no membrane-bound receptor for VEGF or any other of the factors mentioned above could be detected in the thyroid cells. Of the ten proteins belonging to the angiopoietin but not the integrin signaling pathway, four proteins were found in sample 3 (adherent cells) but not in sample 2 (spheroids). One of these was Ras GTPase-activating protein 1 (RASA1), which can bind to paxillin and to PTK2 and supports cell migration and surface ruffling [38,39]. In addition, it forms a complex with survivin [40], which accumulates CHUK protein (Inhibitor of nuclear factor kappa-B kinase subunit alpha) in the nucleus [41]. There, CHUK protein together with IkBKB (Inhibitor of nuclear factor kappa-B kinase subunit beta) regulates nuclear factor kappa B activity [11,42].

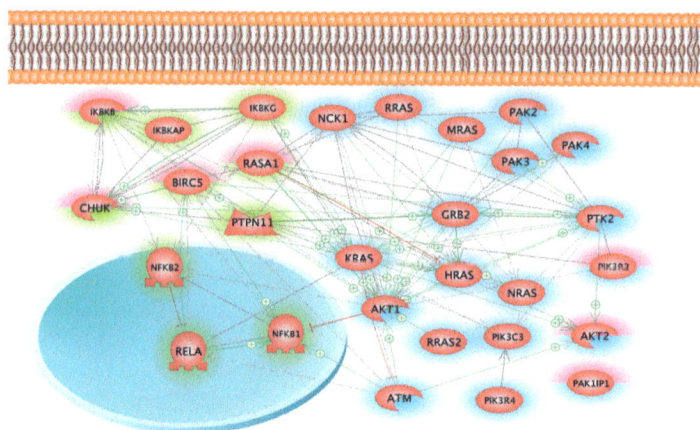

Figure 3. The interaction and localization of proteins, belonging to the angiopoietin signaling pathway and found in samples 2 and/or 3. All proteins marked by a uniform green or blue rim were found in samples 2 and 3, while the proteins of *CHUK, IKBKB, BIRC5, RAS1, PIK3R2, PAK1IP1* and *AKT2*, marked by a upper half red rim were found only in sample 3. Proteins of *CHUK, IKBKB, BIRC5, RAS1, IKBKAP, IKBKG, PTPN11, NFKB1, NFKB2* and *RELA* belong to the angiopoietin pathway only. The rest of the proteins belong to the integrin signaling and the angiopoietin pathway simultaneously. Arrows indicate interaction, while T-bars show inhibition; schemes of a membrane bilayer, a nucleus (circle) and a mitochondrion outline localization.

3. Discussion

The proteome analysis of FTC-133 thyroid cancer cells cultured under various conditions revealed more than 5900 proteins of this cell line. The proteins detected represent about 40% of proteins possibly produced in human cells [13]. Their quantities ranged from 10^7 to 10^{11} LfQ values. Advanced analysis of the detected proteins in regard to their association to the integrin signaling pathway and the angiopoietin pathway challenged the following conclusions: In cells forming spheroids during three days of culture under s-μg, the levels of CAV-1 and p130cas proteins are reduced, but ASAP1 production is enhanced. Under this condition, proteins PXN, VCL, and PTK2 may be positioned within the focal adhesion complex in a way that favors cell detachment from the bottom of a culture flask and mutual attachment. The continuation of adherent growth could be supported by accumulation of p130cas protein in individual cells. In order to prove this hypothesis, we shall investigate the structural changes of the cell adhesion complex applying methods described recently [43].

4. Materials and Methods

4.1. Cell Culture

FTC-133 human follicular thyroid carcinoma cells [44] were cultured in RPMI 1640 (Life Technologies, Naerum, Denmark) medium supplemented with 10% fetal calf serum (Biochrom AG, Berlin, Germany) and 1% penicillin/streptomycin (Life Technologies, Naerum, Denmark) under standard cell culture conditions at 37 °C and 5% CO_2. Prior to culturing under different conditions, 1×10^6 cells were counted and seeded into T25 cm^2 vented cell culture flasks (Sarstedt, Newton, MA, USA). Twelve of these T25 cm^2 culture flasks each containing 10^6 cells were incubated at 37 °C for two days until the cells formed sub-confluent monolayers. Afterwards, three flasks were put nearby the RPM (culture condition 1), while another three were mounted on the RPM (culture conditions 2 and 3). Then, each of these flasks was incubated for another three days prior to harvest. The six remaining flasks continued to be cultured for another three days under normal gravity until the monolayers had reached confluence. Then, again, three of the six flasks were put nearby the RPM (culture condition 4), while the other three were mounted on the RPM (culture conditions 5 and 6). These flasks were also incubated for another three days until harvest (Table 1). Therefore, the main difference between culture conditions 1 and 4 as well as between culture conditions 2 and 3 and culture conditions 5 and 6 is the length of the period of pre-incubation. These cell culture samples were used for MS. For the Western blot analyses, we repeated these experiments (n = 5 per condition).

4.2. Random Positioning Machine

A desktop RPM (Airbus Defense and Space, Leiden, The Netherlands) was placed in a commercially available incubator at 37 °C and 5% CO_2. The RPM was used in real random mode with random speed and random interval and a maximum speed of 75°/s. The flasks were fixed to the central frame, as near as possible to the center of rotation, and were rotated for 3 days. Corresponding static normal gravity controls ($1g$) prepared in parallel were stored next to the device in the same incubator during the time of rotation. Each flask was completely filled with complete medium, taking care that no air bubbles remained in the cell culture flasks in order to minimize shear stress. The mode of action and the effectiveness of this machine at preventing cell sedimentation have been described previously [45,46].

4.3. Cell Harvest

First, the supernatant of each T25 cm^2 culture flask was collected and centrifuged at 4°C for spheroid collection. After centrifugation, the supernatant was carefully aspirated, and the spheroids were collected, washed in phosphate buffered saline (PBS, Gibco, Life Technologies, Naerum, Denmark), and stored in liquid nitrogen. To harvest the adherent cells, 5 mL of ice-cold PBS were carefully added to each T25 cm^2 flask. The supernatant was then aspirated and the cells were scraped off with a scraper. The cell suspension was collected and centrifuged at 4 °C. The PBS was discarded and the dry pellet was washed with PBS and stored in liquid nitrogen.

4.4. Mass Spectometry

Cells were lysed in a buffer containing 6 M guanidium hydrochloride, 20 mM TCEP (tris(2-carboxyethyl)phosphine), and 40 mM chloroacetamide in 25 mM Tris pH 8.0. Lysis buffer, preheated to 95 °C, was added to the cells and sonicated using a Bioruptor plus water bath sonicator (Diagnode, Seraing, Belgium). The lysates were heated again at 95 °C for 2 min, followed by one more sonication step at maximum power settings for ten cycles. Following complete lysis, the sample was diluted 10-fold with 25 mM Tris pH 8.0 and digested overnight at 37 °C with endoproteinase Lys-C (Wako Chemicals GmbH, Neuss, Germany) at a 1:50 protein ratio. The digested peptides were then purified and concentrated on three plugged SDB-XC StageTip [47].

For the liquid chromatography-mass spectrometry analysis, about 2 μg of peptides were loaded onto a 15 cm, 75 μm I.D column packed with 1.9 μm C18 beads (Maisch GmbH, Ammerbuch, Germany) using the Thermo easy n-LC 1000 system (Thermo Scientific, Waltham, MA, USA) and were separated over a 120-min gradient with buffer A (0.1% formic acid) and buffer B (0.1% formic acid and 80% acetonitrile). The LC (liquid chromatography) column was maintained at a constant temperature of 45 °C using a column oven (Sonation, Biberach, Germany). The peptides eluting from the column were directly sprayed into a Q Exactive HF mass spectrometer (Thermo Scientific, Waltham, MA, USA) via a nano-electrospray ionization source (Thermo Scientific, Waltham, MA, USA) [48]. The mass spectrometer was operated in a data-dependent top 15 mode. Survey scans and fragmentation scans were acquired at resolutions of 60,000 and 15,000 respectively (m/z = 200). Fragmentation was performed on precursors isolated within a window of 1.4 m/z with a normalized collision energy setting of 27.

Raw data from the mass spectrometer were processed using MaxQuant computational proteomics platform version 1.5.2.22 (Computational Systems Biochemistry, Max-Planck-Gesellschaft, Munich, Germany) [49] using the standard parameters. Relative protein concentration was performed using the LfQ algorithm (label-free quantitation) as described elsewhere [50].

4.5. Pathway Analysis

To investigate and visualize the original localization and the mutual interactions of detected proteins, we entered relevant UniProtKB entry numbers in an Elsevier Pathway Studio®v.11 software (Elsevier Research Solutions, Amsterdam, The Netherlands). To assign detected proteins to canonical pathways, the Ingenuity Pathway Analysis (IPA) with Advanced Analytics client (CL) (Qiagen GmbH, Hilden, Germany) was applied, also entering relevant UniProtKB entry numbers.

4.6. Western Blot

Western blotting, immunoblotting, and densitometry were performed as described earlier [17]. We used the biorad ChemiDoc XRT+ device. The antibodies used to detect and quantify the antigens are listed in Table 3. The applied secondary antibody, a Horseradish peroxidase (HRP)-linked antibody was utilized at a dilution of 1:4000 (Cell Signaling Technology, Inc., Danvers, MA, USA). In addition, we used glyceraldehyde 3-phosphate dehydrogenase (GAPDH; dilution: 1:1000). Ponceau S red staining was used as an alternative to housekeeping proteins as loading controls. The membranes were analyzed using ImageJ software (U.S. National Institutes of Health, Bethesda, MD, USA; http://rsb.info.nih.gov/ij/), for densitrometric quantification of the bands. Ponceau S was evaluated according to [51]. Statistical analyses were performed as previously published [17].

Table 3. Antibodies applied for Western blot analysis.

Antibody	Dilution	Company	Molecular Weight	Catalog Number
Anti-JNK1	1/1000	Abcam	48 kDA	ab110724
Anti-FAK	1/1000	Abcam	125 KDA (119 kDA)	ab40794
Anti-Vinculin	1/1000	Abcam	130 kDA	ab18058
Anti-Caveolin-1	1/1000	Abcam	22 kDA	ab2910
Anti-NfκB p65	1/1000	Cell-Signaling	65 kDA	#C22B4
Anti-GAPDH	1/1000	Cell-Signaling	37 kDA	#5174

Acknowledgments: This project was supported by the German Space Agency (DLR, grant 50BW1524). The authors would like to thank Nagarjuna Nagaraj, PhD, core facility for mass spectrometry and proteomics, Max-Planck-Institute for Biochemistry, Martinsried, Germany for performing the MS experiments. In addition, we would like to thank Petra Wise, PhD, USC, Los Angeles, CA, USA, for English editing of this manuscript.

Author Contributions: Johann Bauer and Daniela Grimm conceived and designed the experiments; Jayashree Sahana, Stefan Riwaldt, Johann Bauer, and Sascha Kopp performed the experiments; Johann Bauer and Elisabeth Maria Schlagberger analyzed the data; Sascha Kopp and Markus Wehland performed the densitometry; Johann Bauer, Daniela Grimm, Ronald Luetzenberg, Jirka Grosse and Manfred Infanger contributed reagents/materials/analysis tools; Johann Bauer, Daniela Grimm, and Sascha Kopp wrote the paper.

Conflicts of Interest: The authors have declared no conflicts of interest.

Abbreviations

AD	Adherent cells
AKT1	RAC-alpha serine/threonine-protein kinase 1
ARF6	ADP-ribosylation factor 6
ASAP1	Arf-GAP with SH3 domain, ANK repeat and PH domain-containing protein 1
BIRC5	Baculoviral inhibitor of apoptosis repeat-containing 5; Survivin
CAV-1	Caveolin 1
CHUK	Conserved helix-loop-helix ubiquitous kinase; Inhibitor of nuclear factor kappa-B kinase subunit alpha (IKK-α)
GAPDH	Glycerinaldehyd-3-phosphat-Dehydrogenase
HRP	Horseradish peroxidase
IKBKB	Inhibitor of nuclear factor kappa-B kinase subunit beta
JNK1	c-Jun N-terminal kinase 1 or Mitogen-activated protein kinase 8
LfQ	Label-free Quantitation

MAPK8	Mitogen-activated protein kinase 8
MCS	Multicellular tumor spheroids
MS	Mass Spectrometry
NFκB	Nuclear factor kappa B
PAK1IP1	P21-activated protein kinase-interacting protein 1
PIK3R2	Phosphoinositide-3-kinase regulatory subunit 2
PTK2	Protein tyrosine kinase 2, focal adhesion kinase 1 (FAK1)
PXN	Paxillin
Ras1	Ras-like protein 1
r-µ*g*	Real microgravity
RPM	Random positioning machine
s-µ*g*	Simulated microgravity
VCL	Vinculin
VEGF	Vascular endothelial growth factor

References

1. Nath, S.; Devi, G.R. Three-dimensional culture systems in cancer research: Focus on tumor spheroid model. *Pharmacol. Ther.* **2016**, *163*, 94–108. [CrossRef] [PubMed]
2. Chatzinikolaidou, M. Cell spheroids: The new frontiers in in vitro models for cancer drug validation. *Drug Discov. Today* **2016**, *21*, 1553–1560. [CrossRef] [PubMed]
3. Grimm, D.; Wehland, M.; Pietsch, J.; Aleshcheva, G.; Wise, P.; van Loon, J.; Ulbrich, C.; Magnusson, N.E.; Infanger, M.; Bauer, J. Growing tissues in real and simulated microgravity: New methods for tissue engineering. *Tissue Eng. Part B Rev.* **2014**, *20*, 555–566. [CrossRef] [PubMed]
4. Becker, J.L.; Souza, G.R. Using space-based investigations to inform cancer research on Earth. *Nat. Rev. Cancer* **2013**, *13*, 315–327. [CrossRef] [PubMed]
5. Ma, X.; Pietsch, J.; Wehland, M.; Schulz, H.; Saar, K.; Hubner, N.; Bauer, J.; Braun, M.; Schwarzwalder, A.; Segerer, J.; et al. Differential gene expression profile and altered cytokine secretion of thyroid cancer cells in space. *FASEB J.* **2014**, *28*, 813–835. [CrossRef] [PubMed]
6. Warnke, E.; Pietsch, J.; Wehland, M.; Bauer, J.; Infanger, M.; Gorog, M.; Hemmersbach, R.; Braun, M.; Ma, X.; Sahana, J.; et al. Spheroid formation of human thyroid cancer cells under simulated microgravity: A possible role of CTGF and CAV1. *Cell Commun. Signal.* **2014**, *12*, 32. [CrossRef] [PubMed]
7. Riwaldt, S.; Pietsch, J.; Sickmann, A.; Bauer, J.; Braun, M.; Segerer, J.; Schwarzwalder, A.; Aleshcheva, G.; Corydon, T.J.; Infanger, M.; et al. Identification of proteins involved in inhibition of spheroid formation under microgravity. *Proteomics* **2015**, *15*, 2945–2952. [CrossRef] [PubMed]
8. Kopp, S.; Warnke, E.; Wehland, M.; Aleshcheva, G.; Magnusson, N.E.; Hemmersbach, R.; Corydon, T.J.; Bauer, J.; Infanger, M.; Grimm, D. Mechanisms of three-dimensional growth of thyroid cells during long-term simulated microgravity. *Sci. Rep.* **2015**, *5*, 16691. [CrossRef] [PubMed]
9. Bauer, J.; Wehland, M.; Pietsch, J.; Sickmann, A.; Weber, G.; Grimm, D. Annotated Gene and Proteome Data Support Recognition of Interconnections Between the Results of Different Experiments in Space Research. *Microgravity Sci. Technol.* **2016**, *28*, 357–365. [CrossRef]
10. Riwaldt, S.; Bauer, J.; Pietsch, J.; Braun, M.; Segerer, J.; Schwarzwalder, A.; Corydon, T.J.; Infanger, M.; Grimm, D. The Importance of Caveolin-1 as Key-Regulator of Three-Dimensional Growth in Thyroid Cancer Cells Cultured under Real and Simulated Microgravity Conditions. *Int. J. Mol. Sci.* **2015**, *16*, 28296–28310. [CrossRef] [PubMed]
11. Grosse, J.; Wehland, M.; Pietsch, J.; Schulz, H.; Saar, K.; Hubner, N.; Eilles, C.; Bauer, J.; Abou-El-Ardat, K.; Baatout, S.; et al. Gravity-sensitive signaling drives 3-dimensional formation of multicellular thyroid cancer spheroids. *FASEB J.* **2012**, *26*, 5124–5140. [CrossRef] [PubMed]
12. Kossmehl, P.; Shakibaei, M.; Cogoli, A.; Infanger, M.; Curcio, F.; Schonberger, J.; Eilles, C.; Bauer, J.; Pickenhahn, H.; Schulze-Tanzil, G.; et al. Weightlessness induced apoptosis in normal thyroid cells and papillary thyroid carcinoma cells via extrinsic and intrinsic pathways. *Endocrinology* **2003**, *144*, 4172–4179. [CrossRef] [PubMed]

13. Mann, M.; Kulak, N.A.; Nagaraj, N.; Cox, J. The coming age of complete, accurate, and ubiquitous proteomes. *Mol. Cell* **2013**, *49*, 583–590. [CrossRef] [PubMed]

14. Pietsch, J.; Kussian, R.; Sickmann, A.; Bauer, J.; Weber, G.; Nissum, M.; Westphal, K.; Egli, M.; Grosse, J.; Schonberger, J.; et al. Application of free-flow IEF to identify protein candidates changing under microgravity conditions. *Proteomics* **2010**, *10*, 904–913. [CrossRef] [PubMed]

15. Pietsch, J.; Bauer, J.; Weber, G.; Nissum, M.; Westphal, K.; Egli, M.; Grosse, J.; Schönberger, J.; Eilles, C.; Infanger, M.; et al. Proteome Analysis of Thyroid Cancer Cells After Long-Term Exposure to a Random Positioning Machine. *Microgravity Sci. Technol.* **2011**, *23*, 381–390. [CrossRef]

16. Obermaier, C.; Jankowski, V.; Schmutzler, C.; Bauer, J.; Wildgruber, R.; Infanger, M.; Kohrle, J.; Krause, E.; Weber, G.; Grimm, D. Free-flow isoelectric focusing of proteins remaining in cell fragments following sonication of thyroid carcinoma cells. *Electrophoresis* **2005**, *26*, 2109–2116. [CrossRef] [PubMed]

17. Svejgaard, B.; Wehland, M.; Ma, X.; Kopp, S.; Sahana, J.; Warnke, E.; Aleshcheva, G.; Hemmersbach, R.; Hauslage, J.; Grosse, J.; et al. Common Effects on Cancer Cells Exerted by a Random Positioning Machine and a 2D Clinostat. *PLoS ONE* **2015**, *10*, e0135157. [CrossRef] [PubMed]

18. Pietsch, J.; Sickmann, A.; Weber, G.; Bauer, J.; Egli, M.; Wildgruber, R.; Infanger, M.; Grimm, D. A proteomic approach to analysing spheroid formation of two human thyroid cell lines cultured on a random positioning machine. *Proteomics* **2011**, *11*, 2095–2104. [CrossRef] [PubMed]

19. Randazzo, P.A.; Andrade, J.; Miura, K.; Brown, M.T.; Long, Y.Q.; Stauffer, S.; Roller, P.; Cooper, J.A. The Arf GTPase-activating protein ASAP1 regulates the actin cytoskeleton. *Proc. Natl. Acad. Sci. USA* **2000**, *97*, 4011–4016. [CrossRef] [PubMed]

20. Liu, Y.; Yerushalmi, G.M.; Grigera, P.R.; Parsons, J.T. Mislocalization or reduced expression of Arf GTPase-activating protein ASAP1 inhibits cell spreading and migration by influencing Arf1 GTPase cycling. *J. Biol. Chem.* **2005**, *280*, 8884–8892. [CrossRef] [PubMed]

21. Oda, A.; Wada, I.; Miura, K.; Okawa, K.; Kadoya, T.; Kato, T.; Nishihara, H.; Maeda, M.; Tanaka, S.; Nagashima, K.; et al. CrkL directs ASAP1 to peripheral focal adhesions. *J. Biol. Chem.* **2003**, *278*, 6456–6460. [CrossRef] [PubMed]

22. Turner, C.E.; Glenney, J.R., Jr.; Burridge, K. Paxillin: A new vinculin-binding protein present in focal adhesions. *J. Cell Biol.* **1990**, *111*, 1059–1068. [CrossRef] [PubMed]

23. Liu, Y.; Loijens, J.C.; Martin, K.H.; Karginov, A.V.; Parsons, J.T. The association of ASAP1, an ADP ribosylation factor-GTPase activating protein, with focal adhesion kinase contributes to the process of focal adhesion assembly. *Mol. Biol. Cell* **2002**, *13*, 2147–2156. [CrossRef] [PubMed]

24. Randazzo, P.A.; Hirsch, D.S. Arf GAPs: Multifunctional proteins that regulate membrane traffic and actin remodelling. *Cell Signal.* **2004**, *16*, 401–413. [CrossRef] [PubMed]

25. Sabe, H.; Onodera, Y.; Mazaki, Y.; Hashimoto, S. ArfGAP family proteins in cell adhesion, migration and tumor invasion. *Curr. Opin. Cell Biol.* **2006**, *18*, 558–564. [CrossRef] [PubMed]

26. Deakin, N.O.; Turner, C.E. Paxillin comes of age. *J. Cell Sci.* **2008**, *121*, 2435–2444. [CrossRef] [PubMed]

27. Zaidel-Bar, R.; Kam, Z.; Geiger, B. Polarized downregulation of the paxillin-p130CAS-Rac1 pathway induced by shear flow. *J. Cell Sci.* **2005**, *118*, 3997–4007. [CrossRef] [PubMed]

28. Kumbrink, J.; Soni, S.; Laumbacher, B.; Loesch, B.; Kirsch, K.H. Identification of Novel Crk-associated Substrate (p130Cas) Variants with Functionally Distinct Focal Adhesion Kinase Binding Activities. *J. Biol. Chem.* **2015**, *290*, 12247–12255. [CrossRef] [PubMed]

29. Harte, M.T.; Hildebrand, J.D.; Burnham, M.R.; Bouton, A.H.; Parsons, J.T. p130Cas, a substrate associated with v-Src and v-Crk, localizes to focal adhesions and binds to focal adhesion kinase. *J. Biol. Chem.* **1996**, *271*, 13649–13655. [PubMed]

30. Goldmann, W.H. Vinculin-p130Cas interaction is critical for focal adhesion dynamics and mechano-transduction. *Cell Biol. Int.* **2014**, *38*, 283–286. [CrossRef] [PubMed]

31. Goldmann, W.H. Role of vinculin in cellular mechanotransduction. *Cell Biol. Int.* **2016**, *40*, 241–256. [CrossRef] [PubMed]

32. Li, S.; Kim, M.; Hu, Y.L.; Jalali, S.; Schlaepfer, D.D.; Hunter, T.; Chien, S.; Shyy, J.Y. Fluid shear stress activation of focal adhesion kinase. Linking to mitogen-activated protein kinases. *J. Biol. Chem.* **1997**, *272*, 30455–30462. [CrossRef] [PubMed]

33. Bourmoum, M.; Charles, R.; Claing, A. The GTPase ARF6 Controls ROS Production to Mediate Angiotensin II-Induced Vascular Smooth Muscle Cell Proliferation. *PLoS ONE* **2016**, *11*, e0148097. [CrossRef] [PubMed]

34. Pereira, J.; Levy, D.; Ruiz, J.L.; Brocardo, G.A.; Ferreira, K.A.; Costa, R.O.; Queiroz, R.G.; Maria, D.A.; Neto, A.E.; Chamone, D.A.; et al. Azidothymidine is effective against human multiple myeloma: A new use for an old drug? *Anticancer Agents Med. Chem.* **2013**, *13*, 186–192. [CrossRef] [PubMed]

35. Kopp, S.; Slumstrup, L.; Corydon, T.J.; Sahana, J.; Aleshcheva, G.; Islam, T.; Magnusson, N.E.; Wehland, M.; Bauer, J.; Infanger, M.; et al. Identifications of novel mechanisms in breast cancer cells involving duct-like multicellular spheroid formation after exposure to the Random Positioning Machine. *Sci. Rep.* **2016**, *6*, 26887. [CrossRef] [PubMed]

36. Grimm, D.; Infanger, M.; Westphal, K.; Ulbrich, C.; Pietsch, J.; Kossmehl, P.; Vadrucci, S.; Baatout, S.; Flick, B.; Paul, M.; et al. A delayed type of three-dimensional growth of human endothelial cells under simulated weightlessness. *Tissue Eng. Part A* **2009**, *15*, 2267–2275. [CrossRef] [PubMed]

37. Riwaldt, S.; Bauer, J.; Wehland, M.; Slumstrup, L.; Kopp, S.; Warnke, E.; Dittrich, A.; Magnusson, N.E.; Pietsch, J.; Corydon, T.J.; et al. Pathways Regulating Spheroid Formation of Human Follicular Thyroid Cancer Cells under Simulated Microgravity Conditions: A Genetic Approach. *Int. J. Mol. Sci.* **2016**, *17*, 528. [CrossRef] [PubMed]

38. Tsubouchi, A.; Sakakura, J.; Yagi, R.; Mazaki, Y.; Schaefer, E.; Yano, H.; Sabe, H. Localized suppression of RhoA activity by Tyr31/118-phosphorylated paxillin in cell adhesion and migration. *J. Cell Biol.* **2002**, *159*, 673–683. [CrossRef] [PubMed]

39. Yu, J.A.; Deakin, N.O.; Turner, C.E. Emerging role of paxillin-PKL in regulation of cell adhesion, polarity and migration. *Cell Adh. Migr.* **2010**, *4*, 342–347. [CrossRef] [PubMed]

40. Gigoux, V.; L'Hoste, S.; Raynaud, F.; Camonis, J.; Garbay, C. Identification of Aurora kinases as RasGAP Src homology 3 domain-binding proteins. *J. Biol. Chem.* **2002**, *277*, 23742–23746. [CrossRef] [PubMed]

41. Shi, K.; An, J.; Shan, L.; Jiang, Q.; Li, F.; Ci, Y.; Wu, P.; Duan, J.; Hui, K.; Yang, Y.; et al. Survivin-2B promotes autophagy by accumulating IKK α in the nucleus of selenite-treated NB4 cells. *Cell Death Dis.* **2014**, *5*, e1071. [CrossRef] [PubMed]

42. Wu, C.; Ghosh, S. Differential phosphorylation of the signal-responsive domain of IκBα and IκBβ by IκB kinases. *J. Biol. Chem.* **2003**, *278*, 31980–31987. [CrossRef] [PubMed]

43. Corydon, T.J.; Kopp, S.; Wehland, M.; Braun, M.; Schutte, A.; Mayer, T.; Hulsing, T.; Oltmann, H.; Schmitz, B.; Hemmersbach, R.; et al. Alterations of the cytoskeleton in human cells in space proved by life-cell imaging. *Sci. Rep.* **2016**, *6*, 20043. [CrossRef] [PubMed]

44. Goretzki, P.E.; Frilling, A.; Simon, D.; Roeher, H.D. Growth regulation of normal thyroids and thyroid tumors in man. *Recent Results Cancer Res.* **1990**, *118*, 48–63. [PubMed]

45. Borst, A.G.; van Loon, J.J.W.A. Technology and Developments for the Random Positioning Machine, RPM. *Microgravity Sci. Technol.* **2008**, *21*, 287. [CrossRef]

46. Warnke, E.; Kopp, S.; Wehland, M.; Hemmersbach, R.; Bauer, J.; Pietsch, J.; Infanger, M.; Grimm, D. Thyroid Cells Exposed to Simulated Microgravity Conditions—Comparison of the Fast Rotating Clinostat and the Random Positioning Machine. *Microgravity Sci. Technol.* **2016**, *28*, 247–260. [CrossRef]

47. Rappsilber, J.; Mann, M.; Ishihama, Y. Protocol for micro-purification, enrichment, pre-fractionation and storage of peptides for proteomics using StageTips. *Nat. Protoc.* **2007**, *2*, 1896–1906. [CrossRef] [PubMed]

48. Nagaraj, N.; Kulak, N.A.; Cox, J.; Neuhauser, N.; Mayr, K.; Hoerning, O.; Vorm, O.; Mann, M. System-wide perturbation analysis with nearly complete coverage of the yeast proteome by single-shot ultra HPLC runs on a bench top Orbitrap. *Mol. Cell Proteom.* **2012**, *11*. [CrossRef] [PubMed]

49. Cox, J.; Mann, M. MaxQuant enables high peptide identification rates, individualized p.p.b.-range mass accuracies and proteome-wide protein quantification. *Nat. Biotechnol.* **2008**, *26*, 1367–1372. [CrossRef] [PubMed]

50. Cox, J.; Hein, M.Y.; Luber, C.A.; Paron, I.; Nagaraj, N.; Mann, M. Accurate proteome-wide label-free quantification by delayed normalization and maximal peptide ratio extraction, termed MaxLFQ. *Mol. Cell Proteom.* **2014**, *13*, 2513–2526. [CrossRef] [PubMed]

51. Analyzing Gels and Western Blots with ImageJ. Available online: http://lukemiller.org/index.php/2010/11/analyzing-gels-and-western-blots-with-image-j/ (accessed on 26 October 2016).

International Journal of
Molecular Sciences

MDPI

Review

Radiation and Thyroid Cancer

Elisabetta Albi [1,*], Samuela Cataldi [1], Andrea Lazzarini [2], Michela Codini [1], Tommaso Beccari [1], Francesco Saverio Ambesi-Impiombato [3] and Francesco Curcio [3]

[1] Department of Pharmaceutical Science, University of Perugia, 06123 Perugia, Italy;
samuelacataldi@libero.it (S.C.); michela.codini@unipg.it (M.C.); tommaso.beccari@unipg.it (T.B.)

[2] Research Center and Analysis Laboratory CRABION, 06073 Perugia, Italy; andrylazza@gmail.it

[3] Department of Clinical and Biological Sciences, University of Udine, 33100 Udine, Italy;
ambesis@me.com (F.S.A.-I.); francesco.curcio@uniud.it (F.C.)

* Correspondence: elisabetta.albi@unipg.it; Tel.: +39-075-585-7908

Academic Editors: Daniela Gabriele Grimm and Sanjay K. Srivastava
Received: 17 February 2017; Accepted: 24 April 2017; Published: 26 April 2017

Abstract: Radiation-induced damage is a complex network of interlinked signaling pathways, which may result in apoptosis, cell cycle arrest, DNA repair, and cancer. The development of thyroid cancer in response to radiation, from nuclear catastrophes to chemotherapy, has long been an object of study. A basic overview of the ionizing and non-ionizing radiation effects of the sensitivity of the thyroid gland on radiation and cancer development has been provided. In this review, we focus our attention on experiments in cell cultures exposed to ionizing radiation, ultraviolet light, and proton beams. Studies on the involvement of specific genes, proteins, and lipids are also reported. This review also describes how lipids are regulated in response to the radiation-induced damage and how they are involved in thyroid cancer etiology, invasion, and migration and how they can be used as both diagnostic markers and drug targets.

Keywords: radiation; thyroid cancer; cancer genes; lipid metabolism

1. Introduction

Radiation includes ionizing radiation (IR) and non-IR. IR can be distinguished in photon radiation (X- and γ-rays) and particle radiation (such as electrons, protons, neutrons, carbon ions, and alpha and beta particles). IR has enough energy to free electrons from atoms or molecules ionizing them. Non-IR includes ultraviolet (UV), visible light laser, infrared, microwaves, and radio waves.

It is generally accepted that high acute doses of IR may be harmful to living organisms. In radiation accidents, the determination of the radiation dose is a key step for medical decisions and patient prognosis. The estimation of the absorbed dose aids in establishing the risk for acute or chronic health effects, up to months or years after irradiation [1]. The acute radiation syndrome is caused by the exposure to high IR during a short period of time, causing depletion of parenchymal cells in a tissue [2]. Therefore, the doses and duration of radiation exposure are critical for humans. Until now, radiation is largely used in clinical diagnostics and therapy with remarkable clinical benefits for patients. Radiotherapy is essentially based on both X- and γ-rays, which provide photons that are able to specifically penetrate the target and that can be captured on film [3]. Proton therapy uses proton beams that do not traverse the target but stop at an energy-dependent depth within the target with no exit dose [3]. Despite positive diagnostic and therapeutic aspects, the inappropriate use of computed tomography, leading to cancer risk, has been drawing attention for many years [4]. Additionally, the chronic radiation syndrome, ranging from dose-limiting toxicity to the increased risk of secondary cancers following radiation in patients, should always be considered [5]. To this end, adaptive responses to low radiation doses have been widely studied both in vitro and in vivo to ascertain the biological mechanism of radiation action. Radiation-induced signaling pathways in different tissues via EGFR, IGFI-R, PI3K, MAPK, JNK, and p38, as well as via FAS-R, TNF-R, and NFKB, have been reviewed [6].

2. Sensitivity of the Thyroid Gland to Radiation and Cancer Development

Although classically considered resistant to acute effects of radiation [7], the thyroid has actually proved to be particularly sensitive to the long-term effects of radiation exposure as demonstrated in studies of human subjects exposed to sublethal radiation doses [8]. More epidemiological studies were performed. It has been demonstrated by analyzing young adults exposed to radiation during childhood. A screening study of 11,970 residents of Belarus aged $\leq$18 years at the time of the Chernobyl nuclear accident showed a risk for neoplastic nodules significantly higher than for non-neoplastic nodules [9]. High amounts of radiation caused a significant increase in the incidence of thyroid gland carcinoma, as observed in several nuclear catastrophes such as Hiroshima, Nagasaki, Chernobyl, and more recently, Fukushima [10]. The effects of radiation in inducing thyroid nodules have been demonstrated in atomic bomb survivors from 62 to 66 years after exposure during their childhood [11]. The analysis of thyroid consequences of the 2011 Fukushima nuclear reactor accident showed that 35% of the residents developed thyroid nodules and/or cysts [12]. The study of the survivors in Hiroshima and Nagasaki has demonstrated that the risk for thyroid cancer was significantly higher if IR exposure occurred at pediatric ages [9]. Exposure to low or moderate doses of IR appeared to specifically increase the risk of thyroid papillary microcarcinoma, even when exposure occurred during adulthood [13]. Richardson [14] stated that exposure to IR in adulthood was positively associated with thyroid cancer among female survivors from atomic bombs (excess relative rate/Gray (Gy) = 0.70; 90% confidence interval = 0.20–1.46), although the risk seemed to be lower if they were exposed to radiation in their childhood. Ron et al. [15] compared atomic bomb survivors, children treated for tinea capitis, children irradiated for enlarged tonsils, and infants irradiated for an enlarged thymus gland with two case controls of untreated patients with cervical cancer and childhood cancer. The authors reported that the risk to develop thyroid cancer was correlated with age and sex. In fact, in childhood, the pooled excess relative risk per Gy (ERR/Gy) was 7.7 (95% CI = 2.1, 28.7) and the excess absolute risk per 10(4) PY Gy (EAR/10(4) PY Gy) was 4.4 (95% CI = 1.9, 10.1); the excess relative risk was greater (p = 0.07) for females than males. Holm affirmed that, usually, the excess relative risk for thyroid cancer started 5–10 years after radiation exposure and continued until at least 40 years after exposure; it was correlated more to the early age (prior to five years of age) than to the sex [16]. Exposure to ^{131}I during childhood was associated with an increased risk of thyroid cancer and both iodine deficiency and iodine supplementation appeared to modify such risk [17]. Robbins and Schneider confirmed the importance of the age, youth being a risk factor. Although the clinical use of radioiodine has not been reported to cause thyroid cancer, a low number of patients with cancer were young children and the studied cohorts were too small (consisting of 17 to 191 patients) to provide the statistical power to detect such a relatively rare event [18]. Among 585 patients with neck radiation, seven survivors developed papillary thyroid carcinoma (PTC). This indicates that, in adult survivors of cancer during their childhood or young adulthood with a history of radiation therapy to the neck for cancer, an annual physical exam should be considered appropriate as a thyroid cancer screening strategy [19]. Patients with head and neck squamous cell carcinoma showed a strong incidence of a subsequent primary thyroid cancer during the first 5 years after diagnosis and IR-treatment, supporting the concept that continued surveillance of thyroid status is important in this scenario [20]. Molecular mechanisms (genes, proteins, and lipids) that played a role in radiation-induced damages were reported in the following paragraphs.

3. Genes and Proteins Involved in Radiation-Induced Cancer

Advances in biochemistry and molecular biology have allowed the identification of the IR and non-IR molecular events in the thyroid gland (Figure 1).

Both IR and UV induced enhanced production of free oxygen radicals and modified pro-oxidant states [21]. However, the greatest damage to proteins and nucleic acids were with IR.

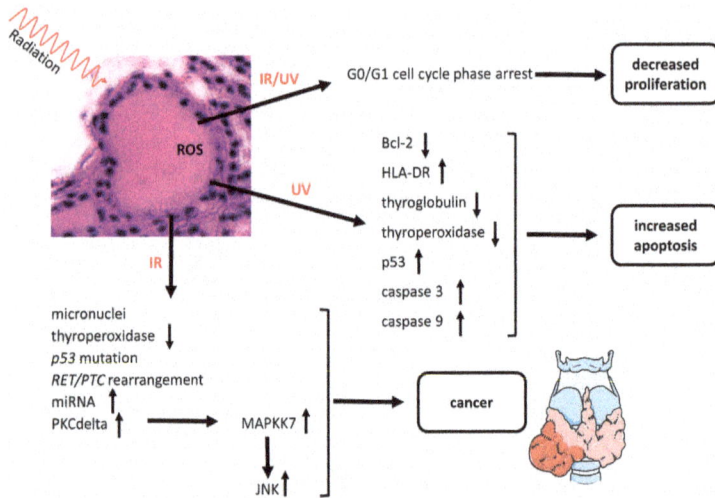

Figure 1. Effects of IR and UV in the thyroid gland. Synopsis of the main literature in the field [22–45]. HLA-DR: human leukocyte antigen-DR; *RET/PTC*: rearranged during transfection/papillary thyroid carcinoma; PKC: protein kinase C; MAPKK7: mitogen-activated protein kinase 7; JNK: c-Jun NH2-terminal kinases; IR: ionizing radiation; UV: ultraviolet rays; ROS: reactive oxygen species. Some graphical elements were taken from the Servier Medical Art Library, available from http://www.servier.com/Powerpoint-image-bank under Creative Commons Attribution 3.0 Unported License. Up-arrows, increase; down-arrows, decrease.

3.1. Ionizing Radiations

IR directly and/or indirectly causes oxidative stresses to the biological systems at the local or systematic level by influencing aging, genetic destabilization and mutagenicity, membrane lysis and cell death, alteration of enzymatic activities and metabolic events, mitochondrial dysfunction, and cancer [22]. The effects of IR in the thyroid gland have been extensively studied. The chronic exposure of mature rats to low-intensity γ-rays between 5 and 50 cGy (dose rates: 25, 400 µGy/h) induced the formation of micronuclei three times higher in irradiated thyrocytes than in thyrocytes of control animals [23]. Furthermore, the residual thyroid of hemi-thyroidectomized rats exposed to acute γ-rays with 2–4 Gy presented micronuclei. Ermakova et al. found that the presence of micronuclei was also a sensitive indicator of radiation-induced genetic damages in the follicular epithelium of thyroid gland [23]. Moreover, IR delayed follicular thyroid cell proliferation [24]. Thyroids of old rats irradiated in the neck region with an X-ray single dose of 3 Gy showed an increase in proliferating follicular cells two days after irradiation, followed by a phase of sharp decrease in cell proliferation between the 2nd and 6th day after irradiation. During the cell proliferation phase, the cell cycle was shortened by approximately 50%, predominantly due to a decrease of the G1-phase duration [24]. ^{131}I was shown to trigger apoptosis in human thyrocytes [25]. The cell viability of human thyroid epithelial cells purified from surgical tissue specimens was not affected by single doses of 5 or 50 Gy IR, and there was no induction of Heat shock proteins (HSP)-72, as an indicator of acute cellular stress. Nevertheless, the expression of thyroperoxidase (TPO), a key enzyme of thyroid hormone synthesis, significantly decreased [26]. The authors hypothesized that the suppression of thyroid hormone synthesis due to TPO reduction could contribute to an early development of thyroid dysfunction following irradiation, and they recommended considering this effect during radiation therapy [26]. On the other hand, the thyroid hormone modulation with X-rays induced neoplastic transformation in vitro [27]. Mizuno et al. [28] indicated that IR caused various oncogene activations, with specificity for early gene alteration uniquely associated with thyroid carcinogenesis. Irradiation of a non-tumorigenic human thyroid epithelial cell line with α-particles or γ-rays stimulated Exons 6 and 7, as well as *p53* mutations in the childhood PTC in Belarus, presumably as a result of radioiodine fallout [29]. In addition, IR exposure of cultured human thyroid cells stimulated the induction of c-Jun NH_2-terminal kinases (JNK) activity, not extracellular signal-regulated kinases (ERK) activity, to a 3.5-fold extent. The effect was specific for thyroid cells as it was absent in fibroblasts [30]. The JNK activation was mediated at least partially through a protein kinase C (PKC)-dependent pathway [30]. Mitsutake et al. [31] reported that among PKC-α, $\beta2$, δ, ε, and ζ isoforms expressed in primary cultured human thyroid cells, only PKC-δ was involved in an IR-induced JNK activation. Moreover, PKC-δ acted via mitogen-activated protein kinase kinase 7 (MAPKK7), not via MAPKK4 [31]. Characteristically, IR was responsible for a dose-dependent REarranged during Transfection/Papillary Thyroid Cance (*RET/PTC*) rearrangement in human thyroid cells [32]. Ameziane et al. [33] demonstrated that this effect was dependent on generated H_2O_2 during irradiation; it was responsible for the breaks of double-strand DNA and facilitated *RET/PTC1* formation. As a consequence, by pretreating the cells with catalase, a scavenger of H_2O_2, *RET/PTC1* rearrangement was decreased. Cells derived from the neural crest, kidney, and enteric nervous system expressed *RET* proto-oncogene [34]. Hamatani et al. [35] reported that in PTC the *RET* proto-oncogene generated a series of chimeric-transforming oncogenes collectively described as *RET/PTCs*. In childhood PTC with a history of radiation exposure, *RET/PTC* rearrangements represented a major event and among atomic bomb survivors, the frequency of rearrangements increased in relation to an increase of radiation dose [35]. In two studies that employed human fetal thyroid tissue xenografts, Mizuno et al. [28] demonstrated that X-ray irradiation generated both *RET/PTC1* and *RET/PTC3* rearrangements, and the *RET/PTC1* type was the most common. Notably, patients exposed to Chernobyl radiation developed PTC, and survivors of the atomic bomb in Japan had a very high frequency of *RET/PTC* chromosomal rearrangement [36]. In addiction, B-Raf proto-oncogene (BRAF) mutation (BRAF V600E) was associated with PTC [37]. Guan et al. [38] demonstrated that high iodine intake was a significant risk factor for BRAF V600E mutation and the development of PTC in the thyroid gland. The prevalence of BRAF V600E mutation in pediatric PTC was significantly lower than that in adults, 54% versus 85% [39]. On the other hand, a clinicopathological study showed that BRAF V600E was associated with older age and larger tumor size [40]. In patients with PTC who were 0–18 years at the time of the Chernobyl accident, BRAF V600E mutation was present, but the percentage was less than that of *RET/PTC1* and *RET/PTC3* rearrangements t [41]. Genomic copy number alterations of PTC of the Ukrainian-American cohort after the Chernobyl accident were associated with BRAF V600E mutation [42]. BRAFV600E mutation was less frequent in the cases of Hiroshima and Nagasaki survivors exposed to higher radiation doses [35]. In atomic bomb survivors in Hiroshima, the median radiation dose able to induce PTC was significantly lower in patients with BRAFV600E mutation than that without the mutation [43]. A screening program of various genetic alterations in children aged 0–18 years old at the time of the Fukushima accident showed that BRAF V600E was highly prevalent in the 63.2% of the population [44]. The difference of the data in various atomic bomb survivors could be due to the different genetic profile of patients, considering that the response to radiation of the thyroid gland was dependent on the genetic profile of the patients [44]. For this reason, Fukushima PTC was completely different from post-Chernobyl radiation-induced PTC [44], indicating the possibility of non-radiogenic etiology of PTC. Significant upregulation of a subset of these miRNAs (miR-187, miR-146b, and miR-155) was found to be more pronounced in PTC carrying *RET/PTC* rearrangements [45]. The association between miRNAs and radiation exposure has been reported in a variety of mouse tissues, including spleen, colon, thymus, and kidney [46]. Acute exposure of thyroid cells to γ-radiation resulted in several specific patterns of miRNA response not directly associated to carcinogenesis [47].

3.2. Non-Ionizing Radiations

UV radiation induced apoptosis in the FRTL-5 rat thyroid cell line [48] by the increase of p53 and caspases 3 and 9, and the decrease of Bcl-2, together with a transient but significant increase in HLA-DR expression [49] and the impairment of genes involved in thyroid hormone production, such as genes for thyroglobulin and TPO [50]. The effect was dependent on TSH that stimulates cell proliferation. Overall, TSH starvation induced virtually all cells to accumulate in the G0/G1 cell cycle phase, blocking cell proliferation, and rendering cells more resistant to UV-C radiation-induced apoptosis [51]. Thus, the effect of UV on FRTL-5 cells in culture was strongly related to the physiological state of the cells. Proliferating cells were more sensitive to radiation treatment than quiescent cells; the cells in a proapoptotic state caused by the lack of trophic support were less sensitive to radiation treatment [52].

4. Lipids as Regulators of Radiation-Induced Cancer

Differences in the responses to IR and non-IR of proliferating, quiescent and proapoptotic thyroid cells were associated with a very complex mechanism of lipid metabolism. A specific cross-talk exists among sphingomyelin (SM), phosphatidylcholine (PC), and phosphatidylinositol (PI) in both cell membrane and nuclei [53,54] (Figure 2).

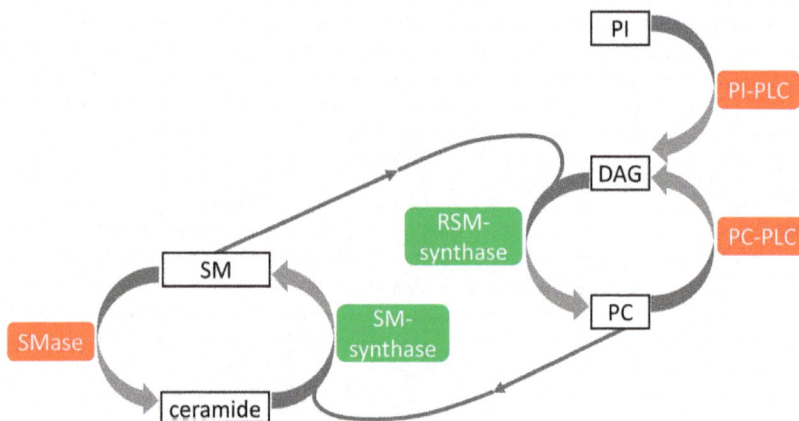

Figure 2. Cross talk among sphingomyelin (SM), phospatidylcholine (PC), and phosphatidylinositol (PI) metabolism. SM is degraded by sphingomyelinase (SMase) to produce ceramide and is restored by sphingomyelin-synthase (SM-synthase) from PC and ceramide. PC is degraded by phosphatidylcholine-specific phospholipase C (PC-PLC) to produce diacylglycerol (DAG) and is restored by reverse sphingomyelin-synthase (RSM-synthase) from SM and DAG. PI is degraded by phosphatidylinositol-specific phospholipase C (PI-PLC). In red, catabolic enzymes; in green, anabolic enzymes; thin arrows, SM for PC synthesis and PC for SM synthesis; thick arrows, the relation of PI and PC with DAG and the relation of SM with ceramide.

4.1. Ionizing Radiation

After IR-exposure, ceramide and diacylglycerol (DAG) acted as second messengers inducing proapoptotic and antiapoptotic signals, respectively [55]. FRTL-5 cells submitted to accelerated proton beams (CERN, Geneva, Switzerland) showed changes of lipid metabolism enzymes [55]. Proton beams induced quiescent thyroid cells towards a proapoptotic state and proliferating thyroid cells towards an initial apoptotic state, by altering the nuclear SM-metabolism. In cell nuclei the strong activation of neutral-sphingomyelinase (N-SMase) reduced SM content that was important for the DNA-stability.

The ceramide produced in the nucleus probably was translocated to the cytoplasm, where it could be metabolized to sphingosine and sphingosine-1-phosphate, lipid mediators involved in apoptosis [56].

4.2. Non-Ionizing Radiation

UV radiation enriched the ceramide pool due to acid-sphingomyelinase (A-SMase) and N-SMase activities and enlarged the DAG pool due to phosphatidylcholine-specific phospholipase C (PC-PLC) and phosphatidylinositol-specific phospholipase C (PI-PLC) in cell membranes of proliferating cells [52]. In purified nuclei, radiation stimulated N-SMase and reverse SM-synthase (RSM-synthase) activities while inhibited PC-PLC, PI-PLC, and SM-synthase activities leading to further ceramide pool enrichment and DAG pool reduction. The effect of UV irradiation on lipid metabolism was higher in the nucleus than in cell membranes [52]. The effect on nuclear lipids was very relevant, because of their role in cell proliferation, differentiation, and apoptosis [57] by acting as a platform for the attachment [58,59] and regulation [60] of active chromatin and for nuclear drug activity [61,62]. The prolonged presence in the stratosphere results in exposure to radiation, so stratospheric balloons were used to expose FRTL-5 cells to radiation present at a 30–40 km altitude for approximately 20 h (BIRBA mission). In proliferating cells, low doses of stratospheric radiation did not induce cell death but only early modifications of nuclear SM and PC metabolism. In purified nuclei, SMase and RSM-synthase activities were increased, whereas PC-PLC and SM-synthase activities were inhibited, leading to an increase of the ceramide/DAG ratio [63]. These studies indicated that nuclear SM metabolism was involved in radiation-induced damage (Table 1). The results were relevant considering the possibility that radiation induced thyroid cancer.

Table 1. The effect of radiation types on nuclear lipid metabolism.

	FRTL-5 Nuclei			
Radiation	Proliferating Cells	Quiescent Cells	Proapoptotic Cells	References
UV	↑ SMase ++ ↑ RSMase ++ ↓ SMsynthase ++ ↓ PCPLC ++ ↓ PIPLC ++	↑ SMase + ↑ RSMase ++ ↓ SMsynthase + ↓ PCPLC + ↓ PIPLC +	↑ SMase ++ ↑ RSMase ++ ↓ SMsynthase ++ ↓ PCPLC ++ ↓ PIPLC ++	[46]
Stratosphere	↑ SMase ++ ↑ RSMase ++ ↓ SMsynthase ++ ↓ PCPLC ++	↑ SMase + ↑ RSMase ++ ↓ SMsynthase ++ ↓ PCPLC ++		[57]
Protons	↑ SMase +++ ↓ SMsynthase =	↑ SMase = ↓ SMsynthase ++		[50]

SMase: sphingomyelinase; RSMase: reverse sphingomyelin-synthase; SMsynthase: sphingomyelin-synthase; PCPLC: phosphatidylcholine-specific phospholipase C; PIPLC: phosphatidylinositol-specific phospholipase C; + low change; ++ medium change; +++ high change. ↑ increased activity; ↓ decreased activity. =: In protons proliferating cell SMsynthase ↓ should be deleted; in protons quiescent cells SMase ↑ should be deleted.

5. Biomarkers of Thyroid Damage

Considering the molecular effects of radiation on the thyroid gland, the analysis of micronuclei frequency in peripheral blood lymphocytes is applicable as a biomarker of chromosomal damage, genome instability, and cancer risk [64]. A negative correlation between micronuclei frequency and the level of platelets without correlation to thyroid-related hormones has been observed in blood of patients suffering from differentiated thyroid cancer and treated with radioactive iodine (^{131}I) [65]. Dom et al. [66] studied children exposed and non-exposed to the Chernobyl radiation and compared them in the transcriptomes of normal contralateral tissues of PTC; in this way, the authors identified a gene expression signature (793 probes) that permits discrimination between both cohorts. To differentiate radiation and no radiation-induced PTC, Port et al. investigated the RNA isolated from 11 post-Chernobyl PTCs and 41 sporadic PTCs [67]. The microarray detected 646 upregulated genes and 677 downregulated genes [67]. The analysis of gene expression can be useful to measure the predisposition to developing cancer after radiation exposure [68]. In particular, the overexpression of the CLIP2 gene is the most promising marker; in fact, it was found in the majority of PTCs from young patients included in the Chernobyl tissue bank [69]. In post-Chernobyl PTC, the expression of CLIP2 gene was radiation dose-dependent [70]. The use of CLIP2 as radiation biomarker was supported by a study indicating its involvement in the fundamental carcinogenic processes including apoptosis, mitogen-activated protein kinase signaling, and genomic instability [71]. In comparison with normal tissues, thyroid carcinoma tissues from patients had a significant increase in lecithin, SM, and cholesterol [72]. Changes of the SM content together with other lipids in the blood plasma of patients with thyroid carcinoma were reported [73]. Serum lipidomic profiling with a panel which included 36:3phosphatidic acid (PA) and 34:1SM can be useful to distinguish between malignant thyroid cancer and benign thyroid tumors [74]. Rath et al. found that glycosylation of ceramide could contribute to the drug-resistance phenotype in thyroid malignancies [75]. Furthermore, it has been suggested that sphingosine kinase 1 (SphK1) and sphingosine-1-phosphate (S1P) may be relevant in the etiology of thyroid cancer, and in the regulation of both invasion and migration of thyroid cancer cells. Therefore, their contents could be useful as specific biomarkers of cancer transformation and progression [76].

6. Conclusions

To date, a number of specific molecular targets have been identified, by which radiation exerts its effects upon the thyroid gland, inducing long-term damages including cancer. Many genes, proteins, and lipids are involved in the mechanism of action, effects, and consequences of radiation, so this field of study is still widely open. It is becoming increasingly evident in the most recent literature that specific genes, proteins, and lipids are important targets of both radiation and cancer, but many points remain obscure. Further studies are required to shed more light on the complexity of interactions among various cellular components.

Acknowledgments: We thank the University of Udine for financial support.

Conflicts of Interest: The authors declare no conflict of interest.

Abbreviations

A-SMase	Acid-sphingomyelinase
DAG	Diacylglycerol
HLA-DR	Human leukocyte antigen-DR
IR	Ionizing radiation
JNK	c-Jun NH2-terminal kinases
MAPKK7	Mitogen-activated protein kinase kinase 7
N-SMase	Neutral-sphingomyelinase
PC	Phosphatidylcholine
PC-PLC	Phosphatidylcholine-specific phospholipase C
PI	Phosphatidylinositol
PI-PLC	Phosphatidylinositol-specific phospholipase C
PKC	Protein kinase C
PTC	Papillary thyroid carcinoma
RET	Rearranged during transfection
RSM-synthase	Reverse sphingomyelin-synthase
SM	Sphingomyelin
SM-synthase	Sphingomyelin-synthase
UV	Ultraviolet rays

References

1. Port, M.; Herodin, F.; Valente, M.; Drouet, M.; Ullmann, R.; Doucha-Senf, S.; Lamkowski, A.; Majewski, M.; Abend, M. MicroRNA expression for early prediction of late occurring hematologic acute radiation syndrome in baboons. *PLoS ONE* **2016**, *11*. [CrossRef] [PubMed]

2. Kazzi, Z.; Buzzell, J.; Bertelli, L.; Christensen, D. Emergency department management of patients internally contaminated with radioactive material. *Emerg. Med. Clin. N. Am.* **2015**, *33*, 179–196. [CrossRef] [PubMed]

3. Yamoah, K.; Johnstone, P.A. Proton beam therapy: Clinical utility and current status in prostate cancer. *OncoTargets Ther.* **2016**, *16*, 5721–5727. [CrossRef]

4. Frush, D. MO-FG-207A-03: Radiation and cancer perspectives from the trenches: Are we providing care or promoting scare? *Med. Phys.* **2016**, *43*, 3714. [CrossRef]

5. Orton, C.; Borras, C.; Carlson, D. Radiation biology for radiation therapy physicists. *Med. Phys.* **2014**, *41*, 532. [CrossRef]

6. Dent, P.; Yacoub, A.; Contessa, J.; Caron, R.; Amorino, G.; Valerie, K.; Hagan, M.P.; Grant, S.; Schmidt-Ullrich, R. Stress and radiation-induced activation of multiple intracellular signaling pathways. *Radiat. Res.* **2003**, *159*, 283–300. [CrossRef]

7. Rubin, P.; Casarett, G.W. Clinical radiation pathology as applied to curative radiotherapy. *Cancer* **1968**, *22*, 767–780.

8. Paro, J.N.; Zavisić, B.K. Iodine and thyroid gland with or without nuclear catastrophe. *Med. Pregl.* **2012**, *65*, 489–495.

9. Cahoon, E.K.; Nadirov, E.A.; Polanskaya, O.N.; Yauseyenka, V.V.; Velalkin, I.V.; Yeudachkova, T.I.; Maskvicheva, T.I.; Minenko, V.F.; Liu, W.; Drozdovitch, V.; et al. Risk of thyroid nodules in residents of Belarus exposed to Chernobyl fallout as children and adolescents. *J. Clin. Endocrinol. Metab.* **2017**. [CrossRef]

10. Imaizumi, M.; Ohishi, W.; Nakashima, E.; Sera, N.; Neriishi, K.; Yamada, M.; Tatsukawa, Y.; Takahashi, I.; Fujiwara, S.; Sugino, K.T.; et al. Association of radiation dose with prevalence of thyroid nodules among atomic bomb survivors exposed in childhood (2007–2011). *JAMA Intern. Med.* **2015**, *175*, 228–236. [CrossRef] [PubMed]

11. Nagataki, S. Thyroid consequences of the Fukushima nuclear reactor accident. *Eur. Thyroid J.* **2012**, *1*, 148–158. [CrossRef] [PubMed]

12. Furukawa, K.; Preston, D.; Funamoto, S.; Yonehara, S.; Ito, M.; Tokuoka, S.; Sugiyama, H.; Soda, M.; Ozasa, K.; Mabuchi, K. Long-term trend of thyroid cancer risk among Japanese atomic-bomb survivors: 60 Years after exposure. *Int. J. Cancer* **2013**, *132*, 1222–1226. [CrossRef] [PubMed]

13. Hayashi, Y.; Lagarde, F.; Tsuda, N.; Funamoto, S.; Preston, D.L.; Koyama, K.; Mabuchi, K.; Ron, E.; Kodama, K.; Tokuoka, S. Papillary microcarcinoma of the thyroid among atomic bomb survivors: Tumor characteristics and radiation risk. *Cancer* **2010**, *116*, 1646–1655. [CrossRef] [PubMed]

14. Richardson, D.B. Exposure to ionizing radiation in adulthood and thyroid cancer incidence. *Epidemiology* **2009**, *20*, 181–187. [CrossRef] [PubMed]

15. Ron, E.; Lubin, J.H.; Shore, R.E.; Mabuchi, K.; Modan, B.; Pottern, L.M.; Schneider, A.B.; Tucker, M.A.; Boice, J.D. Thyroid cancer after exposure to external radiation: A pooled analysis of seven studies. *J. Radiat. Res.* **1995**, *141*, 259–277. [CrossRef]

16. Holm, L.E. Radiation-induced thyroid neoplasia. *Sozial und Präventivmedizin* **1991**, *36*, 266–275. [CrossRef] [PubMed]

17. Cardis, E.; Kesminiene, A.; Ivanov, V.; Malakhova, I.; Shibata, Y.; Khrouch, V.; Drozdovitch, V.; Maceika, E.; Zvonova, I.; Vlassov, O.; et al. Risk of thyroid cancer after exposure to 131I in childhood. *J. Natl. Cancer Inst.* **2005**, *97*, 724–732. [CrossRef] [PubMed]

18. Robbins, J.; Schneider, A.B. Thyroid cancer following exposure to radioactive iodine. *Rev. Endocr. Metab. Disord.* **2000**, *3*, 197–203. [CrossRef]

19. Tonorezos, E.S.; Barnea, D.; Moskowitz, C.S.; Chou, J.F.; Sklar, C.A.; Elkin, E.B.; Wong, R.J.; Li, D.; Tuttle, R.M.; Korenstein, D.; et al. Screening for thyroid cancer in survivors of childhood and young adult cancer treated with neck radiation. *J. Cancer Surviv.* **2016**. [CrossRef] [PubMed]

20. Chan, J.Y.; Gooi, Z.; Mydlarz, W.K.; Agrawal, N. Risk of thyroid malignancy following an index head and neck squamous cell carcinoma: A population-based study. *Ear Nose Throat J.* **2016**, *95*, E7–E11. [PubMed]

21. Borek, C. Molecular mechanisms in cancer induction and prevention. *Environ. Health Perspect.* **1993**, *101*, 237–245. [CrossRef] [PubMed]

22. Islam, M.T. Radiation interactions with biological system. *Int. J. Radiat. Biol.* **2017**, 1–28. [CrossRef] [PubMed]

23. Ermakova, O.V.; Pavlov, A.V.; Korableva, T.V. Cytogenetic effects in follicular epithelium of thyroid gland under prolonged exposure to gamma-radiation at low-doses. *Radiat. Biol. Radioecol.* **2008**, *48*, 160–166.

24. Christov, K. Effect of irradiation on the proliferation kinetics of thyroid follicular cells in infant rats. *Exp. Pathol.* **1982**, *21*, 117–122. [CrossRef]

25. Russo, E.; Guerra, A.; Marotta, V.; Faggiano, A.; Colao, A.; del Vecchio, S.; Tonacchera, M.; Vitale, M. Radioiodide induces apoptosis in human thyroid tissue in culture. *Endocrine* **2013**, *44*, 729–734. [CrossRef] [PubMed]

26. Blasko, I.; Sztankay, A.; Lukas, P.; Grubeck-Loebenstein, B. Decreased thyroid peroxidase expression in cultured thyrocytes after external gamma irradiation. *Exp. Clin. Endocrinol. Diabetes* **2000**, *108*, 138–141. [CrossRef] [PubMed]

27. Guernsey, D.L.; Ong, A.; Borek, C. Thyroid hormone modulation of X ray-induced in vitro neoplastic transformation. *Nature* **1980**, *288*, 591–592. [CrossRef] [PubMed]

28. Mizuno, T.; Kyoizumi, S.; Suzuki, T.; Iwamoto, K.S.; Seyama, T. Continued expression of a tissue specific activated oncogene in the early steps of radiation-induced human thyroid carcinogenesis. *Oncogene* **1997**, *15*, 1455–1460. [CrossRef] [PubMed]

29. Gamble, S.C.; Cook, M.C.; Riches, A.C.; Herceg, Z.; Bryant, P.E.; Arrand, J.E. p53 mutations in tumors derived from irradiated human thyroid epithelial cells. *Mutat. Res.* **1999**, *425*, 231–238. [CrossRef]

30. Hara, T.; Namba, H.; Yang, T.T.; Nagayama, Y.; Fukata, S.; Kuma, K.; Ishikawa, N.; Ito, K.; Yamashita, S. Ionizing radiation activates c-Jun NH_2-terminal kinase (JNK/SAPK) via a PKC-dependent pathway in human thyroid cells. *Biochem. Biophys. Res. Commun.* **1998**, *244*, 41–44. [CrossRef] [PubMed]

31. Mitsutake, N.; Namba, H.; Shklyaev, S.S.; Tsukazaki, T.; Ohtsuru, A.; Ohba, M.; Kuroki, T.; Ayabe, H.; Yamashita, S. PKC delta mediates ionizing radiation-induced activation of c-Jun NH_2-terminal kinase through MKK7 in human thyroid cells. *Oncogene* **2001**, *20*, 989–996. [CrossRef] [PubMed]

32. Caudill, C.M.; Zhu, Z.; Ciampi, R.; Stringer, J.R.; Nikiforov, Y.E. Dose-dependent generation of RET/papillary thyroid carcinoma in human thyroid cells after in vitro exposure to γ-radiation: A model of carcinogenic chromosomal rearrangement induced by ionizing radiation. *J. Clin. Endocrinol. Metab.* **2005**, *90*, 2364–2369. [CrossRef] [PubMed]

33. Ameziane-El-Hassani, R.; Boufraqech, M.; Lagente-Chevallier, O.; Weyemi, U.; Talbot, M.; Métivier, D.; Courtin, F.; Bidart, J.M.; El Mzibri, M.; Schlumberger, M.; et al. Role of H_2O_2 in RET/PTC1 chromosomal rearrangement produced by ionizing radiation in human thyroid cells. *Cancer Res.* **2010**, *70*, 4123–4132. [CrossRef] [PubMed]

34. Schuchardt, A.; D'Agati, V.; Larsson-Blomberg, L.; Costantini, F.; Pachnis, V. Defects in the kidney and enteric nervous system of mice lacking the tyrosine kinase receptor Ret. *Nature* **1994**, *367*, 380–383. [CrossRef] [PubMed]

35. Hamatani, K.; Eguchi, H.; Ito, R.; Mukai, M.; Takahashi, K.; Taga, M.; Imai, K.; Cologne, J.; Soda, M.; Arihiro, K.; et al. RET/PTC rearrangements preferentially occurred in papillary thyroid cancer among atomic bomb survivors exposed to high radiation dose. *Cancer Res.* **2008**, *68*, 7176–7182. [CrossRef] [PubMed]

36. Mizuno, T.; Iwamoto, K.S.; Kyoizumi, S.; Nagamura, H.; Shinohara, T.; Koyama, K.; Seyama, T.; Hamatani, K. Preferential induction of RET/PTC1 rearrangement by X-ray irradiation. *Oncogene* **2000**, *19*, 438–443. [CrossRef] [PubMed]

37. Daliri, M.; Abbaszadegan, M.R.; Bahar, M.M.; Arabi, A.; Yadollahi, M.; Ghafari, A.; Taghehchian, N.; Zakavi, S.R. The role of BRAF V600E mutation as a potential marker for prognostic stratification of papillary thyroid carcinoma: A long-term follow-up study. *Endocr. Res.* **2014**, *39*, 189–193. [CrossRef] [PubMed]

38. Guan, H.; Ji, M.; Bao, R.; Yu, H.; Wang, Y.; Hou, P.; Zhang, Y.; Shan, Z.; Teng, W.; Xing, M. Association of high iodine intake with the T1799A BRAF mutation in papillary thyroid cancer. *J. Clin. Endocrinol. Metab.* **2009**, *94*, 1612–1617. [CrossRef] [PubMed]

39. Oishi, N.; Kondo, T.; Nakazawa, T.; Mochizuki, K.; Inoue, T.; Kasai, K.; Tahara, I.; Yabuta, T.; Hirokawa, M.; Miyauchi, A.; et al. Frequent BRAF V600E and absence of TERT promoter mutations characterize sporadic pediatric papillary thyroid carcinomas in Japan. *Endocr. Pathol.* **2017**. [CrossRef] [PubMed]

40. Cordioli, M.I.; Moraes, L.; Bastos, A.U.; Besson, P.; Alves, M.T.; Delcelo, R.; Monte, O.; Longui, C.; Cury, A.N.; Cerutti, J.M. Fusion Oncogenes are the main genetic events found in sporadic papillary thyroid carcinomas from children. *Thyroid* **2017**, *27*, 182–188. [CrossRef] [PubMed]

41. Tronko, M.; Bogdanova, T.; Voskoboynyk, L.; Zurnadzhy, L.; Shpak, V.; Gulak, L. Radiation induced thyroid cancer: Fundamental and applied aspects. *Exp. Oncol.* **2010**, *32*, 200–204. [PubMed]

42. Selmansberger, M.; Braselmann, H.; Hess, J.; Bogdanova, T.; Abend, M.; Tronko, M.; Brenner, A.; Zitzelsberger, H.; Unger, K. Genomic copy number analysis of Chernobyl papillary thyroid carcinoma in the Ukrainian-American Cohort. *Carcinogenesis* **2015**, *36*, 1381–1387. [CrossRef] [PubMed]

43. Takahashi, K.; Eguchi, H.; Arihiro, K.; Ito, R.; Koyama, K.; Soda, M.; Cologne, J.; Hayashi, Y.; Nakata, Y.; Nakachi, K.; et al. The presence of BRAF point mutation in adult papillary thyroid carcinomas from atomic bomb survivors correlates with radiation dose. *Mol. Carcinog.* **2007**, *46*, 242–248. [CrossRef] [PubMed]

44. Mitsutake, N.; Fukushima, T.; Matsuse, M.; Rogounovitch, T.; Saenko, V.; Uchino, S.; Ito, M.; Suzuki, K.; Suzuki, S.; Yamashita, S. BRAF^{V600E} mutation is highly prevalent in thyroid carcinomas in the young population in Fukushima: A different oncogenic profile from Chernobyl. *Sci. Rep.* **2015**, *5*, 16976. [CrossRef] [PubMed]

45. Nikiforova, M.N.; Tseng, G.C.; Steward, D.; Diorio, D.; Nikiforov, Y.E. MicroRNA expression profiling of thyroid tumors: Biological significance and diagnostic utility. *J. Clin. Endocrinol. Metab.* **2008**, *93*, 1600–1608. [CrossRef] [PubMed]

46. He, L.; He, X.; Lim, L.P.; de Stanchina, E.; Xuan, Z.; Liang, Y.; Xue, W.; Zender, L.; Magnus, J.; Ridzon, D.; et al. A microRNA component of the p53 tumour suppressor network. *Nature* **2007**, *447*, 1130–1134. [CrossRef] [PubMed]

47. Nikiforova, M.N.; Gandhi, M.; Kelly, L.; Nikiforov, Y.E. MicroRNA dysregulation in human thyroidcells following exposure to ionizing radiation. *Thyroid* **2011**, *21*, 261–266. [CrossRef] [PubMed]

48. Del Terra, E.; Francesconi, A.; Meli, A.; Ambesi-Impiombato, F.S. Radiation-dependent apoptosis on cultured thyroid cells. *Phys. Med.* **2001**, *17* (Suppl. S1), 261–263. [PubMed]

49. Kostic, I.; Toffoletto, B.; Toller, M.; Beltrami, C.A.; Ambesi-Impiombato, F.S.; Curcio, F. UVC radiation-induced effect on human primary thyroid cell proliferation and HLA-DR expression. *Horm. Metab. Res.* **2010**, *42*, 846–853. [CrossRef] [PubMed]

50. Baldini, E.; D'Armiento, M.; Sorrenti, S.; del Sordo, M.; Mocini, R.; Morrone, S.; Gnessi, L.; Curcio, F.; Ulisse, S. Effects of ultravioletradiation on FRTL-5 cell growth and thyroid-specific gene expression. *Astrobiology* **2013**, *13*, 536–542. [CrossRef] [PubMed]

51. Del Terra, E.; Francesconi, A.; Donnini, D.; Curcio, F.; Ambesi-Impiombato, F.S. Thyrotropin effects on ultravioletradiation-dependent apoptosis in FRTL-5 cells. *Thyroid* **2003**, *13*, 747–753. [CrossRef] [PubMed]

52. Albi, E.; Cataldi, S.; Rossi, G.; Viola Magni, M.; Toller, M.; Casani, S.; Perrella, G. The nuclear ceramide/diacylglycerol balance depends on the physiological state of thyroidcells and changes during UV-C radiation-induced apoptosis. *Arch. Biochem. Biophys.* **2008**, *478*, 52–58. [CrossRef] [PubMed]

53. Albi, E.; Lazzarini, R.; Viola Magni, M. Phosphatidylcholine/sphingomyelinmetabolismcrosstalk inside the nucleus. *Biochem. J.* **2008**, *410*, 381–389. [CrossRef] [PubMed]

54. Albi, E.; Rossi, G.; Maraldi, N.M.; Viola Magni, M.; Cataldi, S.; Solimando, L.; Zini, N. Involvement of nuclear phosphatidylinositol-dependent phospholipases C in cell cycle progression during rat liver regeneration. *J. Cell. Physiol.* **2003**, *197*, 181–188. [CrossRef] [PubMed]

55. Sautin, Y.; Takamura, N.; Shklyaev, S.; Nagayama, Y.; Ohtsuru, A.; Namba, H.; Yamashita, S. Ceramide-induced apoptosis of human thyroid cancer cells resistant to apoptosis by irradiation. *Thyroid* **2000**, *10*, 733–740. [CrossRef] [PubMed]

56. Albi, E.; Perrella, G.; Lazzarini, A.; Cataldi, S.; Lazzarini, R.; Floridi, A.; Ambesi-Impiombato, F. S.; Curcio, F. Critical role for the protons in FRTL-5 thyroid cells: Nuclear sphingomyelinase induced-damage. *Int. J. Mol. Sci.* **2014**, *15*, 11555–11565. [CrossRef] [PubMed]

57. Albi, E.; Viola Magni, M.P. The role of intranuclear lipids. *Biol. Cell* **2004**, *96*, 657–667. [CrossRef] [PubMed]

58. Cascianelli, G.; Villani, M.; Tosti, M.; Marini, F.; Bartoccini, E.; Magni, M.V.; Albi, E. Lipid microdomains in cell nucleus. *Mol. Biol. Cell* **2008**, *19*, 5289–5295. [CrossRef] [PubMed]

59. Albi, E.; Villani, M. Nuclear lipid microdomains regulate cell function. *Commun. Integr. Biol.* **2009**, *2*, 23–24. [CrossRef] [PubMed]

60. Albi, E.; Lazzarini, A.; Lazzarini, R.; Floridi, A.; Damaskopoulou, E.; Curcio, F.; Cataldi, S. Nuclear lipid microdomain as place of interaction between sphingomyelin and DNA during liver regeneration. *Int. J. Mol. Sci.* **2013**, *14*, 6529–6541. [CrossRef] [PubMed]

61. Bartoccini, E.; Marini, F.; Damaskopoulou, E.; Lazzarini, R.; Cataldi, S.; Cascianelli, G.; Gil Garcia, M.; Albi, E. Nuclear lipid microdomain sregulate nuclear vitamin D3 uptake and influence embryonic hippocampal cell differentiation. *Mol. Biol. Cell* **2011**, *22*, 3022–3031. [CrossRef] [PubMed]

62. Cataldi, S.; Codini, M.; Cascianelli, G.; Tringali, S.; Tringali, A.R.; Lazzarini, A.; Floridi, A.; Bartoccini, E.; Garcia-Gil, M.; Lazzarini, R.; et al. Nuclear lipid microdomain as resting place of dexamethasone to impair cell proliferation. *Int. J. Mol. Sci.* **2014**, *15*, 19832–19846. [CrossRef] [PubMed]

63. Albi, E.; Cataldi, S.; Villani, M.; Perrella, G. Nuclear phosphatidylcholine and sphingomyelin metabolism of thyroid cells changes during stratospheric balloon flight. *J. Biomed. Biotechnol.* **2009**. [CrossRef] [PubMed]

64. Iarmarcovai, G.; Ceppi, M.; Botta, A.; Orsière, T.; Bonassi, S. Micronuclei frequency in peripheral blood lymphocytes of cancer patients: A meta-analysis. *Mutat. Res.* **2008**, *659*, 274–283. [CrossRef] [PubMed]

65. Vrndić, O.B.; Milošević-Djordjević, O.M.; Mijatović Teodorović, L.C.; Jeremić, M.Z.; Stošić, I.M.; Grujicić, D.V.; Zivancević Simonović, S.T. Correlation betweenmicronuclei frequency in peripheral blood lymphocytes and retention of 131-I in thyroid cancer patients. *Tohoku J. Exp. Med.* **2013**, *229*, 115–122. [CrossRef] [PubMed]

66. Dom, G.; Tarabichi, M.; Unger, K.; Thomas, G.; Oczko-Wojciechowska, M.; Bogdanova, T.; Jarzab, B.; Dumont, J.E.; Detours, V.; Maenhaut, C. A gene expression signature distinguishes normal tissues of sporadic and radiation-induced papillary thyroid carcinomas. *Br. J. Cancer* **2012**, *107*, 994–1000. [CrossRef] [PubMed]

67. Port, M.; Boltze, C.; Wang, Y.; Röper, B.; Meineke, V.; Abend, M. A radiation-induced gene signature distinguishes post-Chernobyl from sporadic papillary thyroid cancers. *Radiat. Res.* **2007**, *168*, 639–649. [CrossRef] [PubMed]

68. Maenhaut, C.; Detours, V.; Dom, G.; Handkiewicz-Junak, D.; Oczko-Wojciechowska, M.; Jarzab, B. Gene expression profiles for radiation-induced thyroid cancer. *Clin. Oncol.* **2011**, *23*, 282–288. [CrossRef] [PubMed]

69. Kaiser, J.C.; Meckbach, R.; Eidemüller, M.; Selmansberger, M.; Unger, K.; Shpak, V.; Blettner, M.; Zitzelsberger, H.; Jacob, P. Integration of a radiation biomarker into modeling of thyroid carcinogenesis and post-Chernobyl risk assessment. *Carcinogenesis* **2016**, *37*, 1152–1160. [CrossRef] [PubMed]

70. Selmansberger, M.; Kaiser, J.C.; Hess, J.; Güthlin, D.; Likhtarev, I.; Shpak, V.; Tronko, M.; Brenner, A.; Abend, M.; Blettner, M.; et al. Dose-dependent expression of CLIP2 in post-Chernobyl papillary thyroid carcinomas. *Carcinogenesis* **2015**, *36*, 748–756. [CrossRef] [PubMed]

71. Selmansberger, M.; Feuchtinger, A.; Zurnadzhy, L.; Michna, A.; Kaiser, J.C.; Abend, M.; Brenner, A.; Bogdanova, T.; Walch, A.; Unger, K.; et al. CLIP2 as radiation biomarker in papillary thyroid carcinoma. *Oncogene* **2015**, *34*, 3917–3925. [CrossRef] [PubMed]

72. Das, S.C.; Isichei, U.P. Serum and thyroid tissue lipids in patients with thyroid tumors in euthyroidism. *Indian J. Exp. Biol.* **1989**, *27*, 538–544. [PubMed]

73. Raffelt, K.; Moka, D.; Süllentrop, F.; Dietlein, M.; Hahn, J.; Schicha, H. Systemic alterations in phospholipid concentrations of blood plasma in patients with thyroid carcinoma: An in-vitro ^{31}P high-resolution NMR study. *NMR Biomed.* **2000**, *13*, 8–13. [CrossRef]

74. Guo, S.; Qiu, L.; Wang, Y.; Qin, X.; Liu, H.; He, M.; Zhang, Y.; Li, Z.; Chen, X. Tissue imaging and serum lipidomic profiling for screening potential biomarkers of thyroid tumors by matrix-assisted laser desorption/ionization-Fourier transform ion cyclotron resonance mass spectrometry. *Anal. Bioanal. Chem.* **2014**, *406*, 4357–4370. [CrossRef] [PubMed]

75. Rath, G.; Schneider, C.; Langlois, B.; Sartelet, H.; Morjani, H.; Btaouri, H.E.; Dedieu, S.; Martiny, L. De novo ceramide synthesis is responsible for the anti-tumor properties of camptothecin and doxorubicin in follicular thyroid carcinoma. *Int. J. Biochem. Cell Biol.* **2009**, *41*, 1165–1172. [CrossRef] [PubMed]

76. Törnquist, K. Sphingosine 1-phosphate and cancer: Lessons from thyroid cancer cells. *Biomolecules* **2013**, *3*, 303–315. [CrossRef] [PubMed]

International Journal of
Molecular Sciences

MDPI

Article

TNFSF4 Gene Variations Are Related to Early-Onset Autoimmune Thyroid Diseases and Hypothyroidism of Hashimoto's Thyroiditis

Rong-Hua Song [1,†], Qiong Wang [2,†], Qiu-Ming Yao [1], Xiao-Qing Shao [1], Ling Li [1], Wen Wang [1], Xiao-Fei An [1], Qian Li [1] and Jin-An Zhang [1,*]

[1]   Department of Endocrinology, Jinshan Hospital of Fudan University, No. 1508 Longhang Road, Jinshan District, Shanghai 201508, China; someonesrh66@163.com (R.-H.S.); 14211270001@fudan.edu.cn (Q.-M.Y.); shaoxq2015@163.com (X.-Q.S.); alana3344@126.com (L.L.); wwen910108@163.com (W.W.); anxiaofei2000@163.com (X.-F.A.); ellen38668@163.com (Q.L.)
[2]   The hemodialysis center of Nephropathy Department, Shaanxi Provincial People's Hospital, No. 256 West Youyi Road, Beilin District, Xi'an 710068, China; wq9930@126.com
*   Correspondence: zhangjinan@hotmail.com; Tel.: +86-21-5703-9815; Fax: +86-21-6722-6910
†   These authors contributed equally to this work.

Academic Editor: Daniela Gabriele Grimm
Received: 10 July 2016; Accepted: 16 August 2016; Published: 20 August 2016

Abstract: The aim of the current study was to examine whether the polymorphism loci of the tumor necrosis factor superfamily member 4 (*TNFSF4*) gene increase the risk of susceptibility to autoimmune thyroid diseases (AITDs) in the Han Chinese population, and a case-control study was performed in a set of 1,048 AITDs patients and 909 normal healthy controls in the study. A total of four tagging single nucleotide polymorphisms (SNPs) in the *TNFSF4* region, including rs7514229, rs1234313, rs16845607 and rs3850641, were genotyped using the method of ligase detection reaction. An association between GG genotype of rs3850641 in *TNFSF4* gene and AITDs was found ($p = 0.046$). Additionally, the clinical sub-phenotype analysis revealed a significant association between GG genotype in rs7514229 and AITDs patients who were ≤ 18 years of age. Furthermore, rs3850641 variant allele G was in strong association with hypothyroidism in Hashimoto's thyroiditis (HT) ($p = 0.018$). The polymorphisms of the *TNFSF4* gene may contribute to the susceptibility to AITDs pathogenesis.

Keywords: tumor necrosis factor superfamily member 4 (*TNFSF4*); single nucleotide polymorphism (SNP); autoimmune thyroid diseases (AITDs); Graves' disease (GD); Hashimoto's thyroiditis (HT)

1. Introduction

Autoimmune thyroid diseases (AITDs) are a group of organ-specific and polygenic inherited autoimmune diseases, with an estimated prevalence of up to 1%–5% of the general population [1]. AITDs mainly consist of two clinical subtypes of Graves' disease (GD) and Hashimoto's thyroiditis (HT). GD is predominantly characterized by a variable combination of hyperthyroidism, diffused goiter and high level of thyroid stimulating hormone receptor antibody (TRAb). Meanwhile, some GD patients may present extrathyroidal manifestations, including ophthalmopathy, pretibial myxedema and clubbed fingers. Clinical features of HT include the presence of antibody against thyroid peroxidase (TPOAb) or thyroglobulin (TgAb). Additionally, some patients with HT harbor extensive apoptosis of thyrocytes leading to hypothyroidism. Although there are some common characteristics in GD and HT, such as destruction of thyroid tissue and the existence of circulating thyroid autoantibodies including TRAb, TPOAb and TgAb, the clinical presentations and mechanisms of the two subtypes are different from each other to some extent; for example, our previous studies found that GD, HT or even Graves' ophthalmopathy (GO) have specific genetic backgrounds [2,3]. The pathogenesis of AITDs remains unclear, although there is much evidence demonstrating that the interaction between genetic factors and environmental components may be involved in their etiology [4,5].

More recently, an increasing body of research has confirmed that several specific genes are associated with multiple autoimmune diseases [6,7], implicating that many autoimmune diseases may share some genetic risk factors. For instance, TNFAIP3 has been identified to be related to the genetic etiology of systemic lupus erythematosus (SLE) [8], rheumatoid arthritis (RA) [9], systemic sclerosis (SSc) [10]. Additionally, we also found the relationship between this gene and GD [11]. All these data documented that variants in several genes probably contribute to dysregulation of common immune pathways, and then are involved in the pathological procedure of diverse autoimmune diseases.

The tumor necrosis factor superfamily member 4 (*TNFSF4*) gene encodes a cytokine (OX40L), which is expressed on antigen-presenting cells (APCs) to provide co-stimulatory signals to T cells. In recent years, the *TNFSF4* gene polymorphisms have been reported to be an important predisposition factor to SLE [12], RA [6], SSc [13] and primary Sjogren's syndrome (pSS) [14]. However, to date, whether *TNFSF4* gene variations are associated with AITDs has not been investigated.

In the present study, we evaluated whether mutations in *TNFSF4* gene are genetically predisposed in Han Chinese populations to AITDs via a case-control study. Single nucleotide polymorphisms (SNPs) tagging four independent susceptibility loci were genotyped in a large cohort of AITDs patients and normal healthy controls. We also analyzed the association between each polymorphism locus and the predisposition to different subtypes of AITDs, including GD, HT and ophthalmopathy.

2. Results

2.1. Clinical Phenotype Analysis

The clinical characteristics of the AITDs cohort are displayed in Table 1. Among the investigated 1,048 AITDs patients, 693 were GD patients including 30.736% male and 69.264% female, with mean disease-onset age of 34.010 ± 14.395 years old; 355 were HT patients, including 12.676% male and 87.324% female, with mean disease-onset age of 32.720 ± 13.511 years old. There were 162 (15.458%) teenaged AITDs patients with disease-onset age ≤ 18 years old, 130 (12.405%) AITDs patients with ophthalmopathy, 198 (55.775%) HT patients with hypothyroidism, and 216 (20.611%) AITDs patients, comprised of 143 GD patients and 73 HT patients, with family history.

Table 1. Clinical data of AITDs patients and controls.

Clinical Phenotype	AITDs (%)	GD (%)	HT (%)	Control (%)
Number	1048	693	355	909
Gender	–	–	–	–
Male	260 (24.809)	213 (30.736)	45 (12.676)	314 (34.543)
Female	788 (75.191)	480 (69.264)	310 (87.324)	595 (65.457)
Onset of age	33.570 ± 14.108	34.010 ± 14.395	32.720 ± 13.511	–
≤ 18 years	162 (15.458)	113 (16.306)	49 (13.803)	–
≥ 19 years	886 (84.542)	580 (83.694)	306 (86.197)	–
Ophthalmopathy	–	–	–	–
(+)	130 (12.405)	124 (17.893)	6 (1.690)	–
(−)	918 (87.595)	569 (82.107)	349 (98.310)	–
Family history	–	–	–	–
(+)	216 (20.611)	143 (20.635)	73 (20.563)	–
(−)	832 (79.389)	550 (79.365)	282 (79.437)	–

2.2. Allelic and Genotypic Association

There is a Hardy-Weinberg equilibrium (HWE) in the genotype distributions of *TNFSF4* SNPs in the control group ($p > 0.05$). Additionally, we evaluated the HWE for the loci in our AITDs cases;

the HWE of the four loci for AITDs cases all had *p*-value higher than 0.01. In addition, variant genotype GG of rs3850641 in *TNFSF4* gene is associated with AITDs (*p* = 0.046), as shown in Table 2. Further analysis found that the frequencies of TT genotype in rs7514229 and GG genotype in rs3850641 were lower in the AITDs group than the control group (Table 3), which suggested that people with these genotypes are less susceptible to AITDs (*p* = 0.016, *OR* = 0.236, 95% CI = 0.066–0.850 and *p* = 0.027, *OR* = 0.492, 95% CI = 0.259–0.935, respectively). Those subjects whose genotypes of the four loci failed to be determined were ruled out from the statistical analysis. Moreover, the frequency of genotype GG in rs3850641 was lower in GD patients than the control group in analysis of sub-clinical types of AITDs (GD and HT) although without statistical significance (*p* = 0.086), as shown in Table 4.

Table 2. Allele and genotype distribution of *TNFSF4* in AITDs patients and controls.

SNP ID	Control	AITDs	*p*	*OR* (95% CI)
rs7514229	–	–	–	–
GG	732 (81.424)	838 (81.518)	0.052	–
TT	11 (1.224)	3 (0.292)	–	–
GT	156 (17.353)	187 (18.191)	–	–
G	1620 (90.100)	1863 (90.613)	0.590	1.061 (0.856–1.314)
T	178 (9.900)	193 (9.387)	–	–
rs1234313	–	–	–	–
AA	365 (40.466)	445 (42.871)	0.536	–
GG	110 (12.195)	117 (11.272)	–	–
AG	427 (47.339)	476 (45.857)	–	–
A	1157 (64.135)	1366 (65.800)	0.278	1.076 (0.943–1.228)
G	647 (35.865)	710 (34.200)	–	–
rs16845607	–	–	–	–
AA	4 (0.443)	5 (0.480)	0.993	–
GG	797 (88.359)	920 (88.292)	–	–
AG	101 (11.197)	117 (11.228)	–	–
A	109 (6.042)	127 (6.094)	0.946	1.009 (0.775–1.314)
G	1695 (93.958)	1957 (93.906)	–	–
rs3850641	–	–	–	–
GG	26 (2.879)	15 (1.438)	0.046	–
AA	660 (73.090)	750 (71.908)	–	–
AG	217 (24.031)	278 (26.654)	–	–
G	269 (14.895)	308 (14.765)	0.910	0.990 (0.829–1.182)
A	1537 (85.105)	1778 (85.235)	–	–

Table 3. Genotype frequency of *TNFSF4* loci in AITDs patients and controls.

SNP Name	Genotype	Control (%)	AITDs (%)	*p*	*OR*	95% CI
rs7514229	GG	732 (81.424)	838 (81.518)	0.958	1.006	0.799–1.267
	TT + GT	167 (18.576)	190 (18.482)			
	TT	11 (1.233)	3 (0.292)	0.016	0.236	0.066–0.850
	GG + GT	888 (98.776)	1025 (99.708)			
	GT	156 (17.353)	187 (18.191)	0.631	1.059	0.838–1.339
	GG + TT	743 (82.647)	841 (81.809)			
rs1234313	AA	365 (40.466)	445 (42.871)	0.284	1.104	0.921–1.323
	GG + AG	537 (59.534)	593 (57.129)			
	GG	110 (12.195)	117 (11.272)	0.528	0.915	0.693–1.206
	AA + AG	792 (87.805)	921 (88.728)			
	AG	427 (47.339)	476 (45.857)	0.514	0.943	0.788–1.126
	AA + GG	475 (52.661)	562 (54.143)			

Table 3. *Cont.*

SNP Name	Genotype	Control (%)	AITDs (%)	*p*	OR	95% CI
rs16845607	AA	4 (0.443)	5 (0.480)	0.906	1.082	0.290–4.049
	GG+AG	898 (99.557)	1037 (99.520)			
	GG	797 (88.359)	920 (88.292)	0.963	0.993	0.752–1.311
	AA + AG	105 (11.641)	122 (11.708)			
	AG	101 (11.197)	117 (11.228)	0.983	1.003	0.756–1.330
	AA + GG	801 (88.803)	925 (88.772)			
rs3850641	GG	26 (2.879)	15 (1.438)	0.027	0.492	0.259–0.935
	AA + AG	877 (97.121)	1028 (98.562)			
	AA	660 (73.090)	750 (71.908)	0.561	0.943	0.772–1.151
	GG + AG	243 (26.910)	293 (28.092)			
	AG	217 (24.031)	278 (26.654)	0.185	1.149	0.935–1.410
	AA + GG	686 (75.969)	765 (73.346)			

Table 4. Distribution of genotype and allele of *TNFSF4* gene in sub-clinical types of AITDs patients and controls.

SNP	Control	GD	*p*	OR (95% CI)	HT	*p*	OR (95% CI)
rs7514229	–	–	–	–	–	–	–
GG	732 (81.424)	560 (82.474)	0.128	–	278 (79.656)	0.182	–
TT	11 (1.224)	2 (0.295)	–	–	1 (0.287)	–	–
GT	156 (17.353)	117 (17.231)	–	–	70 (20.057)	–	–
G	1620 (90.100)	1237 (91.090)	0.347	1.123 (0.881–1.432)	626 (89.685)	0.756	0.955 (0.716–1.275)
T	178 (9.900)	121 (8.910)	–	–	72 (10.315)	–	–
rs1234313	–	–	–	–	–	–	–
AA	365 (40.466)	298 (43.504)	0.474	–	147 (41.643)	0.832	–
GG	110 (12.195)	78 (11.387)	–	–	39 (11.048)	–	–
AG	427 (47.339)	309 (45.109)	–	–	167 (47.309)	–	–
A	1157 (64.135)	905 (66.058)	0.261	1.088 (0.939–1.261)	461 (65.297)	0.584	1.052 (0.877–1.263)
G	647 (35.865)	465 (33.942)	–	–	245 (34.703)	–	–
rs16845607	–	–	–	–	–	–	–
AA	4 (0.443)	4 (0.581)	0.897	–	1 (0.283)	0.859	–
GG	797 (88.359)	605 (87.808)	–	–	315 (89.235)	–	–
AG	101 (11.197)	80 (11.611)	–	–	37 (10.482)	–	–
A	109 (6.042)	88 (6.386)	0.69	1.061 (0.794–1.418)	39 (5.524)	0.62	0.909 (0.624–1.325)
G	1695 (93.958)	1290 (93.614)	–	–	667 (94.476)	–	–
rs3850641	–	–	–	–	–	–	–
GG	26 (2.879)	9 (1.306)	0.086	–	6 (1.695)	0.172	–
AA	660 (73.090)	502 (72.859)	–	–	248 (70.056)	–	–
AG	217 (24.031)	178 (25.835)	–	–	100 (28.249)	–	–
G	269 (14.895)	196 (14.224)	0.595	0.948 (0.776–1.156)	112 (15.819)	0.561	1.074 (0.845–1.364)
A	1537 (85.105)	1182 (85.776)	–	–	596 (84.181)	–	–

2.3. Haplotypic Association

Haplotypic analysis using the Haploview software (Whitehead Institute for Biomedical Research, MIT Media Lab, and Broad Institute of Harvard and MIT, Cambridge, MA, USA) revealed that in the HapMap Han Chinese Beijing database, rs7514229 and rs1234313 were in the same block (Figure 1), which contained three haplotypes, namely GA, GG and TG. However, these haplotypes were not associated with AITDs ($p > 0.05$, data not shown).

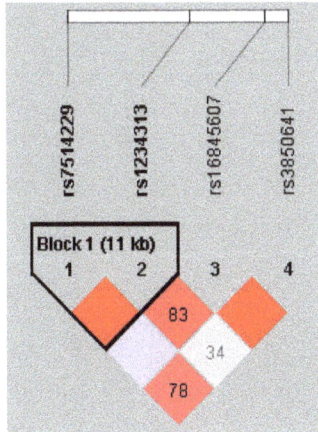

Figure 1. Linkage disequilibrium (LD) block of *TNFSF4* from controls in the Hapmap CHB data.

2.4. Genotyping-Clinical Sub-Phenotype Association

To further investigate the relation between polymorphisms of *TNFSF4* and clinical phenotypes, clinical sub-phenotype analyses were conducted. The results showed that the frequency of genotype TT in rs7514229 marginally declined in AITDs patients with disease-onset age ≥19 years old ($p = 0.049$, as shown in Table 5). However, our present study displayed that *TNFSF4* gene variants were not associated with AITDs patients with ophthalmopathy or family history. Interestingly, the frequency of allele G in rs3850641 was significantly more decreased in HT patients with hypothyroidism than in HT patients without hypothyroidism, suggesting that HT patients with allele G in rs3850641 had increased susceptibility risk to hypothyroidism ($p = 0.018$, Table 6).

Table 5. Allele and genotype distribution of *TNFSF4* in AITDs patients with or without early-onset age.

SNP ID	Onset Age of AITDs Patients		*p*	OR (95% CI)
	≤18	≥19		
rs7514229	–	–	–	–
GG	128 (80.503)	710 (81.703)	0.049	–
TT	2 (1.258)	1 (0.115)	–	–
GT	29 (18.239)	158 (18.182)	–	–
G	285 (89.623)	1578 (90.794)	0.51	1.142 (0.769–1.696)
T	33 (10.377)	160 (9.206)	–	–
rs1234313	–	–	–	–
AA	71 (44.375)	374 (42.597)	0.41	–
GG	22 (13.750)	95 (10.820)	–	–
AG	67 (41.875)	409 (46.583)	–	–
A	209 (65.312)	1157 (65.888)	0.842	1.026 (0.799–1.318)
G	111 (34.688)	599 (34.112)	–	–

Table 5. *Cont.*

SNP ID	Onset Age of AITDs Patients		*p*	OR (95% CI)
	≤18	≥19		
rs16845607	–	–	–	–
AA	1 (0.625)	4 (0.454)	0.246	–
GG	135 (84.375)	785 (89.002)	–	–
AG	24 (15.000)	93 (10.544)	–	–
A	26 (8.125)	101 (5.726)	0.099	0.687 (0.439–1.075)
G	294 (91.875)	1663 (94.274)	–	–
rs3850641	–	–	–	–
GG	2 (1.250)	13 (1.472)	0.639	–
AA	120 (75.000)	630 (71.348)	–	–
AG	38 (23.750)	240 (27.180)	–	–
G	42 (13.125)	266 (15.062)	0.369	1.174 (0.827–1.664)
A	278 (86.875)	1500 (84.938)	–	–

Table 6. *TNFSF4* genotype and allele distribution in clinical sub-phenotype of HT patients.

TNFSF4 SNP	HT		*p*	OR (95% CI)
	Non-Hypothyroidism	Hypothyroidism		
rs7514229	–	–	–	–
GG	107 (80.451)	153 (78.462)	0.668	–
TT	0 (0)	1 (0.513)	–	–
GT	26 (19.549)	41 (21.026)	–	–
G	240 (90.226)	347 (88.974)	0.608	0.874 (0.523–1.462)
T	26 (9.774)	43 (11.026)	–	–
rs1234313	–	–	–	–
AA	50 (37.037)	90 (45.685)	0.287	–
GG	16 (11.852)	19 (9.645)	–	–
AG	69 (51.111)	88 (44.670)	–	–
A	169 (62.593)	268 (68.020)	0.148	1.271 (0.918–1.759)
G	101 (37.407)	126 (31.980)	–	–
rs16845607	–	–	–	–
AA	1 (0.741)	0 (0)	0.136	–
GG	115 (85.185)	180 (91.371)	–	–
AG	19 (14.074)	17 (8.629)	–	–
A	21 (7.778)	17 (4.315)	0.059	0.535 (0.277–1.034)
G	249 (92.22)	377 (95.685)	–	–
rs3850641	–	–	–	–
GG	1 (0.741)	5 (2.525)	0.051	–
AA	104 (77.037)	129 (65.152)	–	–
AG	30 (22.222)	64 (32.323)	–	–
G	32 (11.852)	74 (18.687)	0.018	1.709 (1.096–2.674)
A	238 (88.148)	322 (81.313)	–	–

3. Discussion

The *TNFSF4* gene, also known as the OX40 ligand (OX40L), encodes the OX40L protein which is a co-stimulatory cytokine and belongs to the TNF ligand family. The protein mainly participates in the interaction of T-cell and antigen-presenting cell (APC), T-cell activation and B-cell differentiation, providing CD28-independent co-stimulatory signals for activated CD4[+] T cells [15]. *TNFSF4*, located in chromosome 1 (1q25), contains three exons and two introns (in NCBI database). Previous studies have shown that polymorphisms of *TNFSF4* can confer risk to diverse autoimmune diseases, such as SLE, RA, SSc and pSS, but it remains unknown whether genetic mutations of *TNFSF4* region may induce occurrence of AITDs, which attracts our interest.

AITDs are also regarded as autoimmune diseases targeting the thyroid with a complex genetic and environmental etiology, manifesting mainly as GD and HT. It is notable that genetic factors play a prominent role in the occurrence and persistence of AITDs. Given that autoimmune diseases may share a common genetic predisposition, and that immune dysregulation plays a vital role in AITDs [16,17], we hypothesized that variants within the *TNFSF4* gene, which is a crucial immune regulator, could also elicit abnormal OX40L expression and dysfunction, thus affecting T-cell activation and leading to unbalanced immune regulation and its resultant occurrence of AITDs.

In the present work, we observed the association between four loci of *TNFSF4* gene and AITDs patients in the Han Chinese population. We found that the frequency of genotype GG in rs3850641 was slightly lower in AITDs patients, probably suggesting it could decrease susceptibility to AITDs. In addition, frequencies of GG genotype in rs3850641 and TT genotype in rs7514229 also decreased in AITDs subjects, confirming that variant genotype GG in rs3860541 was indeed a factor protecting people from AITDs, as was variant genotype TT in rs7514229. Our results suggested polymorphisms in the *TNFSF4* gene region, one SNP in 3′UTR (rs7514229) and two intronic SNPs (rs3860541 and rs1234313), may be associated with AITDs susceptibility. To our knowledge, variants in the intron of a gene may influence its expression and regulate its function [18], 3′UTR polymorphisms in the gene region are of important regulation function. We therefore speculated that the molecular action underlying genetic pathology of AITDs is that *TNFSF4* SNPs may affect the expression of *TNFSF4* gene and down-regulate T-cell activation, which requires further in-depth research to confirm.

Further, to investigate the association between genotype and clinical manifestations, we carried out the clinical sub-phenotype analysis. The AITDs occurrence in teenagers (≤18 young patients) may be due to their genetic family history of this disease [19,20], which corresponded with our results showing the frequency of family history was much higher in AITDs patients with disease-onset age ≤18 years old. Meanwhile, marginally significant differences in frequencies of rs7514229 genotype TT and disease-onset age were found between AITDs patients with disease-onset age ≤18 years old and AITDs patients with disease-onset age ≥19 years old. Similar correlations between gene mutations and disease-onset age were reported in RA [21], type 1 diabetes [22] and multiple sclerosis [23]. Furthermore, we revealed that frequency of GG genotype of rs3850641 declined slightly in GD subgroup of AITDs, although without significance. Nevertheless, we observed that *TNFSF4* SNPs were not associated with AITDs patients with ophthalmopathy or family history. Several studies provided clues that thyroid-associated ophthalmopathy (TAO) was correlated with the impact of environmental elements, especially current smoking history [24,25]. Recent studies are suggesting that genetic markers also affect the susceptibility of TAO [26], including genetic variants in the STAT3 [27], TSHR [28] and HLA-DR3 [29] regions. However, our results cannot add the *TNFSF4* gene to the list of the predisposition of thyroid-associated ophthalmopathy (TAO). Moreover, allele A from rs3850641 was associated with the decreased risk for the HT subgroup of hypothyroidism by 41.5%. In HT, hypothyroidism is more associated with a family history of thyroid dysfunction [20]. Our study showed HT hypothyroidism patients with higher ratio of family history, which was consistent with the previous research [20]. To our best knowledge, we were the first to find that genetic factors are also involved in etiology of hypothyroidism in HT. Why do these SNPs not show their susceptibility to GD or TAO? It is possible that thyroid eye disease or TAO is a different disease than Graves' disease and Hashimoto's thyroiditis. In addition, a recent paper found that polymorphisms in calsequestrin (CASQ1) are correlated with HT and GO, but not Graves' hyperthyroidism (GH) [30]. Interestingly, our study found SNPs in *TNFSF4* are associated with hypothyroidism of Hashimoto's thyroiditis, but not thyroid orbitopathy or GD. These two studies do not show contradictory results, and illustrate the complexity of the diseases, GD, HT and TAO or GO. For instance, our previous studies indeed found *UBE2L3* and *CLEC16A* gene polymorphisms to be associated with susceptibility to HT rather than GD and TAO or GO [2,3]. Obviously, the genetic mechanisms of these diseases are still unclear, so more research is needed to reveal the pathomechanism of thyroid ophthalmopathy.

Overall, we provided the first evidence for genetic association between four susceptibility loci in the *TNFSF4* gene in Chinese AITDs patients, with samples exclusively from the Han Chinese population. Nevertheless, considering the validation of a convincing association and discovery of population differences, the importance of replication studies in some different populations should not be overlooked. The statistical power calculated in this research was very strong (larger than 0.8) to detect the association, and it has adequately reached a significant result. Simultaneously, the sample size in this study was large enough with 1,048 cases and 909 controls to effectively reduce the type of errors (type 1 error and type 2 error).

4. Materials and Methods

4.1. Subjects

A total of 1,048 Chinese patients with AITDs (693 GD and 355 HT) and 909 healthy Chinese controls were recruited. All AITDs patients were enrolled from the Out-Patient Department of Endocrinology of Jinshan Hospital of Fudan University. Ethnically and geographically matched and unrelated healthy controls were recruited from the Healthy Check-Up Center of the same hospital.

All AITDs patients were diagnosed as previously described [2,27]. GD patients were diagnosed based on their clinical manifestations and biochemical assessments of hyperthyroidism and the positive circulating TRAb, with or without positive TPOAb or TgAb and diffusive goiter of the thyroid. HT was defined based on the high level of either TPOAb or TgAb, with or without clinical and biochemical hypothyroidism and the presence of an enlarged thyroid. A minority of HT patients were further confirmed by fine needle aspiration biopsies. All the control subjects showed negative thyroid antibodies against TPO. In the current study, TPOAb, TgAb and TRAb were detected with highly specific and sensitive immunochemiluminescence kits from Roche Company (Shanghai, China).

All the subjects, including AITDs patients and controls, were ethnic Han Chinese. Written informed consent was obtained from all participants and the research was approved by the Ethics Committee of Jinshan Hospital of Fudan University (JYLL-2014-06, 2014/2/21), respectively.

4.2. DNA Sample Preparation

Genomic DNA were extracted from 2 mL of peripheral venous blood from each subject using RelaxGene Blood DNA System (Tiangen Biotech Company, Beijing, China), according to the manufacturer's protocol. The concentration and A260/A280 ratio of all DNA samples were measured by NANO DROP 2000 Spectrophotometer (Thermo Scientific Company, Waltham, MA, USA). Finally, the DNA samples with great purity and concentration were used for next genotyping.

4.3. Single Nucleotide Polymorphism (SNP) Selection and Genotyping

Marker-tagging SNPs were chosen from the Hapmap CHB data using the Tagger programme of Haploview software (Whitehead Institute for Biomedical Research, MIT Media Lab, and Broad Institute of Harvard and MIT) to satisfy the following criteria: minor allele frequency (MAF) >0.1, Hardy-Weinberg equilibrium (HWE) with $p > 0.001$ and logarithm of odds (LOD) >3.0. For the *TNFSF4* gene of 23 kb with 42 SNPs in Hapmap CHB population, we selected four loci covering the whole region of the *TNFSF4* gene to capture all the most common variants. Four tag SNPs were selected including rs7514229 located in the 3′ untranslated region (UTR), as well as rs1234313, rs16845607 and rs3850641 in intron 1 of the *TNFSF4* region.

Genotyping of the four SNPs was undertaken using the ligase detection reaction (LDR) platform according to the manufacturer's instructions. Moreover, to ensure detection quality, each reaction was performed in duplicate, and blank samples without DNA were used as negative controls. Furthermore, only SNPs and samples that passed the 95% quality control threshold were subjected to further

Int. J. Mol. Sci. **2016**, *17*, 1369

statistical analysis and SNPs with allele frequencies not meeting Hardy-Weinberg equilibrium (HWE) were removed from the next analysis. The primers specific to the four SNPs at the *TNFSF4* loci are "rs7514229" forward-GATAACACAGAATCATCCAG and reverse-TTGTAGCACATGTTTCCCTG; "rs1234313" forward-ATCTAACACTGGCTCTAGTC and reverse-GCCATTCTGACTAGAATAGG; "rs16845607" forward-AGATATAGCTACCAAGCTCC and reverse-GATGAGAAAACAGAGGCTAC; "rs3850641" forward-GCTGTCACTTTGAAGCTTTG and reverse-TGCCTGATCAAACACATTAC.

4.4. Clinical Sub-Phenotype Analysis

Clinical sub-phenotype stratification analysis was conducted using a case-only approach, in which basic allelic and genotypic examination was performed by comparing minor allele and genotype frequency of cases with a specific sub-phenotype to the whole case group. The clinical sub-phenotypes include: (1) the age of disease onset ($\leq$18 years old versus $\geq$19 years old); (2) presence or absence of ophthalmopathy which was defined as a distinctive disorder characterized by inflammation and swelling of the extraocular muscles, eyelid retraction, periorbital edema, episcleral vascular injection, conjunctive swelling and proptosis; (3) presence or absence of hypothyroidism in HT patients; and (4) presence or absence of AITDs family history, which was defined as the subjects' first-degree relatives including parents, children and siblings or second-degree relatives such as grandparents, uncles and aunts who had AITDs.

4.5. Statistical Analysis

Clinical data were described as $M \pm SD$ (mean $\pm$ standard deviation). Hardy-Weinberg equilibrium (HWE) concordance test in the controls and patient samples, linkage disequilibrium (LD) test and haplotype frequency calculation were performed using HaploView 4.2 (Whitehead Institute for Biomedical Research, MIT Media Lab, and Broad Institute of Harvard and MIT). In order to analyze whether the four predisposing loci are associated with AITDs, allele and genotype frequencies were compared between AITDs cases and healthy controls using the Chi-square test (χ^2-test) or Fisher's exact test. LD among the selected SNPs was measured using the pairwise LD measures D' and r^2. All data were statistically calculated with the SPSS 18.0 software (International Business Machines Corporation, Armonk, NY, USA). A p value of less than 0.05 was considered statistically significant. Odds ratio (OR) and 95% confidence interval (95% CI) were applied to assess the association between each genotype and AITDs.

4.6. Power Calculation

Power calculations for AITDs in this research considered allele frequency of SNPs from 0.05 to 0.5, a population prevalence of 1%–5% for AITDs, and OR of 0.2–0.5 at a 0.05 significant level. As a result, this study had sufficient power (larger than 0.8) to detect the association of OR of 0.2 or above with 1,048 cases and 909 controls.

5. Conclusions

In conclusion, the preliminary findings of our present study are the first to indicate the association of novel genetic susceptibility loci of the *TNFSF4* region with the predisposition to AITDs. Additionally, our results support the importance of T cells in the pathology of AITDs, and reveal the frequency overlap of risk loci in immune pathways between AITDs and other autoimmune diseases.

Acknowledgments: This project is supported by grants from the National Natural Science Foundation of China (No. 81471004) and the Key Disciplines Development of Shanghai Jinshan District (No. JSZK2015A02). The authors would like to thank all of the people who took part in the study.

Author Contributions: Rong-Hua Song and Qiong Wang carried out the work and contributed equally to the work. Qiu-Ming Yao, Xiao-Qing Shao, Ling Li, Wen Wang, Xiao-Fei An and Qian Li helped specimens collection. Rong-Hua Song conducted the data analysis and wrote the manuscript. Jin-An Zhang designed the study and coordinated the research team.

Conflicts of Interest: The authors declare no conflict of interest.

References

1. Tomer, Y. Mechanisms of autoimmune thyroid diseases: From genetics to epigenetics. *Annu. Rev. Pathol.* **2014**, *9*, 147–156. [CrossRef] [PubMed]
2. Muhali, F.S.; Cai, T.T.; Zhu, J.L.; Qin, Q.; Xu, J.; He, S.T.; Shi, X.H.; Jiang, W.J.; Xiao, L.; Li, D.F.; et al. Polymorphisms of CLEC16A region and autoimmune thyroid diseases. *G3 (Bethesda)* **2014**, *4*, 973–977. [CrossRef] [PubMed]
3. Wang, Y.; Zhu, Y.F.; Wang, Q.; Xu, J.; Yan, N.; Xu, J.; Shi, L.F.; He, S.T.; Zhang, J.A. The haplotype of UBE2L3 gene is associated with Hashimoto's thyroiditis in a Chinese Han population. *BMC Endocr. Disord.* **2016**, *16*, 18–23. [CrossRef] [PubMed]
4. Balazs, C. The role of hereditary and environmental factors in autoimmune thyroid diseases. *Orv. Hetil.* **2012**, *153*, 1013–1022. [CrossRef] [PubMed]
5. Effraimidis, G.; Wiersinga, W.M. Mechanisms in endocrinology: Autoimmune thyroid disease: Old and new players. *Eur. J. Endocrinol.* **2014**, *170*, 241–252. [CrossRef] [PubMed]
6. Orozco, G.; Eyre, S.; Hinks, A.; Bowes, J.; Morgan, A.W.; Wilson, A.G.; Wordsworth, P.; Steer, S.; Hocking, L.; Thomson, W.; et al. Study of the common genetic background for rheumatoid arthritis and systemic lupus erythematosus. *Ann. Rheum. Dis.* **2011**, *70*, 463–468. [CrossRef] [PubMed]
7. Gourh, P.; Arnett, F.C.; Tan, F.K.; Assassi, S.; Divecha, D.; Paz, G.; McNearney, T.; Draeger, H.; Reveille, J.D.; Mayes, M.D.; et al. Association of TNFSF4 (OX40L) polymorphisms with susceptibility to systemic sclerosis. *Ann. Rheum. Dis.* **2010**, *69*, 550–555. [CrossRef] [PubMed]
8. Graham, R.R.; Cotsapas, C.; Davies, L.; Hackett, R.; Lessard, C.J.; Leon, J.M.; Burtt, N.P.; Guiducci, C.; Parkin, M.; Gates, C.; et al. Genetic variants near TNFAIP3 on 6q23 are associated with systemic lupus erythematosus. *Nat. Genet.* **2008**, *40*, 1059–1061. [CrossRef] [PubMed]
9. Bowes, J.; Lawrence, R.; Eyre, S.; Panoutsopoulou, K.; Orozco, G.; Elliott, K.S.; Ke, X.; Morris, A.P.; Thomson, W.; Worthington, J.; et al. Rare variation at the TNFAIP3 locus and susceptibility to rheumatoid arthritis. *Hum. Genet.* **2010**, *128*, 627–633. [CrossRef] [PubMed]
10. Dieude, P.; Guedj, M.; Wipff, J.; Ruiz, B.; Riemekasten, G.; Matucci-Cerinic, M.; Melchers, I.; Hachulla, E.; Airo, P.; Diot, E.; et al. Association of the TNFAIP3 rs5029939 variant with systemic sclerosis in the european caucasian population. *Ann. Rheum. Dis.* **2010**, *69*, 1958–1964. [CrossRef] [PubMed]
11. Song, R.H.; Yu, Z.Y.; Wang, Q.; Muhali, F.S.; Jiang, W.J.; Xiao, L.; Shi, X.H.; He, S.T.; Xu, J.; Zhang, J.A. Polymorphisms of the TNFAIP3 region and Graves' disease. *Autoimmunity* **2014**, *47*, 459–465. [CrossRef] [PubMed]
12. Zhou, X.J.; Lu, X.L.; Nath, S.K.; Lv, J.C.; Zhu, S.N.; Yang, H.Z.; Qin, L.X.; Zhao, M.H.; Su, Y.; Shen, N.; et al. Gene–gene interaction of BLK, TNFSF4, TRAF1, TNFAIP3, and REL in systemic lupus erythematosus. *Arthritis Rheum.* **2012**, *64*, 222–231. [CrossRef] [PubMed]
13. Coustet, B.; Bouaziz, M.; Dieude, P.; Guedj, M.; Bossini-Castillo, L.; Agarwal, S.; Radstake, T.; Martin, J.; Gourh, P.; Elhai, M.; et al. Independent replication and meta analysis of association studies establish TNFSF4 as a susceptibility gene preferentially associated with the subset of anticentromere-positive patients with systemic sclerosis. *J. Rheumatol.* **2012**, *39*, 997–1003. [CrossRef] [PubMed]
14. Sun, F.; Li, P.; Chen, H.; Wu, Z.; Xu, J.; Shen, M.; Leng, X.; Shi, Q.; Zhang, W.; Tian, X.; et al. Association studies of TNFSF4, TNFAIP3 and FAM167A-BLK polymorphisms with primary Sjogren's syndrome in Han Chinese. *J. Hum. Genet.* **2013**, *58*, 475–479. [CrossRef] [PubMed]
15. Kinnear, G.; Wood, K.J.; Fallah-Arani, F.; Jones, N.D. A diametric role for OX40 in the response of effector/memory CD4+ T cells and regulatory T cells to alloantigen. *J. Immunol.* **2013**, *191*, 1465–1475. [CrossRef] [PubMed]
16. Simmonds, M.J. GWAS in autoimmune thyroid disease: Redefining our understanding of pathogenesis. *Nat. Rev. Endocrinol.* **2013**, *9*, 277–287. [CrossRef] [PubMed]
17. Marique, L.; Van Regemorter, V.; Gerard, A.C.; Craps, J.; Senou, M.; Marbaix, E.; Rahier, J.; Daumerie, C.; Mourad, M.; Lengele, B.; et al. The expression of dual oxidase, thyroid peroxidase, and caveolin-1 differs according to the type of immune response (Th1/Th2) involved in thyroid autoimmune disorders. *J. Clin. Endocrinol. Metab.* **2014**, *99*, 1722–1732. [CrossRef] [PubMed]

18. Alberobello, A.T.; Congedo, V.; Liu, H.; Cochran, C.; Skarulis, M.C.; Forrest, D.; Celi, F.S. An intronic SNP in the thyroid hormone receptor β gene is associated with pituitary cell-specific over-expression of a mutant thyroid hormone receptor β2 (R338W) in the index case of pituitary-selective resistance to thyroid hormone. *J. Transl. Med.* **2011**, *9*, 144–151. [CrossRef] [PubMed]

19. Brix, T.H.; Petersen, H.C.; Iachine, I.; Hegedus, L. Preliminary evidence of genetic anticipation in Graves' disease. *Thyroid* **2003**, *13*, 447–451. [CrossRef] [PubMed]

20. Manji, N.; Carr-Smith, J.D.; Boelaert, K.; Allahabadia, A.; Armitage, M.; Chatterjee, V.K.; Lazarus, J.H.; Pearce, S.H.; Vaidya, B.; Gough, S.C.; et al. Influences of age, gender, smoking, and family history on autoimmune thyroid disease phenotype. *J. Clin. Endocrinol. Metab.* **2006**, *91*, 4873–4880. [CrossRef] [PubMed]

21. Radstake, T.R.; Barrera, P.; Albers, M.J.; Swinkels, H.L.; van de Putte, L.B.; van Riel, P.L. Genetic anticipation in rheumatoid arthritis in Europe. European consortium on rheumatoid arthritis families. *J. Rheumatol.* **2001**, *28*, 962–967. [PubMed]

22. Paterson, A.D.; Kennedy, J.L.; Petronis, A. Evidence for genetic anticipation in non-mendelian diseases. *Am. J. Hum. Genet.* **1996**, *59*, 264–268. [PubMed]

23. Cocco, E.; Sardu, C.; Lai, M.; Spinicci, G.; Contu, P.; Marrosu, M.G. Anticipation of age at onset in multiple sclerosis: A sardinian cohort study. *Neurology* **2004**, *62*, 1794–1798. [CrossRef] [PubMed]

24. Villanueva, R.; Inzerillo, A.M.; Tomer, Y.; Barbesino, G.; Meltzer, M.; Concepcion, E.S.; Greenberg, D.A.; MacLaren, N.; Sun, Z.S.; Zhang, D.M.; et al. Limited genetic susceptibility to severe Graves' ophthalmopathy: No role for CTLA-4 but evidence for an environmental etiology. *Thyroid* **2000**, *10*, 791–798. [CrossRef] [PubMed]

25. Melcescu, E.; Horton, W.B.; Kim, D.; Vijayakumar, V.; Corbett, J.J.; Crowder, K.W.; Pitman, K.T.; Uwaifo, G.I.; Koch, C.A. Graves orbitopathy: Update on diagnosis and therapy. *South. Med. J.* **2014**, *107*, 34–43. [CrossRef] [PubMed]

26. Li, H.; Chen, Q. Genetic susceptibility to Grave's disease. *Front. Biosci.* **2013**, *18*, 1080–1087. [CrossRef]

27. Xiao, L.; Muhali, F.S.; Cai, T.T.; Song, R.H.; Hu, R.; Shi, X.H.; Jiang, W.J.; Li, D.F.; He, S.T.; Xu, J.; et al. Association of single-nucleotide polymorphisms in the *STAT3* gene with autoimmune thyroid disease in chinese individuals. *Funct. Integr. Genom.* **2013**, *13*, 455–461. [CrossRef] [PubMed]

28. Liu, L.; Wu, H.Q.; Wang, Q.; Zhu, Y.F.; Zhang, W.; Guan, L.J.; Zhang, J.A. Association between thyroid stimulating hormone receptor gene intron polymorphisms and autoimmune thyroid disease in a Chinese Han population. *Endocr. J.* **2012**, *59*, 717–723. [CrossRef] [PubMed]

29. Jurecka-Lubieniecka, B.; Ploski, R.; Kula, D.; Krol, A.; Bednarczuk, T.; Kolosza, Z.; Tukiendorf, A.; Szpak-Ulczok, S.; Stanjek-Cichoracka, A.; Polanska, J.; et al. Association between age at diagnosis of Graves' disease and variants in genes involved in immune response. *PLoS ONE* **2013**, *8*, e59349.

30. Lahooti, H.; Cultrone, D.; Edirimanne, S.; Walsh, J.P.; Delbridge, L.; Cregan, P.; Champion, B.; Wall, J.R. Novel single-nucleotide polymorphisms in the calsequestrin-1 gene are associated with Graves' ophthalmopathy and Hashimoto's thyroiditis. *Clin. Ophthalmol.* **2015**, *9*, 1731–1740. [CrossRef] [PubMed]

MDPI AG

St. Alban-Anlage 66

4052 Basel, Switzerland

Tel. +41 61 683 77 34

Fax +41 61 302 89 18

http://www.mdpi.com

IJMS Editorial Office

E-mail: ijms@mdpi.com

http://www.mdpi.com/journal/ijms

www.ingramcontent.com/pod-product-compliance
Lightning Source LLC
Chambersburg PA
CBHW051842210326
41597CB00033B/5748